THINKING FOR OURSELVES

HOW SCIENTISTS AND DETECTIVES THINK

by

Michael Disney ;   2020

The age of the expert is fading as expert knowledge migrates on to the web. But for any of us to exploit that newly available knowledge we need to evaluate evidence, think straight, judge soundly, and gamble wisely. But these survival skills – Common Sense in other words – are the very same skills required by successful scientists. The author, who is a Space scientist, has therefore analysed successful scientific practice – to discover it relies on 5 key tools: Bayes' Rule, Ockham's Razor, the Principle of Animal Wisdom, The Detective's Equation and The Inference Table – none of which are taught in the current educational system because scholars appear to despise Common Sense – which is far more like detective work than academic logic. He uses numerous examples to show how we can all become wise, even if we can't become "clever" (i.e. pass exams with high marks – a singularly futile skill). Can anyone afford not to become expert in Common Sense Thinking?

COPYRIGHT PAGE

© 2020 Michael Disney

The moral rights of the author have been established. All rights reserved: no part of this publication may be reproduced, or transmitted in any form, or by any means, without the prior permission, in writing, of the author at:

mjd@astro.cf.ac.uk

Published by Kindle Direct Publishing.

ISBN – 978 -1-71250556-4

A DOZEN OUTRAGEOUS BUT TRUE STATEMENTS ABOUT THINKING AND LEARNING.   TFO Disney (draft 8/9/18)

One might imagine that thinking about thinking would be a pretty dry business. But nothing could be further from the truth, as the following statements illustrate. In the course of the book I hope to persuade you that they all make sense.

1) Time and again, from Aristotle to Popper, philosophers have misunderstood Thinking, most often because they didn't ask 'How do animals think?' [Ch.1&14]

2) Like all our other vital powers, we must have inherited Common Sense Thinking [CST] from our animal forbears by gradual Evolution. Therefore, *without artificial aids*, we cannot expect to think very differently from them – or much more effectively. Our blinding conceit is to mistake culture for intellect. [Ch.1&14]

3) Since we all need to think to survive, Nature couldn't afford to hand out significantly unequal capacities to do so to different individuals. Brilliance, IQ, the capacity to pass exams, call it what you will – is of little use when it comes to serious thinking. That requires other qualities including curiosity, breadth, relevant knowledge, judgement, imagination, flexibility, integrity and – above all doggedness. Don't be impressed by, or defer to, superficially clever people. [Ch. 8,10 &13].

4) Serious Thinking is almost always based on Induction, which is, to some extent, guesswork – and indeed a gamble. We can't live without constantly gambling [Ch. 2, 4 & 14]

5) Beware of very strong arguments – they could well be the product of the Systematic Errors, the 'Elephants in the room'. Far better to rely on a network of weaker clues which generally cohere with one another. [Ch.10]

6) The decisiveness of evidence rises virally with the number of clues involved. Thus doubling of evidence can easily increase ones power to reach a wise decision by 5,000 per cent ! (16:3)

7) The spectacular Ascent of Mankind is the result of learning to write and so to overcome the limitation in our power to recall and combine more clues, and thus to accomplish far more complex tasks successfully. That ascent is NOT due to greatly superior brains. [(16:9) and Appendix 9].

8) Much of conventional Education is a waste of time and money – because it doesn't comprehend Common Sense. While Education is a good idea in principle, in practice much of it has been, and remains, designed more to benefit the teachers than their students – let's be honest about it. How much of your young life and your future savings can you afford to spend in class-rooms? Never let your scholastic experiences, however bruising, deter you from aiming high or becoming wise.[Ch. 16]

9) Because we desperately need to understand a subject does not mean we can – *or ever will*! God, the Mind, Consciousness .... if the tools of CST cannot get a grip on a subject then we are helpless to progress. For instance The Principal of Limited Variety reveals that neither Economics

nor Psych***y are sciences – and are mostly irreproducible hocus-pocus. [Ch. 13]

10) The fundamental truth we all need to understand is that *almost all serious arguments about the real world can only reach provisional conclusions.* The best any of us can hope for are good betting odds deriving from a wide variety of evidence. Once we recognize that single truth we might be less hostile to folk who don't believe the same things as we do. Provisionality is the only soil in which any progressive civilization can germinate or flourish. A craving for Certainty is childish, futile and sometimes destructive. Had we rejected it we might have had the jet engine before the birth of Christ. [Ch. 4,7,10,11&13]

11) You can assume that Statistics, where it amounts to anything more than simple Common Sense – is either wrong or irrelevant.[Ch. 11]

12) Once a subject has acquired a sufficient following it can be kept alive in face of all evidence to the contrary, by attaching more and more prostheses (free parameters) to it. Greek Medicine, the Communist Manifesto, Psychoanalysis and Big Bang Cosmology come to mind. [Ch. 7 & 13]

'THINKING FOR OURSELVES' (140 kw)
by MICHAEL DISNEY

CONCISE OUTLINE BY CHAPTERS
Draft (29/1/19)

THEME

When we explore the secrets of scientific discovery we find they are very largely based on Common Sense and Observation, not, as commonly supposed, on Experiment, Deduction and leaps of genius. But how does Common Sense Thinking work? We start by consulting a successful gambler, then a detective, then a research astronomer before going into a jury room to find out how Common Sense works in coming to a verdict on a complicated murder. Example by example, piece by piece the secrets of Common Sense emerge from the deep shadows in which they must have evolved over hundreds of millions of years as our key survival mechanism. It's bloody clever, and a lot more effective then Logic or Statistics.

All but Chapter 1 have a brief summary at the end.

A DOZEN OUTRAGEOUS STATEMENTS ABOUT THINKING AND LEARNING. P2

CHAPTER 1: CAN WE LEARN TO THINK BETTER? P 15

Here we lay out the promise of the book in outline.

(1:1) The Aim of this book.
(1:2) The strange history of thinking.
(1:3) Who this book is for.
(1:4) The big picture.
(1:4) Why learn to think straight now?

CHAPTER 2 : **DIFFERENT KINDS OF THINKING** p 25

Deduction, Induction and Common Sense compared.

(2:1) Deduction, Induction and Common Sense.

(2:2) Induction versus Deduction.

(2:3) Exact and Comparative logic; one man-made, the other of animal vintage.

(2:4) Which kind of thinking is more persuasive? Why have we been so confused?

(2:5) Theories and hypotheses: the real business of a thinking life is deciding between them.

(2:6) History again.

(2:7) Other Thinkers and their work.

CHAPTER 3: **HOW DO SCIENTISTS THINK?** P 46

Nineteen short stories from the history of discovery reveal how scientists think and, more particularly, how they don't. Stories include : Why the sky is dark at night; Hunting down the origins of both Cholera and Malaria; How studying storms at sea led to the unification of Electricity and Magnetism; tripping over Continental Drift in a library; a fogged photograph and the discovery of Deep Time; how a rotten theory of starvation led to the discovery of Evolution; how Relativity forced itself upon us…….

Section (3:20) then looks at the above stories and asks 'So how do Scientists think?' We conclude that Common Sense forms the basis of most scientific thinking, not scientific genius. But how does Common Sense Work? That will become our quest.

CHAPTER 4: **NATURAL THINKING AND BAYES' RULE** p 95

From a successful gambler we learn how to incorporate new evidence into sound decision-taking and hence to win bets.

(4:1) Induction.

(4:2) Probability and Odds.

(4:3) Thinking like a Gambler.
(4:4) The Reverend Bayes and Burglars.
(4:5) The case of the Missing Galaxies.
(4:6) The case of the Dreaded Cancer Test.
(4:7) Did they see Flying Saucers on Jura?
(4:8) The case of The Miraculous Cure.

CHAPTER 5: **DECISIVE THINKING AND THE DETECTIVE'S EQUATION**. P 132

How can different clues be combined together to reach Odds on which one can reasonably decide between one idea and another? We stumble over The Detective's Equation, the $E = mc^2$ of Common Sense Thinking.

(5:1) Detectives and Astronomers.
(5:2) Combining Evidence.
(5:3) Murder in the Library.
(5:4) The Case of the Missing Wife.
(5:5) Animal Thinking.
(5:6) Thinking like Stone Agers.
(5:7) When should we use the Detective's Equation?

CHAPTER 6: **NUMBERS AND THINKING** p 154

It's all too easy to be bamboozled by numbers because we have no instincts for dealing with them. Here we find out how to tame them.

(6:1) Introduction.
(6:2) Looking at Numbers.
(6:3) Is the Universe Expanding?
(6:4) Problems with Finding Weights.

CHAPTER 7: **WOOLLY THINKING AND OCKHAM'S RAZOR** (p170)

Why are some ideas progressive while others lead into a maze? It's vital to know. The great philosophical principle – Ockham's Razor or 'Parsimony' – has so often proved to be the beacon light of scientific success.

(7:1) The Expanding Universe and Ockham's Razor.

(7:2) The complexity of hypotheses.

(7:3) Ockham's Razor in action. Some great triumphs and tragedies of Science.

(7:4) Parsimony, Noise and Prediction.

(7:5) Examples of Woolly thinking in Science.

CHAPTER 8: **COMMON SENSE** p198

Now we have assembled 4 of the 5 tools of Common Sense we try them out in the real world of the jury room and the science lab.

(8:1) Reaching the watershed.

(8:2) Inference Tables

(8:3) The Logic of Common Sense Thinking

(8:4) Tipping Weights

(8:5) Setting Priors

(8:6) Jury Duty: you must reach your verdict on a murder.

(8:7) Common Sense Thinking in the real world.

(8:8) Do you believe in the Big Bang?

CHAPTER 9 : **ERROR ANALYSIS** p 236

Successful thinkers need to recognize and avoid some nasty pitfalls en route towards a wise decision.

(9:1) Weighing Evidence.

(9:2) Counting-Errors and Scatter.

(9:3) Small Number Statistics.

(9:4) Correcting Inference Tables for Scatter.

9

(9:5) Weighing Numbers.

(9:6) Measurement Errors: a looming disaster.

CHAPTER 10: **SYSTEMATIC ERRORS, THE ELEPHANTS IN THE ROOM.** P 268

How can we prevent plausible, but wildly wrong ideas from leading us too far astray?

(10:1) Introduction

(10:2) A Gallery of Rogue Elephants.

(10:3) Elephant Lessons.

(10:4) The Principle of Animal Wisdom (PAW).

(10:5) Unscientific Elephants.

(10:6) Wisdom and Weights

CHAPTER 11: **STATISTICS – OR TERROR ANALYSIS.** P 294

Modern thinking is bedazzled by Mathematical Statistics. We show why much of it is so wrong and how Common Sense, properly used, will do the job instead.

(11:1) Introduction

(11:2) The Magic of Statistics. The perils of smoking.

(11:3) The Gambler's Secret.

(11:4) Probability Wars.

(11:5) The Magic Fades Away. Why Statistics became a disaster.

(11:6) Good-simple versus Bad-simple.

(11:7) Death in the Sky: bomber casualties over Nazi Germany: more statistical disasters.

(11:8) An Aside to the Angry Reader who still believes in Statistics.

(11:9) Death in the Cot. The Sally Clark tragedy. How hard thinking can be.

CHAPTER 12: **PERSUASION.**   P342

How can Common Sense Thinking (CST) – which is private – lead to more general agreement on issues of wide interest?

(12:1) Introduction.

(12:2) Beyond Reasonable Doubt.

(12:3) Scientific debate in practice; Peer Review.

(12:4) A hesitant guide to Persuasion.

(12:5) Afterword.

CHAPTER 13: **POOR THINKING.**   P 357

(13:1) We need to identify the main causes of poor thinking, both in our own head and in others'. The Internet is changing everything.

(13:2) Misunderstanding Common Sense : Popperism.

(13:3) The Principle of Limited Variety: where Common Sense Thinking won't work.

(13:4) When a science is not a science: Economics.

(13:5) Psy****y: why it can't help us to think better.

(13:6) Tunnel Vision I; dismissing alternative hypotheses.

(13:7) Tunnel Vision II: Favouring pet clues. The case for Animal Wisdom.

(13:8) Prejudicial Thinking in general.

(13:9) The Burden of Proof.

(13:10) Misplaced Deference: how to screw yourself.

(13:11) Careerism: conforming for promotion.

(13:12) Groupthink.

(13:13) Resurrective Thinking; forgetting Ockham.

(13:14) Uncritical use of Analogy.

(13:15) Forgetting the Tarantula.

(13:16) The Pygmalion Complex.

(13:17) Blinkered Thinking

CHAPTER 14: **THE EXTRAORDINARY HISTORY OF THINKING** p 407
 Why scholars and educationalists have been so confused for so long, and why most of them still are.
(14:1) Introduction. You're not going to believe this.
(14:2) The Geometer's World.
(14:3) The Instrumental World.
(14:4) The struggle for Induction.
(14:5) The appearance and disappearance of the Reverend Bayes.
(14:6) Subjective Bayesianism.
(14:7) Objective Bayesianism.
(14:8) Why the Detectives Equation remained hidden.
(14:9) Bayesians and Detectives.
(14:10) Ockham's Razor: its history.
(14:11) The Detective's Equation versus Bayes' Theorem.
(14:12) The obstacle course. How we got held up for 2000 years.

CHAPTER 15: **THE PECULIARITIES OF SCIENCE.** P 451
and why scientific thinking can't always be applied in other spheres.
(15:1) Introduction.
(15:2) Systematic Exploration.
(15:3) Devising new instruments.
(15:4) Posing fruitful questions.
(15:5) Measuring and quantifying.
(15:6) Experimenting.
(15:7) Repeating.
(15:8) Relating.

(15:9) Explaining.

(15:10) A cautionary tale.

(15:11) Publishing.

(15:12) What about Mathematics?

CHAPTER 16: **CONSEQUENCES: THE METEORIC ASCENT OF MANKIND.** P 476

Now we know how to think we take wing and make some extraordinary discoveries in the fields of learning, education, time, cunning, government, religion, ethics, civilization and the emergence of homo sapiens as the dominant species. The prospects ahead look dazzling.

(16:1) Where could our newfound knowledge of Common Sense Thinking (CST) lead?

(16:2) The value of learning – each new thing you learn is valuable in proportion to what you *already* know.

(16:3) Decisiveness goes viral. The Exponential value of Evidence.

(16:4) Mental Laziness: to be expected; to be exploited.

(16:5) Education. The questionable value of a modern-day education lasting 20 years. It's way too long: it's for growing big babies, not wise men and women.

(16:6) Common Sense and Governance. The impact of CS ideas on governance at all levels. Well ordered committees are likely to be more decisive – and wiser – than strong CEOs.

(16:7) Religion and Common Sense. The possible bearing of CST on atheism, religion and tolerance in general.

(16:8) The Meteoric Ascent of Mankind: How come a zoological failure took over the world in a mere 5000 years? It's Common Sense – but with a new twist. And we haven't started yet.

(16:9) Honesty

(16:10) Priestliness, Holiness and Certainty.

(16:11) Last Thoughts.

FINAL SUMMARY OF THE BOOK. (1.5 kw)   p 503

Epilogue   'THROTTLE UP' (0.6 kw)   p 510

GLOSSARY   p 513

REFERENCES (and discussion of), by chapter. (7.4 kw) p 526

ACKNOWLEDGEMENTS   p 546
APPENDICES.
Appendix A1* : A Table of Normal and PAW Weights.   p 547
Appendix A2* : Murder in a University Town: your verdict.   p 551
Appendix A3:   The Expected Scatter.   p 553
Appendix A4:   The Boy/ Girl problem.   p 557
Appendix A5:   Where the Normal Law of Error comes from. p 559
Appendix A6*:   The Sally Clark Case.   p 562
Appendix A7:   Philosophy and our Book.   p 566
Appendix A8**: Certainty, Falsifiability and Common Sense. p 572
Appendix A9**: Categorical Inference and Animal Wisdom   p 581
Appendix A10:   Hidden Galaxies netted at last   p 592
Appendix A11:   How this book got written   p 595
Appendix A12:   Can machines be trained to Think ?   p 598
Appendix A13:   A sketch-map of Thinking   p 604
INDEX   p 612
Where * means 'Indispensable Reading' and ** 'Absolutely Spinal'.
   TOTAL WORD COUNT 185 k-words

14

THE APPENDICES: We have placed some key material in appendices. It is there not because it is less significant, on the contrary, but so it can be easily consulted from many places in the main text. A single asterisk* attached to the appendix title means that it is indispensable reading, two ** that it is absolutely spinal.

THE UP TO DATE WEBSITE FOR THE BOOK

and the author is now up and running at **mjdisney.org** and there is a special post under the 'My Books' Category called **TFO&&&** dealing entirely with this book. There you will find Exercises with Answers, corrections, additions, updates, modifications and much else besides, including a mechanism for readers to post comments.

EXERCISES: There are a few sets of simple Exercises put in to familiarize readers with concepts such as gambling-odds with which they might not be acquainted. They are designed to speed up progress: as such are exercises rather than hard problems – which are labelled as such where such problems do appear. To aid with feedback the actual questions, together with Hints and answers, are not in the text but on the books' website: [https://tinyurl.com/yj94ffwf]. Forty years of teaching students has taught me that such feedback is vital to *both* sides of the learning process

THE OLD WEBSITE still contain the Exercises with Answers; some additional appendices; many extra references brought up to date; a place for feedback between readers and author, and between readers ; many extra figures and pieces of evidence; short bibliographies and caricatures of some of the main heroes and villains. It will be interactive and accretive, and be maintained by the author.

For latest news on that old website go to
https://tinyurl.com/yj94ffwf

15

THINKING FOR OURSELVES (Disney)
CHAPTER ONE (14/9/18)
CAN WE LEARN TO THINK BETTER? (3.4 k-words)

"It ain't what a man don't know as makes him a fool; it's what he do know as just ain't so." Josh Billings

(1:1) THE AIM OF THIS BOOK

To think effectively for ourselves we require the right information and the right tools. At last the Internet is bringing an increasing flood of the kind of information we need to make crucial decisions for ourselves. But do we have the right mental tools to make use of it? I will argue that we do and we don't. We do because deep inside us Nature has embedded the sophisticated mental mechanisms needed to survive in the perilous stone-age world out of which we evolved. We don't, in the sense that we don't know how those tools work and so we cannot adapt them for best use in the modern world with all its culture and technology. The aim of this book is to open Nature's tool-bag and demonstrate her mental tools working one by one. In other words we will be discovering how Common Sense Thinking (CST) works; it's bloody clever. Reassuringly it has very little to do with 'brilliance', mathematics or the capacity to pass academic exams. It turns out that Science, and indeed serious thinking in general, is far more like detective work than anything else. We have suspicions, we follow up clues, hoping to eventually accumulate Odds on this or that suspect high enough to make an arrest and convince a jury. Thus we employ symbols like $O(S|C)$ which stands for our betting Odds on the guilt of some suspect S, after discovering clue C. And we change our mind, our Odds, when a new clue N comes in, according to the notorious gamblers' rule $O(S|N) =$

W(N|S) × O(S), which works just fine[1]. So, in a sense, this book is a series of detective stories in which we learn how to work out rough Odds on the various suspects (ideas) being guilty. The symbols are necessary to distinguish one suspect from another, and to keep a rough account of our accumulating Odds because 2 to 1 won't convince anybody, 10 to 1 might, while 50 to 1 ought to carry the day.

I bet you haven't learned to use all these gambling tricks at school or university because scholars like to pretend that there are more 'logical' ways to think. But to judge from science they are quite wrong. I hope to convince you, using many examples, that when you do learn to adapt those tricks to your world, you can become a far wiser thinker than you are now

(1:2) THE STRANGE HISTORY OF THINKING

How do we think and can we learn to think better? What passes through our minds when we try to reach conclusions and make decisions?

Over the centuries many scholars have pondered deeply upon the matter; philosophers, mathematicians, scientists, divines; latterly psychologists and statisticians. Indeed the biggest industry in the West, it's called Education, is largely based on the proposition that we can be taught to think better. At the same time there is a general feeling that some individuals are imbued with almost magical powers of thought, Einstein for instance, or Sherlock Holmes. There's a divine spark somewhere that the rest of us must lack.

But wait a moment. We are all, or nearly all Evolutionists nowadays. That means that if we can think then our ancestors must have been able to think before us, hominids, apes, monkeys, lemurs and so on even further back on the trail. And if you've had anything to do with animals you will know that they can think

---

[1] Most of the 'formulae' are like ' O(S|C)' . All it stands for is 'The Odds on Suspect S given clue C' and it will be a number like 4 meaning '4 to 1 on', or ½ meaning '2 to 1 against'. That's hardly maths. I have put in plenty of simple exercises so that you can become familiar with it. We all need to be.

too, that is to say reach conclusions and act upon them much as you and I. They may not do it as well, or in exactly the same fashion as us, but an animal unable to make sensible decisions wouldn't survive for long.

Thus thinking, in the sense of concluding and deciding, must be an ability buried in all of us, going back tens of millions of years and shared to some extent with our furred and feathered relatives. It is no divine spark but a survival mechanism presumably encoded in our genes. The great question is "How does thinking work?" If we could find out we might be able to exercise the faculty in such a way as to greatly improve our performance in life. We might.

I intend to approach the question from the point of view of a working scientist, one who has been struggling to do astronomy and space research for fifty years. That is why my subtitle is 'How do scientists and detectives think?' Scientists spend a great deal of time thinking and outsiders, particularly philosophers, are inclined to imagine that we must have, because of our evident successes, developed superior methods of thought (see below). Alas and alack! Scientists are no better than other people at articulating what goes on in their heads. This was wittily summed up by the distinguished zoologist Peter Medawar who wrote in 1972: "Ask a scientist what he considers the scientific method to be, and he will adopt an expression that is at once solemn and shifty eyed: solemn because he feels he ought to declare an opinion; shifty eyed because he is wondering how to conceal the fact that he has no opinion to declare." Ironically Medawar, who won the Nobel prize for discovering graft-rejection, went on to advocate a method of scientific thinking called "Popperism" which we recognise as completely unsound today.

What science does have, that most other disciplines lack, is progress. It may be halting but modern scientists can usually look back to see the misconceptions of their predecessors and identify the weaknesses in the arguments which led to them. Both as fallible individuals and as members of a progressive craft we, more than most others, become aware of the snakes and ladders of mental combat. We can learn from scientific history and thus may have something to teach other no less intelligent individuals who are not so luckily situated in the vanguard of human exploration. The historian Isiah Berlin put it thus: "The most successful method of identifying, discovering and inferring facts is that of the natural sciences. This is the only region of human experience, at any rate in modern times, in which progress has indubitably been made. It is natural to wish to apply methods successful and authoritative in one sphere to another, where there is far less agreement among specialists."

('The Proper Study of Mankind', 1997, p 18).

Some few of us, perhaps as the result of a personal experience, become obsessed by the thinking processes underlying scientific research. We plunge into the literature desperately seeking for clever tactics that will either get us out of a deep hole in our own research or make us wiser and more successful. Such was my experience. I was at a conference when young Dr Z got up and demonstrated, apparently beyond question, that my beautiful theory about Hidden Galaxies (more about them later) to which I'd devoted a life's work, was totally and disastrously wrong. And yet. What about all that positive if circumstantial evidence I had so carefully gathered over thirty years? Could it all be a delusion, a figment of the biased mind? Despite being a professor of 20 years standing I had no idea how to react or proceed. His evidence was damning, unanswerable. But mine wasn't easily dismissed either. I went to the library, took out several learned tomes and went down to the beach to think. Little did I realize that I was embarked on a quest that would last twenty years and range across two dozen centuries.

When I finally emerged from the maze I was clutching not the Holy Grail but something a good deal more useful – a boy-scout's penknife for the thinking mind. It comprised five extremely powerful implements: Bayes' Rule, Ockham's Razor, The Principle of Animal Wisdom (PAW), and the Inference Table, all auxiliary to the central tool which I call 'The Detective's Equation'. Once grasped I couldn't do without my wonder tool – either as a scientist or as an ordinary citizen. For instance

I applied it to Dr Z's evidence – to find that it was fishy, and almost certainly flawed.

The Detective's Equation is the $E = mc^2$ of Common Sense Thinking. Ever since stumbling over it in 2010 I have been asking myself why I should have picked up a jewel that has evaded so many greater minds. The answer is two-fold. First of all I asked a rather different question from my eminent predecessors: I asked 'How do *animals* think?' I did so because, in his achievements, *homo sapiens* has really only risen above his fellow creatures in the last 10,000 years or so, a mere blink in the long cavalcade of Evolution. If we can think, it must be largely because they can think too. And second, through a piece of blind luck. As we shall see Inductive thinking involves gambling and gambling has been discussed in terms of two rather different vocabularies: 'Odds' and 'Probabilities'. Serious thinkers about Thinking have almost exclusively preferred 'Probabilities' in which, as it happens, the Detective's Equation is a tangled mess of ropes and knots: hard to spot, hard to remember, almost impossible to use. In Odds though, as I found by accident, it is utterly transparent, indeed, with a few qualifications, completely obvious. So much so that when you first point it out to people their initial reaction is; "So what; everybody knows that!"

The truth is 'They do and they don't.' They do because they've been using Common Sense Thinking (CST henceforth) all their lives; they don't because they can't explain how CST actually works. We need to know exactly how it works to realise its full power – and its limitations. That is what this book is about.

How could a mere change of vocabulary make such a profound difference? There is a precedent. The ancient Greeks, despite all their brilliance, couldn't solve Zeno's Paradox – the argument that the hare could never catch the tortoise. They couldn't do so because of their clumsy arithmetical vocabulary – I, II, III, IV ,X…and so on. Using Arabic numerals and decimals, Zeno's Paradox evaporates. We now realize that to understand arithmetic you need to understand both fractions and decimals. Likewise to understand Common Sense Thinking you need to use both Probabilities and Odds. That was by no means obvious. It was my luck to find out.

In any case CST is so vital to survival that Nature could not afford to leave it to parents to teach it to their offspring. It has to be an innate part of our being – like our liver. Most of us have no idea how our liver works but use it, without thinking, every moment of the day. So I believe it is with CST. To work, it didn't have to be self-explanatory.

Because the great thinkers of the past didn't realize that we are descended from ape-like creatures they didn't ask the simple question 'How do animals think?' They thought that Man was somehow related to the gods and that his thinking process was the highest evocation of his immortal spirit. It is not surprising therefore that they were led astray into realms of Philosophy and Mathematics which have little to do with the real case. For instance there is a widespread delusion that Logical Deduction is especially useful in the higher realms of thought – notably Science. We shall discover that this is not the case and that in order to think straight we will have to put aside many other such delusions left behind by the detritus of history and education. (Chapt.3)

The history of Common Sense Thinking [CST], or at least Man's understanding of it, is extraordinary, indeed barely credible, and will have to wait for a chapter of its own (14). The ancient Greeks buried it in Geometry; the Romans and their successor church discouraged serious thinking of any kind; peoples everywhere were blind to Evolution, yet even so glimpses of the truth were spotted by Christiann Huyghens (1690) and Thomas Bayes (1761) during the Enlightenment. But then mathematicians, who didn't like the look of it, successfully reburied it under a formidable looking theorem (which was actually wrong, or at least irrelevant). Statisticians, who felt they knew how science ought to work, made the situation far far worse in the twentieth century so that scientists of my generation could only go about science by ignoring the self appointed experts on The Scientific Method – whether they came from Mathematics, Statistics or Philosophy. We stuck to our Common Sense Thinking, even though we couldn't explain it, because it worked in practice. It is possible to argue (Ch.14) that if Mankind hadn't been waylaid by all this philosophizing we might have reached the jet engine before the birth of Christ; we might. Yes it is an extraordinary and unlikely story.

(1:3) WHO THE BOOK IS FOR

The book is aimed at two kinds of readers: those who feel they are educated – and those who feel they are not. The educated ones will, I believe, discover that they have not learned Common Sense Thinking at school or university – and badly need to – while the uneducated ones will, I hope, find that dropping out of the current educational system is no handicap if they now decide that they really do want to think seriously and well. After all, thinking seriously has been studiously avoided in class-rooms and lecture halls for over two millennia. Greek philosophy, with its misplaced emphasis on deductive logic, and Abrahamic religion with its delusion that we are the chosen vessels of God, lured scholars away from the real question – how did we, and our animal forbears, learn to use our brains in order to survive for millions of years before either Socrates or Christ were born. It is only when we ask "How could animals think?" that the subtle tools of Common Sense come to light. And when they do we will find that there can never be a more effective way of thinking than Common Sense – one that Nature has evolved within us over uncountable generations. We don't need to change it, but we urgently need to understand it well enough to exploit it best in a modern world overwhelmed by data, culture and technology.

In short if you are over fourteen, and not thoroughly familiar with Bayes' Rule, Ockham's Razor, PAW (The Principle of Animal Wisdom), PLV (the Principle of Limited Variety) and The Detective's Equation, then you probably need to read this book – no matter how many degrees, or how few GCSEs, you have got.

(1:4) WHY LEARN TO THINK STRAIGHT *NOW*?

It is sad but true that even if you stay on in full time education until you are 25 you will be taught almost nothing of Common Sense Thinking at present – not even if you study Science. And yet there never was a time when we needed Common Sense more. In the past CST was often crippled by lack of specialised knowledge. For instance what was the point of thinking about your disease if you had no access to a medical library? We became reliant on experts, all too often flawed. The Internet is upending, capsizing, overturning the whole balance between experts and the rest of us. From now on we will be limited not so much by our access to information as by our ability to make use of it wisely, in other words by our ability to think straight about it. It is going to take Education, by far the biggest business in the Western world, generations to catch up with this awkward fact, awkward from the point of view of its 'business model'. Big Education is planning, indeed expecting to occupy a third of your life and drain away much of your savings. But that is already a redundant assumption. If you have access to the Internet, and you really know how Common Sense Thinking works, I can't see why you need to spend so many of those invaluable years between 15 and 25 sitting in class-rooms, piling up debts, not learning CS, and therefore not learning to think straight (Chapter 16).

What this book aims to do is enable the reader to become thoroughly conscious of, and proficient in the use of, the main tools of Common Sense Thinking. And I claim that anyone who uses the Internet to look for information and think about it will benefit from reading us. The approach is to work through numerous stories and examples, some serious, some light hearted, all hopefully interesting and relevant. A couple of hours devoted to each chapter should give most readers a working skill in using those main tools. After all you were born with Common Sense. Yes there are formulae in the book, and lots of simple arithmetic, but nothing more than you would encounter in the bank or betting shop. I'm not saying Common Sense Thinking is easy – on the contrary – not if you want to do it well. But even if it looks a bit numerical at times the book doesn't need a mathematical background. If ducks use common sense, as they certainly do, how could it?

The book should help you optimise your powers of judgement, given that you have evidence and have weighed that evidence yourself. It cannot weight the evidence for you; that obviously depends on specialised knowledge and experience of the application in question. It might make a great plumber out of a good one. It won't make even a mediocre plumber out of an ignorant astrophysicist; I speak from experience. There's is no substitute for knowledge, for curiosity, for keen observation, for breadth of learning, for openness of mind and yes, experience. But all of these admirable qualities are nothing without Common Sense.

(1:5) THE BIG PICTURE

Our book is largely taken up with how we might be able to think better as individuals. However *homo* is a social animal so if we were able to think more wisely as a species the consequences might be sensational. To see how this could be so let us briefly touch on four topics that will arise in detail later:

(a) Political discussion very often centres around economic issues (e.g. Brexit in the UK just now). The protagonists talk as if they knew just what the economic consequences of this or that policy will be. But how can they know? Does Economics have a sound conceptual foundation upon which reliable forecasts can be based? Our analysis of CST will lead us to prove that it does not, *and never can have* such a foundation – because of 'The Principle of Limited Variety'. According to J.K. Galbraith, a distinguished historian of the subject: "Economics was invented to make Astrology look respectable."

(b) Many of the catastrophes of history are based on the proposition that "We are certainly right, and they are certainly wrong." But, as we shall discover, all serious thinking has to be based on Induction – which can never deliver Certainty – only Provisional truth. It is, and has to be, open to new evidence. This is the very secret and heart of Progress. Only false gods can promise Certainty in the real world. Priests, philosophers, demagogues, logicians, dictators, scientists, and statisticians have all exploited a widespread craving for Certainty – which we need to understand is both childish – and generally impossible of fulfilment [Sect (16:9); Appendix 8].

(c)   The Earth is flat; witches cause diphtheria; the world was formed in 4004 BC; black people are sub-human; the Pharaoh is a god; you will go to Hell; animals can't think; the Sun orbits the Earth; the pope is infallible; women are inferior…… were all disastrously wrong beliefs that held mankind up for centuries. How do we know that some parts of our present belief system are not equally treacherous 'Systematic Errors' (Elephants in the room)? And, more to the point, how do we prevent them from leading us into grand follies? Animals, it will appear, employ 'The Principle of Animal Wisdom' (PAW) as a very powerful self-preservative strategy, one which we need to disinter from beneath the encrustations of human culture and education. It's a vital – absolutely vital strategy for avoiding extinction.[Sect. (10:4)]

(d)   10,000 years ago we were just another mammal, gathering, hunting, scavenging, unobtrusive, rare. Then something happened that led to a spectacular transformation, the like of which the Earth had not seen in a billion years of Evolution. What was it? If it wasn't magic from Outer Space [viz the film '*2001*'] it could only have come about from the way we thought. When we come to pin down CST we shall see that a single trick would have enabled *homo* to increase his and her thinking power by over a million, build Venice, plan D-Day and go to the Moon. But if a single modification of CST is capable of that – what about a second leap? [Chapt. 16 'The Ascent of Mankind']

So let's start out on our quest for the secrets of Common Sense by looking at the different kinds of Thinking that have been described (Chapter 2).

THINKING FOR OURSELVES (Disney)
CHAPTER 2 (draft 14/9/18)
DIFFERENT KINDS OF THINKING (6.7 kw.)

"...although the arguing from Experiments and Observations be no Demonstrations of general Conclusions, it is the best way of arguing which the Nature of Things admits of." Isaac Newton

(2:1) INTRODUCTION

There are many kinds of thinking ranging from day-dreaming to mathematical logic. The only kind we shall be concerned with is that required to reach conclusions and make decisions. Even so there is more than one kind of thinking employed to do even that. In this short, rather superficial chapter we try to distinguish Common Sense Thinking (CST), which will be our main concern, from its rivals and alternatives.

The poshest kind of thinking is called 'Deduction'. It's what mythically clever people like Sherlock Holmes and Albert Einstein are supposed to do – but actually don't most of the time. It's posh because Philosophers talk about it incessantly and because it is indeed capable of reaching absolutely certain conclusions. The problem is that it's mostly not applicable to real life situations, including science; take a trivial example of what is called a 'Logical Syllogism':

1 James is bigger than George
2 Ben is bigger than James
3 Therefore Ben is bigger than George.

1 and 2 are called 'premises' and 3 is a 'deduction'. The deduction is logically inescapable, absolutely valid, but whether it is true in real life depends on the correctness and precision of the two premises. Bigness is a rather vague concept and not easy to define exactly; James may look bigger than George because he dresses rather differently.

Deduction is most useful when the premises are established by fiat (law), i.e. are unarguable by definition. For instance in games the rules are laid down arbitrarily and unambiguously; you can't argue with them. Thus deduction is highly useful when you are playing chess or doing Sodoku. Likewise in Mathematics where the premises, or 'axioms' as they are then called, must be assumed as true before you can even start. Whether those rules apply to the real world is another matter entirely.

So Deductions are all very well but only as reliable as the premises on which they rest. In the real world you might only be able to say:

1 I think it's 70 per cent certain that James is bigger than George.

2 And I think its 60 per cent certain that Ben is bigger than James.

3 So I can't deduce that Ben is *for certain* bigger than George. But what can I say?

There *are* areas where the general laws are so well established that they can be used as reliable premises on which to base many useful deductions. In designing bridges for instance the laws of Statics were established centuries ago and the properties of building materials can be accurately controlled. Thus the designer can deduce the stresses and strains in every component of the bridge and so ensure that it is safe. Likewise the behaviour of complex electronic circuits can be deduced from the component characteristics and the well-established laws of electro-magnetism. Engineers and technologists use deduction extensively – but pure scientists much less so. After all, they are trying to establish new laws of Nature, not to exploit those already understood. Likewise in everyday life we have to deal with components – for instance people – who do *not* move on predictable tramlines. Some other kind of thinking has to be looked for.

The main other kind of thinking is called 'Induction'. Instead of working down from premises assumed certain towards conclusions equally so, Induction works in the opposite direction. It moves upwards from specific instances to try and infer broader generalisations. Suppose for instance you've seen hundreds of swans – and all of them were white – you might be inclined to infer that 'All swans are white'. That would be Induction. The problem of course is that no such inference is logically warranted; no matter how many white swans you've seen there can be no certainty that the next one won't be black – or orange. In a logical sense the conclusion that 'All swans are white" is not valid. This is the infamous 'Problem of Induction' – general laws cannot be inferred from specific instances, no matter how numerous they are.

But we can't leave the matter there. We have to think, and think effectively. But how?

(2:2) DEDUCTION, INDUCTION AND PLAUSIBLE INFERENCE COMPARED

As an illustration of the contrast between the first two types of thinking let us apply each by turn to the same famous problem: the well known assertion that the angles inside *any* triangle add up to 180 degrees or two right angles. How might you convince yourself of its truth (or otherwise)? {1, 2}

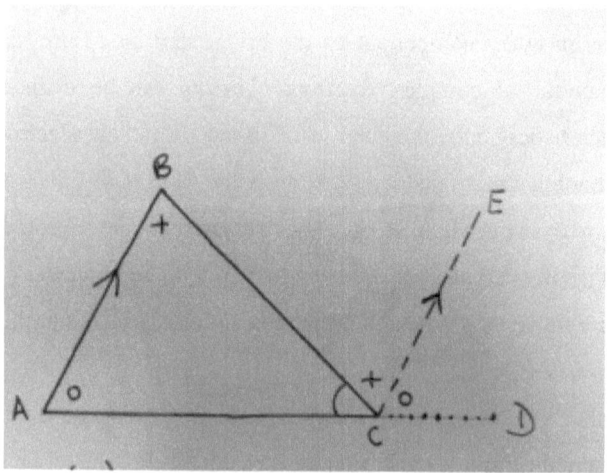

Fig (2:1) The Triangle ABC has two artificial "constructions" added to help make the proof that its 3 interior angles add up to 180 degrees or 2 right-angles. CD is simply an extension of the line AC while CE is drawn deliberately parallel to AB. With these constructions it becomes obvious that the labelled angles are equal because of their relationship to the two parallel lines. Thus the sum of all three such angles is, at C, obviously equivalent to a 'straight' angle i.e. two right angles.

Fig (2:1) outlines the well known deductive 'proof' that is taught at the beginning of every school course on Geometry and which goes back to Classical Greece. It involves attaching two not-obvious 'constructions' or extra lines to the triangle and then recognising that certain pairs of (marked) angles in the resulting drawing must be equal. With that established the three angles interior to the triangle can be seen as equal to the three adjacent angles at C which make up a straight line and which therefore must add up to two right angles – which by definition also make up a straight line. QED, or '*Quod Erat Demonstrandum*' as they say.

This deductive proof is very ingenious and far from obvious – until the necessary constructions are pointed out to us. But it is artificial and therefore, to many minds, too bloody clever by half – and repugnant on that count alone (One can see why many good minds immediately get turned off by Geometry while others are enthralled). Its great virtue is that it is completely general i.e. applies to all triangles, because the argument nowhere makes use of the particular shape of the triangle referred to.

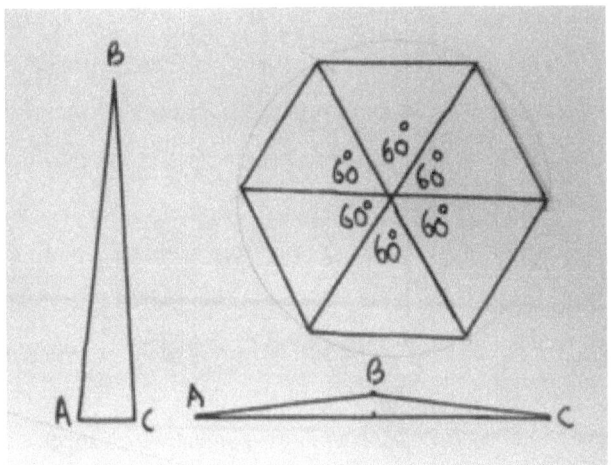

Fig (2:2) Three very different triangles and yet a little reflection will convince you that each must contain internal angles totalling 2 right angles. Take the tall one to an extreme and its top angle vanishes while the bottom 2 become right angles. Squash the flat one even further and its bottom angles vanish leaving a single 'straight' angle at the top. As for the equilateral triangle it could obviously tile the entire plane and six such tiles would meet at every vertex in the form of a hexagon as shown. Six internal angles coming to a whole circle, or 360 degrees, implies 60 degrees each.

Fig (2:2) illustrates the alternative Inductive argument. I have drawn three extreme kinds of triangle; one tall, one squat, and one maximally symmetric or 'equilateral' which lies midway in its properties between the other two. It's not difficult to see from inspection that in their limits both the first two will contain exactly two right angles in all because some angles will then shrink to zero. As for the Equilateral, imagine the whole plane tiled with such triangles. I've drawn only six of them meeting at a point. Its obvious that the six equal angles at the intersection come to a full circle or 360 degrees. Thus each internal angle must equal 60 degrees so three together, as in a triangle. will amount to 180 degrees or two right angles once again. We cannot write QED this time however because it is always possible that some other odd-shaped triangle will not conform with these three. But that's becoming to seem unlikely, at least to me. Surely there's some sort of continuity or uniformity to be expected among all the intermediate types of triangle? Although this Inductive argument is merely probable, and not conclusive, in fact I'd be prepared to bet money on it. And if I can find some more kinds of triangles which conform to the rule (I can) my conviction would quickly move towards, if never quite reach, absolute conviction.

I find this Inductive argument, although imperfect, comes far more naturally to me (and to other scientists I know), than the Deductive one. Indeed I'm sure the ancient Greeks used this argument long before they found their ingenious constructions and deductive certainty.

There is however a finality about a Deductive argument which appeals to a certain kind of mind – the mathematical and philosophical mind in particular. Unfortunately there are at least two fatal limitations to Deduction as a means for advancing thought. First; how can one be absolutely sure of ones' premises – which one has to be for the system to work in the real world (as opposed to in a game). For instance in arguing about those equal angles in the deductive proof above we unconsciously had to assume that parallel lines never meet. That may be true of an imaginary 'flat' world, it will in fact not be exactly true of parallel lines drawn on the surface of the Earth, because it's spherical. You might respond: "But *real* Space is 'flat' – or 'straight' in 3 dimensions." But how do you know? That's merely an assumption which might not survive a precise survey (According to eclipse measurements Space near the Sun, i.e. near us, appears to be very slightly curved). Second, and far worse, Deduction applies only to a *closed* system. All it can do is move the existing furniture around to find out which propositions are consistent with its premises, and which are not. In the real world however one must always be able to introduce *new* information into the argument, information which *may* conflict with ones existing premises (for instance in the Triangle argument above I needed to introduce new kinds of triangles to provide more evidence). On its own Deduction simply cannot do the job of advancing Science – or indeed any progressive field of thought. However unhappy philosophers and mathematicians may be with The Problem of Induction they cannot escape it by appealing to Deduction instead. Alas, as we shall see, they try all the time, disguising their misunderstanding from themselves, as much as from us, in a smokescreen of jargon. This is particularly true of Statisticians, whose entire subject we will find very good reasons to question. No we can't avoid using Induction in some form. As Isaac Newton put it (1704), having got the matter wrong himself several times; "…..although the arguing from Experiments and Observations be no Demonstrations of General Conclusions, it is the best way of arguing which the Nature of things admits of."

In real life, and in science, we have to make decisions, based on such incomplete information as we have, life and death decisions sometimes. How we make such decisions based on imperfect or incomplete evidence is technically called 'Plausible Inference' or 'Common Sense' in our parlance and is obviously a form of Induction. It differs only in attaching some kind of probability to each of its clues and each of its conclusions. We say things like "The discovery that *equilateral* triangles have internal angles summing to two right angles is a clue with a Weight of 4 in my mind – that is to say, on its own it quadruples my belief in the hypothesis that *all* triangles contain two right angles." Or I might say "My Odds on the idea that men are related to monkeys are now ten to one on." Or "I think it 99 per cent certain that Man is responsible for global warming." So Plausible Inference is really Induction with some kind of Weights (Likelihoods) attached. The Likelihood might be rather precise ['95%' for example] or it might be very crude – 'High', 'Medium' or 'Low' say. In its cruder version Plausible Inference tends to be called 'Common Sense' and is a form of reasoning we (probably) share with animals because they too can recognize when danger say is 'High', 'Medium' or 'Low' without having to put a number to it. They, and Plausible Inference, may share logic of the 'More-or-Less' kind rather than the 'Yes-or-No' kind employed by Deduction. We cannot assume it is less reliable just because it is less precise. On the contrary. Our quest will be to try and find out how that Common Sense Thinking (CST) works, and to optimise our use of it. A particular challenge will be to find a way to handle conflicting evidence.

Whenever I have needed practical guidance I have tended to look towards astronomy first – because that is the field I know by far the best. There is a certain danger in doing so because astronomy may not be typical of science in general. I admit to that. But I feel it is better to look at data whose weaknesses I know than to look towards some other field for 'better' data – whose weaknesses I don't. Additionally astronomy is a simple subject compared to say genetics so that its stories are easier told, not yet masked in the arcane language that is bound to accrue in a more sophisticated field. Some sciences have become very esoteric these days. For instance I can't even understand the *titles* of many genetics papers let alone the arguments therein contained.

I go further and introduce my own research topic 'Hidden Galaxies' from time to time. I do so because I know intimately all its blunders and controversies – which are an essential part of any living field of research or debate. I don't so know the equivalent blunders and controversies in other fields because we scientists tend to be pretty reticent about our follies – particularly with outsiders.

I also spend time on Hidden Galaxies because it gives the chance to tell stories. We all like a good story – not least because we can learn from it more pleasurably than we can from an abstract disquisition. I've tried to pack the book with good stories.

(2:3) EXACT AND COMPARATIVE LOGIC.

The mathematics we are taught at school is all about the (=) equals sign; $2+2=4$ while $2+3 \neq 4$. It is exact. Arithmetic statements are either right or wrong. Likewise in geometry: two angles are either equal or they are not. If you are to claim that they are equal then prove your claim by applying exact logic to the known facts and/or to previously established theorems.

Animals, so far as we know, don't use the above kind of logic – and nor do we until we are taught to do so at school. There is nothing natural about such exact logic but it can be very useful—for instance in dealing with money, or in quantities of goods to barter.

Animals have to calculate too, but it is a different *kind* of calculation. It is more a question of *comparison* than equality. What are the risks of fighting compared to the risks of fleeing? Is it better to try and snatch that food out in the open or to remain here starving in the bush? No very exact estimates of the Odds are then usually possible (perhaps to no better than a factor of 2) but the consequences of poor judgement could be fatal.

Exact Logic and Comparative Logic are both valid. Neither is better than the other; they are useful in different *contexts*. If you pay the taxman roughly the right amount of money he won't be pleased. If you decide to have an operation only if the risks can be guaranteed to within a per-cent you will probably die unnecessarily. You need to choose your logic, Exact or Comparative, according to the circumstances.

We shall find the Common Sense Thinking, the kind that we share with other animals, is almost entirely based on Comparative logic, if only because the risks and advantages of action or inaction can usually be judged only to within rough margins. But there's nothing wrong with that. We are looking for a *sound* final decision – which can be reached by compounding several clues with rough and ready odds together. If the final Odds so reached are roughly 100 to 1 on who cares if they 'should' be exactly 93 or 107 to 1 on instead? What matters is their overall reliability, not their exact value.

From the context it is usually clear which kind of logic to use. The one serious exception, I claim, is the importation of exact logic into hypothesis-testing (decision-making) by Statisticians. Statisticians are mathematically trained types who often confuse exactitude for reliability. They strain for 4-figure precision when they should instead be looking for more, and more reliable clues with single-figure accuracy. Alas they have often been far too dogmatic about the matter, considering themselves to be experts on scientific methodology when they are not; on the contrary. For good reasons mathematicians actually have Induction and guesswork examined out of them at an early age; that makes them better mathematicians because the job of a mathematician is to establish arguments exactly. But it completely handicaps them when it comes to doing science where Induction and guesswork are absolutely fundamental. Indeed I shall argue in Chapter 11 that Statistics has become a menace to Science, and to sound thinking in general. This is a controversial claim but I believe, along with John Maynard Keynes, "Better to be roughly right than exactly wrong."

(2:4) PERSUASION.

At the back of one's mind while thinking must be the issue of convincing others. That is one reason why philosophers and statisticians favour deduction because a rock-solid deductive proof ought to convince everyone. Actually, if it is couched in abstruse mathematics, or logical jargon, it may do nothing of the kind because we cannot easily follow it. A few practical examples may prove far more convincing. Nevertheless, thinking that cannot explain itself clearly is at a severe disadvantage. Judgement, intuition, hunches may well be invaluable – but not persuasive to those who do not share them. That is why there is a premium, not to say a fetish about 'objective' evidence.

One such kind of supposedly objective evidence is numerical. It may be the result of measurement, or of counting, or from theory. Imagine you have a theory of the Solar System that predicts that the Sun will rise in the morning – and it does. That's not very convincing evidence that your theory is right; many other theories could predict likewise. But imagine your theory predicts that it will rise at 05 22 30 GMT and it actually rises at 05 22 45; that sounds significant. For a wrong theory to get such a closely corresponding answer would be more than something of a coincidence. But what about the 15 second discrepancy? Is it significant because it is so small – or significant because it is so large? This highlights the crucial nature of 'Error Analysis'. Your prediction will not have been perfect, neither will the recording of sunrise – one can easily imagine a slop of two minutes or more in either direction. The correspondence between theory and measurement is only useable as evidence once the errors (or uncertainties) in the two numbers are well understood. A scientist would be forced to write something like; " 05 22 45 plus or minus 2 minutes (say)" implying that the odds are about 2 to 1 of that measurement lying within the two limits.

Once numbers enter the discussion so may Mathematics, the language of numbers. Some sciences like Physics are dominated by measurements (and their errors) and so are densely mathematical; others such as Zoology much less so. Mathematics thus has an important role in some sciences, but it shouldn't be exaggerated. Some of the greatest scientists, even physicists like Michael Faraday, were mathematical dunces. It's possible to think without mathematics just as its sometimes possible indeed necessary to decide without measurements. Measurements and Mathematics are both powerful tools, but they are not substitutes for Common Sense Thinking. Alas they are too often used to bamboozle or intimidate opposition. Take the two British eclipse expeditions of 1919 which 'verified' The General Theory of Relativity and so shot Albert Einstein into the firmament as an intellectual superstar. They were both to remote parts of the tropics – one to Africa, one to South America. They were so important because, during total eclipse that year both Sun and Moon would lie in front of the Hyades star-cluster, the perfect and unique place to make the crucial measurements – an opportunity that wouldn't recur for centuries. Things didn't go well; clouds partially obscured the eclipse at both sites. One of the telescopes became warped in the heat. Photographic plates were taken but were very far from ideal and, worse still, disagreed with one another. And yet at the press conference held afterwards at The Royal Society in London one of the astronomers, the one least qualified to do so, turned the performance into a triumph for science and a triumph for Relativity. Professor Eddington's more knowledgeable but less famous colleagues remained notably taciturn. My point is that numerical evidence is not always as objective as it may seem. Eddington appears to have shamelessly highlighted those numbers which favoured Einstein's theory, ignored or down-played those which did not. I'd like to think one wouldn't get away with such chicanery today – but I'm not so sure. It's a human weakness to sometimes admire, even revere those we least understand. As a boy and a young scientist my heroes were all mathematical 'wizards' like Eddington, wizards precisely because they'd apparently achieved profound insights by manipulating hieroglyphics in ways I couldn't then comprehend. They seemed like prophets who had gone out into the Wilderness and there seen the face of Almighty God. But as my knowledge and experience grew so my reverence – or most of it – melted away. They'd botched their equations together using trial and error and Common Sense, just like the rest of us. Their equations' validity, or otherwise, depended entirely on subsequent measurements. Some got lucky, most did not.

We are not solitary animals. Coming to a decision is all very well but if one cannot persuade others it is often of little consequence. If our partners, our colleagues and our fellow citizens can't follow us they won't follow us. That is why thinking and persuasion are inevitably linked in a human world. One of the great virtues of being able to think clearly, and to be conscious of how exactly one is doing so, is its power of persuasion. Conversely, if one is going to be persuaded, one needs to understand just how good the arguments are. We are surrounded on all sides by persuaders whose arguments are sometimes sound and sometimes not, sometimes honest and sometimes not. Only a fool can be content with arguments he is in no position to understand. The modern world is precisely the world in which we have to largely deal with people we do not know (We even marry people we scarcely know who may have gone to elaborate lengths to disguise who they actually are, as much from themselves as from us). This is a very different world from the one in which we evolved. The fellow members of our hunter-gatherer band were too well known to us to fool us, or even to try. And anyway there were obvious consequences. Today that is simply not true. Today's world is a world of impersonal, usually transient transactions. Why should we buy, sell, believe, work with or for, employ, move to, spend on, be advised by, trust, listen to, agree with, respect, read, oppose, love, accede to, pay......why, why, why? It's all about arguments. With choice comes the need to decide, and with the need to decide comes the overwhelming need to understand sound thinking: our thinking and theirs. ...... It is no good relying on expertise or even eminence. Physicians, judges and yes, famous scientists are, as we shall see time and again, prone to making silly mistakes in their thinking – just like the rest of us. Sir Peter Medawar, mentioned earlier, was one of the cleverest and wittiest writers about science who ever lived. But then he trespassed into a field he evidently didn't understand – philosophy – and emerged advocating principles on how to think scientifically which are, in important respects, completely wrong. Thinking is hard – even for the smartest of us.

But why? Why has it been so hard for humans to understand thinking when animals have been doing it successfully, though not writing books about it, for ages? Here briefly are some of the confusions which have bedevilled the subject and which we will encounter in specific examples later on:

(a) The confusion between 'deciding' and 'concluding". We can decide on the balance of the evidence without waiting for conclusive proof.

(b) The confusion between 'Induction' and 'Deduction' both as words and as thinking processes. Thus Sherlock Holmes' superhuman 'powers of deduction' were most often Induction allied to acute observation and specialised knowledge (of cigar ash for example). Remember that deduction is thinking perfectly downwards from the general to the particular. Induction tries to think imperfectly upwards from the particular to the more general. It draws plausible inferences rather than indubitable conclusions.

(c) Confusion between Exact and Comparative logic, a fault particularly of Statisticians.

(d) The confusion between mathematical thinking and the scientific variety has caused and is still causing enormous puzzlement today. It is very difficult for those brought up with a strong mathematical training to realize that deduction and conclusiveness must necessarily play minor roles in science. Thus statisticians have time and again tried to import into science methods of deductive reasoning that are wholly illegitimate. Perhaps because of their very dogmatism this has given them an influence out of all proportion to their contribution. Having learned, used and then taught Statistics over 50 years I have to say I am less and less impressed. As Rutherford opined, if you need elaborate statistics to interpret your data then the data isn't good enough.

(e) The confusion between models and reality. The real world is complex with a mixture of phenomena interacting and confusing the onlooker. The way to understanding is often to simplify; to extract out of the messy reality a simpler model that embodies the main features while ignoring the rest. That's fine and indeed has often been the secret of much great science. But sometimes models are taken too seriously. Take mathematical models of the atmosphere. They're idealizations of the real thing, but very useful for weather forecasting. But they do have serious limitations simply because they are, by design, incomplete. They can tell you whether it's likely to rain tomorrow, but not whether it will rain most of the summer because they cannot presently handle the Jet Stream. Some commentators believe that the global financial crisis of 2007 was a direct consequence of taking economic models too seriously. (Sect 13:2).

(f) As we have seen Inductive thinking deals in Probability not certainty. But what is Probability? Alas there has been a three hundred year confusion between Subjective Probability and Objective Probability. The first is a measure of the conviction in your own mind about some proposition. The second is a number that can be calculated from the mechanics of the situation; what for instance is the Probability that I will draw

2 aces in a hand of 13 cards? Probability Theory is a very useful tool for clear thinking – but not if the two kinds of Probability get confused with one another – as they have been time and again over the past three centuries. Both kinds are fine – in their respective provinces. But using the same word for both still causes endless trouble. Imagine the same word being used to describe both passports and driving licences. Contemplate the resulting mayhem at Heathrow airport every day – then triple it. Whenever you encounter or use the word 'Probability' be crystal clear which kind, Subjective or Objective, is implied. 'Probability Wars' are bitter, lasting and unnecessary.

Having mentioned these special kinds of thinking, and the frequent confusions between them, it is worth emphasising that most scientists don't use them so much as Common Sense Thinking. So we won't need to use them much either. We will use quantitative evidence when we can but we won't need to massage it with elaborate mathematics or statistics. We will however emphasise the importance of Error Analysis. (Chapter 9)

(2:5) THEORIES AND HYPOTHESES.
The main task of CST is to weigh up the evidence for or against hypotheses, and if neccessary come to a decision as to whether to act on them.

What is an hypothesis? It has to be more than a vague idea before it can attract evidence either for or against. "The weather will be OK tomorrow" won't do. Either "It won't rain tomorrow" or "The temperature will reach 30 degrees" will.

Where do hypotheses come from? From everywhere. We observe the world around and explanations for what we see bubble continuously to the surface of the mind. Most of them will turn out to be wrong; our job is more to separate the grain from the chaff than to gather more of them. People, particularly silly people, will quickly find an explanation (hypothesis) for almost anything that happens to them. And we love to share our ideas – so we are bombarded by hypotheses from every side. The task of CST is to winnow them. And note that the more plausible an hypothesis sounds the more wary one should be of accepting it because it has probably undergone less critical scrutiny. Here's a plausible piece of nonsense that I was unintentionally responsible for. I had arrived with a party of two dozen other divers at a remote island in the Andaman Sea. I was standing outside my grass hut when a hefty coconut thudded onto the beach not far away. 'Dangerous place' I mused, and wandering about the village I noted that there were indeed very few elders but hoards of children. I mentioned my 'hypothesis' over supper and by next day all the European divers who had not been to a Third World country before were stepping gingerly in between the cocoanut palms, keeping a wary eye overhead.

Scientists tend to label hypotheses according to their respectability using a rough scale ascending something like this;

A tentative idea

A conjecture,

An hypothesis,

A working hypothesis,

A model,

The theory of….,

The Law of ……

Some hypotheses remain theories for a very long time because it is impossible or impractical to test them. Thus The Theory of Evolution remained as such for a century because Natural Evolution is so slow that it couldn't be observed at work until it was spotted in rapidly breeding bacteria.

The nature of scientific theories has changed over time. Originally they were meant to answer the question "Why?" – why did the gods order so and so. Then Galileo recognised that a far more fruitful question was 'How'…How does the cannon ball fly through the air – a question of practical moment and one susceptible to measurement and experiment. At first theories tended to be modelled on the familiar world – thus Newton's model of the Solar System was labelled 'clockwork'. But generally models became more abstract when it was found that Nature on very much larger and smaller scales didn't resemble the familiar one around us. Finally the pretence of familiarity was abandoned altogether and replaced by mathematical equations which seemed to work – but for no very obvious reasons. Thus James Clerk Maxwell constructed his wildly successful equations of Electromagnetism using a model of the aether with fluid properties such as vortices and idler wheels. Nobody believes his model any more, if they ever did, but his equations live on as the indispensable scaffolding of the modern world.

Some people feel that this process of abstraction has gone too far – Einstein for one – but history has not been kind to them. John Bell showed that the Quantum Theory in its weirdest and most abstract sense, is in fact correct.

What does one demand of a theory? Is it to be a conduit into the mind of God – or is it merely a jingle, a short and easily recalled mnemonic for describing an otherwise unwieldy heap of information? Theoreticians and commentators on science tend to profundify theories after the event, are prone to see great insights where none was originally intended. Thus Max Planck, the father of Quantum Theory, bodged up a mathematical expression to describe the experimentally determined spectrum of heat radiation. Afterwards it was realized that the same expression could be arrived at by assuming that energy had a particulate nature – a wholly revolutionary concept. Profundity makes for a much better story than bodging – but it often ain't, very often ain't necessarily so. Theoretical Physics is much more akin to plumbing than some of its most ardent fans would like us to believe. There ain't no such thing as magic – not even mathematical magic.

Likewise Statistics. Statistics can be a useful tool for describing otherwise unwieldy amounts of data. What for instance would we do without the concept of 'average'? The problems start when such statistics are used to draw inferences: that such and such an hypothesis is true or false. A statistic is, of its very nature, a simplification of the complete picture; a thousand numbers say are being represented by their single average value. Most of the original information is being thrown away. For instance how do the real numbers scatter about that average? Certainly a second statistic can be devised to describe that scatter – the 'Standard Deviation' (Chapter 9). Now a thousand numbers can be represented, after a fashion, by only two – the Average and the Standard Deviation. But that's still a gross simplification and thus a distortion of the truth. Not long ago statistical thinking was a necessary evil: there was no other way of handling extensive data. The discipline of Statistics grew enormously in influence: became a very weighty church of its own and like most churches it wielded power by means of terror. What better way to terrorize the peons than to thunder about 'The Kolmogorov-Smirnov Test' or 'The Central Limit Theorem'? I belonged to that terrorized generation who felt unable to think or do science without genuflecting to the God of Statistics. Unfortunately that church divided against itself, as churches tend to do, and eventually some of us peons were able to see that the high priests had feet of clay – and dirty toenails to boot. Today computers can handle large data-sets without gross simplifications. Deriving inferences from that data still requires careful thought but we don't have to genuflect to Statistics as we used to do in the past. Most of it is wrong anyway. (Chapter 11)

For reasons I would like to understand, but don't, we humans have a weakness for prophets, especially if they look and sound the part. The emaciated cheek, the lofty brow, the wild Einsteinian hair, the eccentric garb, the equations rattling like a witch doctor's bones, all bespeak profundity – almost certainly undeserved. Just as there is no such thing as magic, there are no wizards either, much to our childish regret. Is there not a part in all of us that would like to believe otherwise? Wouldn't it be marvellous if some wizard could stir his equations in the cauldron and out of the steam would emerge a Theory of Everything, all-wise, all-wonderful, the djinni of Eternal Truth? It could happen of course, it just might. But it's never happened so far – outside the pages of a child's story book.

So: Deduction won't do – unless its premises are certain, known and complete. Mathematics is only a tool. Statistics is a fallen god. Scientists are hopeless at explaining their own thinking. Wizards belong in fairy tales. There is no such thing as magic. We are left with good old Common Sense. Yes – but how does it work?

(2:6) HISTORY AGAIN

Much of this book will be about unlearning what we have been taught, or have picked up from our surroundings. That will mean being critical of many great minds from the past, not to say many in the present who may have taught us at school or university. How could they have been so deluded – if indeed deluded they were? It is certainly not because they were or are fools, any more foolish than you or I. No it is because we are all the playthings of history. The Greeks began seriously thinking about Thinking about 500 BC and in particular defined Logic and Mathematics. Then the Abrahamic religions came along and saw Man ascending to the right hand of God, with Thinking seen as the highest evocation of his immortal spirit. So for more than two thousand years Man puzzled over Thinking without having any idea that he was descended from apes. No wonder we got so many of our conceptions so wrong; it is a miracle that we got anything much important right at all. After Darwin's 'Origin of Species' (1859) we should, in an ideal world, have torn up the entire book of Learning and started again from scratch. Of course we didn't do that, we couldn't do that. We are thus the inheritors of a ragbag of inconsistent culture which has come down to us from Erasmus, from China, from St. Thomas Aquinas, the Koran, from Ptolemy, from Aristotle, from the glory-hole of history and from goodness knows where else. It was always going to take generations to sweep the cobwebs away. And if we scientists have taken an honourable part in that it is because Nature won't allow us to fool ourselves – or not for long. When we use daft ideas we quickly come into bruising contact with her, and learn to retreat. That is one reason why scientists have generally been sceptical of religion, indeed of philosophy in general. So if we seem to be tearing up the past now, it is not because we are smarter than our predecessors but because, every day, we are forced to learn a little more.

We shall devote a whole chapter (14) to the fascinating History of Thinking later on. It deserves a whole library.

(2:7) OTHER THINKERS AND THEIR WORK

When I originally drafted this book every paragraph, almost every sentence contained references to previous authors' work. Non-academic readers however found this tedious, indeed off-putting. So I've removed most of the references in the text. That may give the impression that most of my thinking is original to myself. It isn't – it simply isn't. I learned most of it by listening, by asking, by arguing, by making mistakes and above all by reading. So, at the end of the book, I have put back in a list of key references for each chapter which the reader can consult if desired, but they are not essential to CST. And of course there's the website full of more extensive material. For instance you might now like to consult there 'Our Jack' the story of a young jackdaw who came to live with us for a few weeks when I was a boy. He taught me the quite invaluable lesson that animals can think too. It's a foolish conceit to imagine otherwise.

SUMMARY

In this very superficial chapter I have listed various kinds of thinking including Deduction, Induction, Exact logic and Comparative logic, Plausible Inference (Common Sense) and the Mathematical and Statistical modes of thought. We've also briefly discussed evidence, particularly of the objective and/or quantitative kind and underlined the need for Error Analysis. We've mentioned some of the widespread confusions which have bedevilled thinking generally and finally I've claimed, without evidence, that scientists don't use the more esoteric approaches enumerated above so much as Plausible Inference or good old Common Sense Thinking (CST). The next chapter contains the evidence for this important, and I believe, comforting claim.

44

45

THINKING FOR OURSELVES (Disney)
CHAPTER THREE (14/9/18)
HOW SCIENTISTS THINK (12.7 kw)

'Science has several rewards, but the greatest is that it is the most interesting, difficult, pitiless, exciting and beautiful pursuit that we have yet found. Science is our century's art."

(Horace Freeland Judson: "The Search for Solutions"; 1987; Johns Hopkins University Press, abridged version, p12)

(3:1) INTRODUCTION

One might suppose we scientists are extensively taught how to think. We aren't. In 50 years as a professional I have never received or given a single lesson on the subject and my colleagues report likewise. Quite why this is so would make a fascinating inquiry in itself (Chapter 14) but the fact is that we are supposed either to know how to think already, or to learn on the job. In my experience both expectations are wildly optimistic and help to explain why scientists are reluctant to talk about thinking, or inclined to become incoherent when they do. Our witty zoologist Peter Medawar said: "You must admit that this adds up to an extraordinary state of affairs. Science, broadly considered, is incomparably the most successful enterprise human beings have engaged upon; yet the methodology that has presumably made it so, when propounded by learned laymen, is not attended to by scientists, and when propounded by scientists is a misrepresentation of what they do. Only a minority of scientists have received instruction in scientific methodology, and those that have done so seem no better off." {1}

Part of the problem of course is the sheer diversity of scientific research. What modes of thought could a naturalist studying gorillas in the cloud forest share with a particle physicist measuring bubble tracks from an accelerator, or with a theoretical astrophysicist manipulating equations, or with a microbiologist looking for bacteria in a culture dish? And if we do seek for guidance we generally run into books written either by philosophers – which seem irrelevant, or mathematical treatises written by statisticians who disagree violently with one another. So what we learn to do when we are young is to think like our elders and betters and when we are old to think like our peers. No it doesn't sound healthy does it. We rely on Nature to set us straight when the thinking goes wildly astray. That's all very well, but it's often a sluggish process because scientists are just as reluctant to change their minds as other people so that sometimes the subject can only advance 'funeral by funeral'.

The general result is most confusing, as much among the experts as laymen. In such an atmosphere myths arise which tend to exaggerate the differences between scientific thinking and the everyday variety. For instance it is 'more deductive', 'more objective', 'more logical', 'based on experiment', 'statistically significant' etcetera and it is mainly advanced by scientific geniuses like Newton and Einstein whose methods of thought are utterly incomprehensible to the rest of us.

Before we could use science as an exemplar for thinking in general we need to find out whether all or any of the above myths are true. We aim to do so in this chapter by looking at 19 short stories taken from the history of scientific discovery. I hope you will find each episode absorbing in itself, though it is not the particular discovery, nor the individual protagonist, nor even its significance to mankind, that will interest us, but the underlying mental tactics employed. Are those tactics familiar to us, or extraordinary? Were they deductive, or inferential? Were they mathematical or indeed quantitative at all? Was the discoverer a lone genius – or did he or she have rivals or precursors? Could the discovery have been made long before, or were the preconditions for it of recent origin? At the end of the chapter we look back to see if history has supplied answers to these questions. We shall conclude, with Einstein, that "The whole of science is nothing more than a refinement of everyday thinking". If that is so then we shall be ready to start out on our quest to find out how such thinking actually works.

At the end of each episode you might like to pause and ask yourself two questions:
"Have I thought along the same lines, but in a different context, myself?" and "If I have not, then could I usefully add this tactic to my box of mental tools for future use?"

There is a list of references on the history of discovery in the References to Chapter 3.

So let us begin, heading each episode with the mental tactic employed.

(3:2) RECOGNISING THERE IS A PROBLEM

"Why is the sky dark at night?" To most of us it would never occur to ask this question. We might suppose, "It just is – so what's the problem?" or if we thought a little harder we might come up with "Darkness is an absence of light, surely it is the presence of something, not its absence, that we have to explain". Johannes Kepler thought otherwise, and so founded the subject of scientific cosmology – the study of the heavens in their entirety. He was stimulated to think about the matter after reading "The Starry Messenger" a pamphlet put out by Galileo in 1610 describing the sensational discoveries he had just made with his early astronomical telescope. Galileo reported that the instrument "…set distinctly before the eyes other stars in myriads which have never been seen before……." And of the Milky Way he said "…….it.is nothing else but a mass of innumerable stars plastered together in clusters. Upon whatever part of it you direct the telescope straight way a vast crowd of stars presents itself to view……….."

Struck by this last remark Kepler noted that nevertheless the sky remained dark – very dark. If it were true that wherever you cast an eye or telescope in the heavens it were to fall upon a star, then the whole heavens should burn with the furnace heat of a typical stellar-surface – such as the Sun's. That it does not, implied to Kepler that the Milky Way, and indeed the entire firmament of stars must somewhere, somehow come to an end. Whatever else it might be, the Universe could not be infinite.

This was a profound thought – and not one that man has found it easy to explain. Or rather it was one we found all too easy to explain – using unsound arguments.

First it was supposed that there was absorbing smoke in interstellar space that obscures the distant stars in an infinite universe. The snag with that explanation is that any conceivable smoke particles would evaporate after absorbing so much luminous energy. Next theoretical cosmologists came up with two elaborate calculations to prove that either a spatially finite cosmos, or later an expanding one, would explain Kepler's so called "Dark Sky Paradox". Both very complex calculations turned out eventually (1970) to be wrong. Today we suppose the way out is connected with "The Big Bang" i.e. the idea that the Universe began in a titanic explosion thirteen billion years ago. For the sky to be bright the universe would have to be vastly older than that, old enough for the light of truly distant stars to reach us and fill up our local sky.

The important point here is not whether there is a solution, but to recognise that there is a problem. Kepler did, whereas lesser men either did not, or dismissed it with some facile explanation. When the ancients saw the mast of a distant vessel sink below the horizon, or picked up the fossil of an extinct creature, only a very rare few were prepared to confront the issues being raised.

It is fascinating to wonder if there are other problems as fundamental as Kepler's still staring us in the face but being ignored today. The Origin of Inertia may be one such. How does a massive body know when it is being accelerated, and thus resist the forces acting on it? Think of Foucault's pendulum, a massive weight swaying back and forth on a long wire suspended from a ceiling far above. Foucault hung his giant pendulum from the dome of the Pantheon in Paris in 1851. Too immense to be affected much by air resistance, such a pendulum will beat back and forth for days. If you suspend it above a large sand-tray the pendulum will etch a mark on the sand each time it swings by. As the Earth turns, with the sand-tray resting upon it, so the pendulum cleaves to a fixed constant plane with regard to the distant stars. Over the course of 23 hours and 56 minutes, precisely the time it takes the Earth to turn once in relation to those distant stars, the marks in the sand tray make one exact revolution.[Because of the curvature of the Earth the period is different at different latitudes being 23hours and 56 minutes only at the poles]. The cathedral has turned, the continent upon which it is built has turned, the massive Earth has turned but the pendulum has ignored them all, preferring to mark its beat by the distant Universe. Why, or rather how? It is apparently an effect without a cause. What forces unknown connect the pendulum, and indeed all accelerated masses to the distant constellations? Most scientists do not appear to care, do not see it as more than a "philosophical problem", but a few of us do. Einstein certainly did and was bitterly disappointed when his Theory of General Relativity, designed specifically to solve The Problem of Inertia, failed to do so.

(3:2) RECOGNISING A COINCIDENCE

Over the history of mankind no individual has saved as many lives as Dr. John Snow. In 1848 he was called in to a cholera epidemic raging at Soho in the heart of London. Within 10 days 500 people died in a very localised area around Golden Square and Broad Street. In some households everybody died, whereas in the ones to either side no one was affected. For instance only 5 out of the 535 inmates of the Poland Street Workhouse were affected whereas the neighbouring buildings were devastated. From the odd pattern of the disease Snow began to suspect the water from the Broad Street pump (there was no piped water in those days). The workhouse for instance had its own well while the inhabitants in unaffected houses drew their water from elsewhere. His suspicions were definitely confirmed by the sporadic cases that occurred in distant parts of London, all of which could be traced to drinking some water from the Broad Street Pump. After Snow had the pump-handle removed the epidemic died out.

Snow's observations led soon afterwards to a huge public-works program, to the laying of sewers and the supplying of truly clean water across the civilized world. Not only cholera was banished but a host of other agues and distempers which dragged down those living in urban stews everywhere. Mankind's enormous population growth and increased longevity owes more to the discovery of John Snow's coincidence than to anything else.

(3:4) SPOTTING A CONNECTION BETWEEN TWO PREVIOUSLY UNRELATED PHENOMENA.

Hans Christian Oersted (1820) was employed by the Danish government to study meteorological records. He was given the task of looking through ships' logbooks to find out about storms at sea. Time and again seamen reported that during electrical storms their compasses went haywire. This was inexplicable because at the time no one suspected that electricity and magnetism were in any way related. But Oersted could hardly avoid that inference, and so decided to do a simple experiment. Luckily for him the battery had just been invented. Passing a current through a wire he was able to map its indubitable effect upon any compass placed nearby.

News of this discovery spread like wildfire through the laboratories and academies of Europe. In particular Ampere in Paris and Faraday in London were able to tease out the precise relationship between two hitherto unrelated phenomena. Electromagnetism was born, and with a small addition from James Clark Maxwell 30 years later, so was the bulk of the modern world. After Oersted's discovery, Faraday, Marconi and Einstein were inevitable. The radio demagogues, television, code-breaking, computers, Relativity and mobile phones were all waiting in the wings of history.

Associating two previously unrelated observations has been one of the most widely productive tactics across all of science. Perhaps it explains why so many great scientists have exhibited omnivorous curiosity. If you know only 10 things there are only (10 times 9)/2 = 45 possible connections between them [the 2 enters because otherwise you will count the same connection twice]. But if you know a hundred there are (100 times 99)/2 or nearly 5000 possible connections to explore. [This insight should help to make all that homework seem worthwhile because every extra thing you learn is made valuable in proportion *to what you already know!*]

(3:5) NOTICING A NUMERICAL COINCIDENCE

From the sight of a falling apple – in conjunction with the Moon seen in the sky above the apple-tree – the young Isaac Newton was, so he later said, led in 1665 to suspect that the force of the Earth's gravity extended so far as the Moon. But how? Years later, with an accurate distance to the Moon before him, Newton was able to calculate its acceleration toward the Earth, the acceleration that keeps it in its orbit instead of straying off to infinity. He found the acceleration to be 3,600 times less than the falling apple's. Why 3,600 he wondered. Newton noticed that 3,600 is the square of 60 precisely [i.e. 60 times 60] and that 60 is the ratio of the Moon's distance from the centre of the Earth, to the apple's. He had his famous 'Law of Universal Gravitation' . He guessed that gravitation, an attractive force between any two bodies in the Universe, must fall away as the inverse square of the distance between them – at twice the distance it is four times as weak When he tried it out on the planets he found he could explain exactly all three of Kepler's Laws of Planetary Motion, to say nothing of the comets, the tides and the complex trajectory of the Moon.[2]

---

[2] Some historians consider that Newton was a liar and that he pinched the idea of the inverse square law from his rival Robert Hooke much later. It's certainly possible because Newton was not notably scrupulous and Hooke was his declared enemy. But he certainly *could* have come by the idea under the apple tree as described. Newton's

Another singular numerical coincidence astonished the Scottish physicist James Clerk Maxwell when he was working on electromagnetism at Kings College London in 1862. He was trying to cast into differential equations the laboratory work of men like Oersted and Faraday. To his consternation he found that his equations allowed for electric charge to disappear, something never observed in Nature. Realizing that some effect must be missing from his equations he guessed it must be the "displacement current", the transient pulse of current observed in open circuits when they are first switched on. The addition of this new current term sufficed to conserve electric charge in his equations. Satisfied, he went on to explore the other consequences of his new equations.

In perhaps the greatest moment of epiphany in the history of science Maxwell discovered that his new equations predicted that electromagnetic waves would propagate through empty space. Furthermore they predicted the speed of these waves in terms of two obscure electric and magnetic constants of nature, previously measured in the laboratory. Filling in their numbers Maxwell found that the wave speed would be 300,000 kilometres a second. But that was precisely the speed of light, long known to science from observing the moons of Jupiter. Maxwell must have been flabbergasted. Light, Electricity and Magnetism must be one and the same phenomenon, otherwise the numerical coincidence in their velocities would be inexplicable. He said: " The velocity of transverse undulations in our hypothetical medium, calculated from the (previous) electromagnetic experiments of Kohlrausch and Weber, agree so exactly with the velocity of light calculated from the optical experiments of Fizeau

---

insight here principally lay in assuming gravitation was universal, an attraction between *all* masses in the universe.

that we can scarcely avoid the inference that light consists in the transverse undulations of the same medium, which is the cause of electric and magnetic phenomena". Intriguingly other such numerical coincidences have been found in physics, which are still totally unexplained. For instance the ratio of the electric to the gravitational forces between elementary particles appears to be equal to the square root of the number of protons in the observable Universe.

(3:6) RECOGNISING A PATTERN

The apparent fit between the opposite shore-lines of the Atlantic Ocean had long intrigued map-makers. Had the two sides once been parts of a super-continent which had split, leaving the halves to drift apart? The first man to seriously entertain this hypothesis of 'Continental Drift' was the German meteorologist Alfred Wegener. He reasoned (1912) that if there was anything to it then the detailed geology on opposing sides of the ocean ought to correspond. When he looked into the geological libraries he indeed found an uncannily good correspondence. Rocks thousands of miles apart matched so well that it seemed there could be no doubt about the matter.

Unfortunately Wegener, a man of many parts, died a heroic death out on the Arctic ice in 1930. This enabled the theoretical geophysicists, led by Harold Jeffreys at Cambridge, of whom more later, to dismiss the awkward patterns and to prove, with the aid of a string of equations, that the idea of solid continents drifting about was preposterous.

It took another 30 years to discover even more striking patterns in geomagnetic measurements of the ocean floor. The bed of the sea was indubitably spreading outwards on both sides from the Mid-Atlantic Ridge at several centimetres a year. Rock was far more plastic than Jeffreys and the theoreticians had been able to imagine. Wegener's pattern-recognition argument, albeit in a modified form, was far too strong to be gainsaid by any number of partial differential equations. Some solids, such as the andesite rock underlying the continents, are rigid in the short term but can flow plastically over geological time. Likewise the glass in ancient Egyptian tombs had visibly 'dripped' by the time it was unearthed.

(3:7) FOLLOWING UP A CHANCE OBSERVATION

– is one of the classic high-roads to discovery. Thus in 1899 Henri Becquerel was experimenting with a compound called potassium uranyl sulphate which fluoresces, that is to say continues to glow after it has been exposed to sunlight. It is the kind of substance that enables you to see the hands of your watch in the dark. Unfortunately, in the middle of his experiments, the sky clouded over and Becquerel put away in a light-tight drawer the packet of unexposed photographic plates that he was using to detect the fluorescence. When, several days later, he resumed his experiments he was disconcerted to find out that the plates were all fogged over – as if they had been exposed to light. Instead of throwing them away as defective – which is what most of us would probably have done, Becquerel tried to find out how the plates could have become blackened in an apparently light- tight package within a light-tight drawer.

He traced the source of the problem to a dusting of the aforesaid potassium uranyl sulphate compound left by accident on the surface of the package in the course of his earlier experiment. Some strange radiation, capable of penetrating light-tight cardboard, was emerging from his fluorescent compound. Becquerel had discovered "radioactivity" whose source was eventually traced to the single uranium atom in every molecule of the compound.

Becquerel's accidental, and apparently trivial discovery, was to turn physics, chemistry and geology entirely on their heads. Atoms must contain energy, as it happens vast amounts of energy, which they can occasionally emit. But if they can emit energy they must be changing inside, which implies they are not entirely indestructible, as chemists had supposed. And if the rocks of the Earth contain some radioactive atoms, then the Earth's interior may actually be heating up instead of cooling down as everyone had supposed. But if it was heating up then the age of the Earth, calculated from its supposed rate of cooling to be in tens of millions of years only, must be entirely wrong. The vast aeons of time required by biologists to accommodate evolution, and by geologists to explain the gradual erosion of mountains and canyons, was now a possibility. Becquerel's fortuitous discovery catapulted us into the previously unimaginable abyss of Deep Time.

Serendipitous, fortuitous discoveries are the sign of a science in birth – or about to be reborn: X-rays, cosmic radio-emissions, penicillin, Big-Bang radiation, blood-sugar, galvanism, small-pox immunity, bacterial disease agents, anaphylaxis, helium………….the list of chance scientific discoveries goes on and on and on…………

(3:8) DOING A CAREFUL EXPERIMENT
One of the most fertile ideas in science is 'The Conservation of Energy' – the recognition that while energy can switch back and forth from one kind to another, it can neither be created nor destroyed. The pendulum for instance works by continually switching its energy back and forth between

the gravitational form – height, and the kinetic form - speed . Unfortunately this conservation law appeared to be only partially and sporadically true. The pendulum for instance gradually slows down until it stops altogether. And if you throw a bucket of water over a cliff into a pool far below, the gravitational energy of the water appears to vanish completely. James Joule a brewer from Manchester (Joule's Ales still exist) wondered whether the missing energy had turned into heat. To find out he constructed ever more precise thermometers. He took two of those on his honeymoon to Switzerland in 1848. He and his wife Amelia visited the highest waterfalls in the Alps. The one of them at the top would measure the temperature of the water before it plunged over the edge, while the one at the bottom would simultaneously measure the temperature in the pool below. They found no systematic differences.

Undeterred Joule constructed ever more sensitive thermometers and used them in carefully controlled experiments in his laboratory at home. Eventually he succeeded in getting definite and reproducible results. Water, when violently stirred for long enough, heated up by a minute but measurable amount, and when it was stirred for twice as long its' temperature rose by twice as much. Joule's experiment proved what had long been suspected, that mechanical energy can turn into heat and that it does so in a systematic way: in modern parlance 4.18 joules of mechanical energy turns into 1 calorie of heat. Energy is indeed conserved – so long as you include the heat component.

Nowadays children repeat Joules' beautiful experiment with modern thermometers at school, and they memorize 'the mechanical equivalent of heat' as '4.18 joules per calorie'. Unfortunately this last litany disguises a wonderful truth – for it hopelessly mixes two schemes of units together when they truly differ by a factor of ten million. Think of Joule and his new wife out in Switzerland. Why did they fail?

You can put the question in another way, as I once put it to four distinguished professors of physics round a dining-table: "If you cool a bucket of water by one degree centigrade, and use the heat-energy so extracted at one hundred per cent efficiency to raise the bucket, how high would it go? I asked them all to make a rough guess. Their answers varied between "a few millimetres" to "not more than a centimetre or two".

The true answer is an astonishing 473 meters. Or put in other terms the heat energy implied in "one degree centigrade" would suffice to accelerate that same bucket to no less than 172 mph. Joule and his wife failed because a vast amount of mechanical energy amounts to very little in the way of heat. Reverse this last statement and you have the entire secret of the Industrial Revolution: a very little amount of heat – say one calorie, is equivalent to 42 million ergs! In other words the heat from comparatively small amounts of coal (or oil) can pump out mines, drive locomotives, propel battleships and power factories. But of course the inventors had discovered all this long before the scientists caught up.

Experiments enjoy a hallowed place in the legend of science, although they are much less common than you might suppose. In many sciences they are either unethical, impossible or uneconomical. Astronomers obviously can't carry out experiments at all, while it has been said that to do an interesting experiment in fundamental physics today costs at least half a billion dollars. Nor were many of the iconic experiments beloved by historians of science actually carried out. Galileo for instance seems to have executed most of his famous experiments only "in his head";

there is no evidence that he actually dropped canon-balls and musket-balls side by side from the Leaning Tower of Pisa. Nevertheless, when they can be done, experiments are the way for Nature to speak.

(3:9) CONSTRUCTING MORE ACUTE INSTRUMENTS

Long ago in classical Greece Aristarchus first suggested that the Earth orbits the Sun, and not vice-versa. He predicted that from our orbital viewpoint we ought to see the nearer stars oscillate back and forth once a year against the background of the distant stars. But astronomers, despite the most minute observation, were unable to detect any such 'parallactic motion' and Aristarchus's Heliocentric hypothesis was dismissed as fanciful.

It took more than 2000 years before Fraunhoffer was able to construct a telescope acute enough for Bessel in 1848 to detect the parallax of the star 61 Cygni. Poor Aristarchus had complained that the stars might turn out to be too far away for their parallaxes to be detectable with the technology of his day (the Mark 1 eyeball). And so it proved. In 1989 the European Space Agency launched Hipparcos a satellite able to measure parallaxes for over a hundred thousand stars in the course of a decade.

Much of science is based on improving the technologies of observation and measurement. Anthony van Leeuwenhoek (1670) secretly constructed over 500 ever more powerful microscopes to enable him to see deeper and deeper into the fascinating world of tiny animalcules and cells. Very often the smaller the target the larger the instrument has to be. To see quarks and bosons CERN (Centre for European Nuclear Research) has built an heroic particle accelerator 27 kilometres in circumference under the Jura mountains.

(3:10) CONCIOUSLY LOOKING UNDER DIFFERENT STONES

Theoretical astrophysicists convinced NASA that there was no point in looking for cosmic X-rays because there would be none – except perhaps from the Moon. So Riccardo Giacconi in 1962 launched a small probe on a sounding rocket to look for X-rays from the Moon. The probe, which was designed to 'accidently' look about in all directions, failed to pick up the Moon but it did see a powerful X-ray source in the constellation of Scorpio, as well as a general X-ray glow from all over the sky. [This is a recurrent warning that observation and experiment must always take precedence over theory in science. Man's imagination isn't up to nature's creativity.]

That was the start of a continuing crusade to open up the entire electromagnetic spectrum to astronomy from space, to explore the universe through a multiplicity of windows, stretching from gamma-rays, through X-rays, through the ultraviolet, infra red and microwave bands. Each window has yielded a cornucopia of cosmic surprises which between them have entirely altered our perspective on the cosmos. For instance we now know that most of the detectable matter is invisible. It is extremely hot and radiates most of its energy as X-rays, not as light. In a like manner every experimental scientist is on the lookout for ways to open up new sectors of 'parameter-space', or in other words to look under new stones.{2}

(3:11) ARGUING BY ANALOGY

In 1798 the Revd. Thomas Malthus published his highly influential "Essay on the Principle of Population". In it he argued that an unrestrained population would always multiply faster than its food resources, leading inevitably to starvation, to misery, and to a "struggle for existence". In his own words this struggle entailed "…..every cause, whether arising from vice or misery, which in any degree contributes to shorten the natural duration of human life. Under this head, therefore, may be enumerated all unwholesome occupations, severe labour and exposure to the seasons, extreme poverty, bad nursing of children, great towns, excesses of all kinds, the whole train of common diseases and epidemics, wars, plagues and famine". The two naturalists who later (1858), separately but simultaneously, first explained the "origin of species" by "evolution and natural selection", that is to say Alfred Russell Wallace and Charles Darwin, had both read and were heavily influenced by Malthus. The analogy between the "struggle for existence" among men, and a like struggle between plants and animals led them directly to their seminal idea. To quote Darwin: "Hence, as more individuals are produced than can possibly survive, there must in every case be a struggle for existence, either one individual with another of the same species, or with the individuals of distinct species, or with the physical conditions of life. It is the doctrine of Malthus applied with manifold force to the whole animal and vegetable kingdom…."

Wallace said: "The life of wild animals is a struggle for existence. The full exertion of all their faculties and all their energies is required to preserve their own existence and provide for that of their infant offspring……The numbers that die annually must be immense, and as the individual existence of each animal depends upon itself, those that die must be the weakest – the very young, the aged, and the diseased, - while those that prolong their existence can only be the most perfect in health and vigour – those that are able to obtain food regularly, and avoid their numerous enemies. It is, as we commenced by remarking 'a struggle for existence', in which the weakest and least perfectly organised must always succumb".

Arguing thus by analogy has always been a fertile, if not always a sound way to think. Numerous examples can be quoted from all walks of science but here is one with which I am personally acquainted.

Radar scientists in World War II noticed that the radar 'returns' from an aircraft approaching a ship often went haywire. This was caused by the direct return being interfered with by a second return from the aircraft bouncing off the sea-surface into the antenna. The sea was acting like a giant mirror for radio waves. Immediately after the war radio scientists in Sydney decided to use this phenomenon to track down the previously mysterious sources of radio radiation coming from the cosmos. Using an antenna on a cliff-top looking out over the Pacific Joe Pawsey, Ruby Scott-Paine and their colleagues used the interference between the two incoming signals, one direct and one reflected off the ocean, to pinpoint cosmic radio-sources far more accurately on the sky. This allowed them to identify the radio signals for the first time with objects already known to optical astronomy: the radio source labelled Taurus A with the exploding star called 'the Crab Nebula', Centaurus A with a strange galaxy millions of light-years away. Arguing by analogy had advanced astronomy by a giant leap.

Much earlier in 1727 James Bradley, another astronomer, was able to explain his own discovery of stellar aberration – the apparent annual oscillations of stars in the sky – by means of another analogy which came to him in the course of a picnic on the river Thames. Noticing that the burgee at the mast-head of his sailing vessel was pointing in a different direction from the flags ashore (because of the vessel's motion) he realised in a flash that stars would *appear* to annually shift their positions slightly on the sky (to 'aberrate') due to the motion of the Earth around the sun. Bradley's aberration was, by the way, the first *direct* proof that the Earth does indeed orbit the Sun.

(3:12) FOLLOWING A MISTAKEN PATHWAY

None of the true pioneers of radio – Maxwell, Hertz and Lodge for instance – could see any practical use for it. After all, at a time when transoceanic telegraph cables were already in service (1894), radio could never hope to defeat the curvature of the Earth. But the young Guiglielmo Marconi didn't know that – in fact he didn't know much at all for he was a bit of a dunce. However, when he read in a newspaper of the death of Heinrich Hertz – the first man to have conclusively demonstrated the existence of radio-waves, the 20 year old Marconi was seized with the idea of building a transmitter-receiver system for himself. He succeeded and he and his brother would wander about their father's large estate in Italy sending dots and dashes to one another. By trial and error Marconi discovered two vital things: that large grounded aerials were necessary for long- range communication; and that waves could apparently pass through the large hill in the middle of the estate.

When Marconi senior tried to put a stop to these un-gentlemanly activities Guiglielmo escaped with his very rich mother to her native Britain. There he was exceedingly fortunate in being introduced to Sir William Preece – chief engineer of the Post-Office. Preece was still simmering with rage from a public humiliation he had suffered at the hands of Oliver Lodge, a brilliant radio pioneer who was trying to set up a public radio company. Preece saw the young Marconi as his way of exacting revenge. Were Marconi to succeed in sending the first long-distance messages the prize would be snatched from under the nose of his enemy Lodge.

And so it came to pass. With all the resources of the Post Office behind him, including huge aerials on the cliffs at Cardiff (1897), Marconi succeeded in sending messages across the Bristol Channel – a distance of about ten miles. In 1901, using aerials suspended from giant kites, he picked up signals in Newfoundland from his transmitter in Cornwall.

Apart from his huge aerials Marconi's apparatus was exceedingly crude and 'unscientific'. He had very little imagination – for instance he couldn't see any future for broadcasting, thinking of radio merely as a way of undercutting the monopolistic cable companies – who did everything they could to prevent his success.

What Marconi did have, apart from a rich mother, dogged persistence, luck, and a gift for publicity, was a useful degree of ignorance. Unlike his far better educated rivals such as Lodge, he was in no position to realise that sending trans-Atlantic radio must be impossible because radio waves, which are merely longer wavelength light-waves, cannot travel round corners or indeed through the solid Earth. So in ignorance he went ahead – and succeeded. It wasn't until 1924 that Edward Appleton demonstrated that the ionosphere far above us can reflect long-wavelength radio waves around the globe. For Marconi it was a case of ignorance proving bliss. He was, in the annals of science, far from alone. It has been said that as a scientist's learning increases so he or she learns more and more reasons for why any unconventional idea will not work.

(3:13) RECOGNISING UNEXPECTED SIMILARITIES

Nothing so holds up progress, not even ignorance, as belief in a plausible but mistaken "Truth". Thus chemistry was held entangled in the silken web of Aristotelian thought not for centuries but millennia. In 350 BC Aristotle had argued that all substances are made up in different proportions out of his Four Elements: earth, water, fire and air.

Escape from this web was only made possible by improving the precision of scientific instruments such as the chemical balance (Joseph Black, Glasgow, 1750). When burned in air some substances such as potassium actually gained rather than lost weight, an affect which the Aristotelian sophists cold hardly explain – though they tried.

With Aristotle's hold finally broken progress was rapid. Between 1790 and 1870 no less than 60 true elements were identified, as well as the exact ratios (valences) in which they combined to form compounds. This proliferation was however rather too much of a good thing. It threatened chaos – or rather such a superabundance of unrelated facts that no chemist could hope to memorise but a tiny fraction of them.

So patterns were sought for, and in tantalising fragments here and there found. If you arranged the elements by their atomic weight compared to the lightest of them (Hydrogen said to have a weight of "one" by definition), then John Newlands (1864) noticed that the eighth in his table (Fluorine) had properties similar to the first (Hydrogen) and the fifteenth (Chlorine) while the twelfth (silicon) resembled the fifth (Carbon). Newlands was tempted into suggesting that the elements were organized in 'octaves' like the notes in music. Unfortunately there were major discords in Newland's harmonious scheme, for instance Sulphur in no way resembled Iron.

Dmitri Mendeleev, born in Siberia the youngest of 17 children, seems to have been a cross between a dynamo and an encyclopaedia. In 1869 he discerned a periodicity of his own among the elements. His "Periodic Table" ranged elements with similar properties together in lines of increasing Atomic Numbers (in brackets), for instance: Lithium (19) with Sodium (23), Potassium (39), Rubidium (37), and Caesium (133).

Mendeleev read everything, summarising the latest properties of each element on the cards that he pinned to his wall. Fond of playing solitaire he shuffled the cards about, seeking for the order that had evaded his fellow chemists.

Eventually it nearly all made sense, so much sense that he was prepared to bet that those elements which didn't quite fit his scheme, such as Chlorine, had been wrongly measured. Subsequent experiments, often made by himself, showed that he was right. And where there were gaps in his pattern he daringly suggested that new elements would in time be found. He even named these hypothetical elements – eka silicon, eka aluminium and predicted their properties. Sure enough eka aluminium was found and renamed gallium, and eka-silicon named germanium.

Mendeleev's Periodic Table is the classical example of unexpected order discovered in science. Although no one was to understand the physical basis of it for another 60 years, the Table proved to be an indispensable map for generations of chemists. A century later physicists looked for and found similar periodic patterns among the strange fundamental particles that emerge from the atomic nucleus.

As we now understand it chemistry deals with interactions between the outer electronic shell of one atom and the outer shells of its fellows. In his search for order Mendeleev had discovered groups of atoms with equal numbers of electrons in their outermost shells and therefore equivalent chemical properties.

Note that you don't have to explain something in Nature in order to realize its significance. The failure to understand this principle on the part of theoreticians has often been the cause of long delays in science (viz Continental Drift, see {3/11}.

(3:14) SYSTEMATICALLY ELIMINATING ALTERNATIVES

Malaria, or 'million murdering death' was the scourge of mankind – and not only in the tropics. The lower lying marshy areas of Europe were rendered more or less uninhabitable by the deadly disease. For instance East Anglia contains no large cities to this day because, before its wetlands were drained, 'ague' as the pandemic was known in Britain, ravaged the population.

What caused it and why was it prevalent in some areas but absent in others? Its very name "malaria" or "bad air" enshrined the idea that poisonous vapours stole from foetid marshes at night to overwhelm its victims. In afflicted countries such as Italy towns were built on hilltops to escape the murdering miasma.

The Frenchman Charles Laveran established in 1889 that a parasite was involved by examining the blood of victims under the microscope. But where did it come from – and how was it transmitted from man to man?

Some suspicion fell on the mosquito for mosquitoes were always present where malaria thrived. For instance Manson in Britain suspected that victims were infected by drinking water contaminated by mosquito-eggs. He inspired Ronald Ross to go out and look for evidence in malaria-ridden India.

In 1898, after years of frustration and failure Ross managed to show that birds contracted malaria after being bitten by certain brown mosquitoes and that the malaria parasite passed part of it life-cycle in the mosquito's gut before swimming up to its salivary glands ready to pass into the next avian victims bloodstream. Unfortunately Ross was unable to show that the same unlikely process affected man. In fact, as we now know, bird malaria is caused by an entirely different parasite.

In the very same month another man in another continent, Giovanni Grassi in Italy, set out to confirm his own suspicions that mosquitoes, or at least a specific variety of mosquito, was involved. Grassi was a skilled zoologist with systematic habits of thought. He knew of areas of Italy where mosquitoes swarmed yet malaria was unknown. But he also knew that malaria was never found where mosquitoes were entirely absent . Of the forty or so varieties of mosquito in Italy perhaps only one or two were the deadly agents?

Grassi spent the summer of 1898 tramping round the mosquito-ridden areas of his country collecting the insects and talking to locals. One by one the different sub-species were eliminated until his suspicion centred on a single variety dubbed 'zanzarone' by the peasants, but *Anopheles claviger* by entomologists. Concentrating on these he was able to prove the connection beyond doubt by means of number of simple experiments. Healthy volunteers from unaffected areas invariably came down with the disease when bitten by parasite-carrying Anopheles. But healthy *Anopheles*, and biting mosquitoes of other kinds, had no such affect on their victims.

In a not uncommon travesty of Nobel justice Ross got the prize while Grassi is forgotten.

(3:15) FINDING CORRELATIONS

Once the parallax (distance) of one star was measured by Bessel, the dam burst. Nineteenth century astronomers measuring many more distances were astonished to discover the vast range they found in stellar properties. Our own Sun turned out to be relatively sub-luminous, some stars giving out more than 100,000 times as much energy. Some were red (cool) and some were blue (hot), some were tiny – no bigger than the Earth, some so titanic that they would drown our entire inner Solar system below their bloated surfaces. How they worked, and where they got their energy from, were total mysteries.

Then two astronomers, plotting stellar properties on a graph, noticed an unexpected correlation (Fig. 3:1). In 1913 Ejnar Hertzpung and Henry Russel separately found that the luminous stars were blue whilst the faint ones were red, with the yellow ones, like our Sun, in between. This is odd because one would expect the luminosity of a star to depend on both its temperature(colour) *and* its size. If there are both big and small red stars, there ought to be both luminous and faint ones. But there were not. The overwhelming majority of red stars were under-luminous, and therefore small.

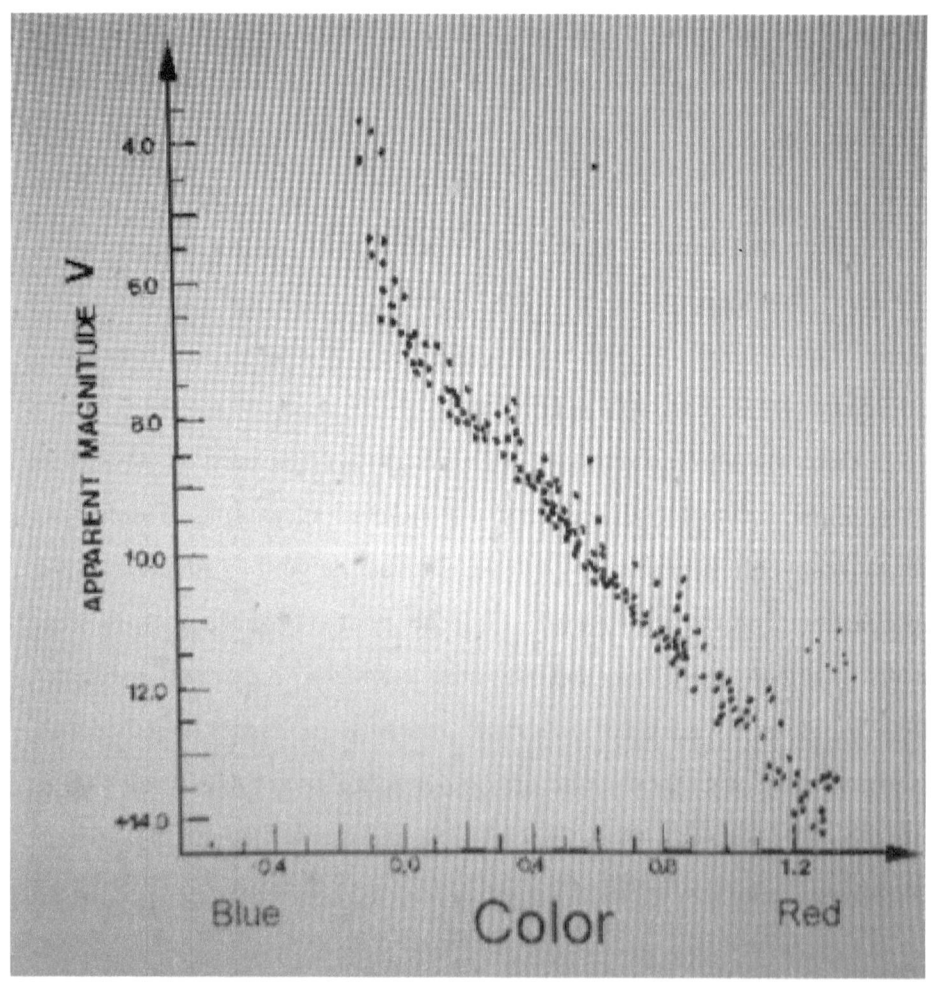

Fig (3:1). The Colour/Luminosity correlation for stars in a sample all at the same distance but fairly close to the Sun. The vertical axis plots their Luminosities in a logarithmic fashion. Thus the Sun itself on this diagram would have a magnitude of 10 and a colour (yellow) of about 0.6. Stars near the top have a Luminosity 250 times greater than the Sun, stars near the bottom are 40 times fainter than it is. What was totally unexpected was the tight correlation between luminosity and colour. Note the scatter, which is typical of real data; maybe it is a reflection of real astrophysics, or maybe the measurements are imperfect, or 'noisy' as we say.

Explicable or not this correlation was immensely useful. Knowing only the colour of a star the astronomer could immediately read off its intrinsic luminosity from the Herzprung-Russel graph. Comparing this with the apparent brightness of that same star, as seen through a telescope, instantly yielded its distance. Thus vast distances could be probed, and very quickly. The flat sky of the ancients became a 3-dimensional ocean of space inhabited by clusters and nebulae almost overnight.

Usefulness aside though, the unexpected correlation cried out for an explanation. Within a generation this was to be found in the thermostatic properties of the nuclear reactions that must power the stars.

Whenever scientists do not understand, they measure and compare – looking for correlations which may suggest theories – and which can certainly be used to test them. They have to beware though because not all correlations imply a causal relationship viz. the unquestionable correlation between annual US steel production and the average size of Chinese women's feet in the first half of the 20th century. {3/2} [See Appx. 12 for another surprising, and thus far totally unexplained set of correlations.]

(3:16) REFUSING TO ABANDON A CHERISHED IDEA

The study of radioactive decay in the 1920's and 30's revealed that within radioactive nuclei a neutron would frequently decay into a proton plus an electron. This so called "Beta-Decay" had however some very worrying consequences. Careful measurement showed that the decay products contained, in total, less energy, less momentum and less spin than the mother neutron from which they had descended. In other words energy, momentum and spin were not being conserved.

Unwilling to abandon some corner-stone ideas of Physics –The Laws of Conservation of Energy, Momentum and Spin – Wolfgang Pauli proposed the existence of an extraordinary hypothetical particle, later to be named the 'neutrino', expressly for the purpose of balancing up the measurements. This neutrino could have no mass, or very little of it, and no electric charge, but it could carry away the energy, momentum and spin apparently missing from Beta decays. Because of its bizarre hypothetical properties the neutrino would be virtually undetectable and yet was likely to be the commonest particle in the universe, with billions of them passing unnoticed through the human body every second. Less substantial than moonshine, more elusive than fairies, these products of the megalomaniac theoretical mind could, in their hypothetical billions, dominate the entire cosmos.

But Pauli had his reasons. He knew that abandoning the Conservation Laws meant abandoning deeper principles still, symmetry principles without which it might be impossible to construct science at all . In 1917 Emmy Noether had shown that each Conservation Law was the consequence of a corresponding Symmetry Principle. For instance to abandon The Conservation of Energy would be to abandon the notion that scientific laws could be framed which could withstand forever the ravages of time. Pauli revolted from such a hideous possibility and, in preference to chaos, opted for a multitude of fairy particles .

Subsequent events proved that he was right. The nuclear reactor, unknown in Pauli's time, simply glows with neutrinos, particles which cannot be confined within any man-made containment vessel. Their properties, measured in 1956, saved theory from chaos.

In general though Pauli's argument is a dangerous one for it puts theory above experiment – the Platonist fallacy which held up science for more than 2000 years. In this instance Pauli was right, and perhaps he deserved to be because he argued from theoretical principles too strong to be lightly gainsaid.

But take a more modern instance of the Pauli argument, one which has led to the 'discovery' of overwhelming amounts of 'Dark Matter'. The outermost planets orbit the sun more slowly than those closer in. Neptune for instance orbits at 6 kilometres a second as compared to the 30 kilometres a second of our Earth. This is a consequence of Newton's Law of Gravitation which sees the gravitational effect of a massive body, in this case the Sun, falling off as the inverse square of your distance away from it. Were Neptune to orbit any faster the sun's weaker gravitational pull out there would be unable to hold it in its present orbit.

When the first radio telescope was built capable of looking inside spiral-galaxies [at Westerbork in Holland in the 1970s], it was naturally expected that the same 'slowing-down' law would be found inside them, with the outermost skirt of a spiral orbiting more slowly than the innermost – which would be closer to the main seat of gravitational mass. In fact this seemed so obviously trivial that the task of checking it was left to a graduate student at the University of Groningen called Albert Bosma.

When Bosma's observations of some 20 spiral galaxies showed nothing of the expected sort it was naturally supposed that Albert must have made a mistake. After all, students often do. But Albert stubbornly stuck to his guns. There were rows. Albert's eminent supervisors resigned. Others were brought in to check through his results. No errors could be found so, reluctantly, Bosma was allowed to publish his PhD thesis in 1978.

Bosma himself, fed up perhaps with all the rowing, left his results to speak for themselves – and speak they did. Galaxies spinning in their outmost parts as fast as Albert had found them to be, would fly apart unless they were held together by far more matter than one can see in the form of their visible stars. The following year two distinguished American reviewers were to say '…..we think it likely that the discovery of invisible matter will endure as one of the major conclusions of modern astronomy.'

This was Pauli's argument all over again. This time, as I write in 2017, there is no *direct* evidence that this Dark Matter actually exists. No one has captured an elusive Dark Matter particle. Nevertheless the majority of astronomers seem to believe in it because they are unwilling to abandon Newton's Law of Gravitation, or at least its modern form as slightly modified by Einstein. The Gravitation Law has none of the theoretical pedigree of Pauli's Conservation Laws, but nonetheless it does have 300 years of respectable usage behind it, to say nothing of a great deal of elaborate and rather beautiful mathematical structure. It works magnificently at short range within the solar system. It never seems to work at long range in the largest structures of the universe, not without invoking the presence of overwhelming amount of the presently mysterious Dark Matter. Do you remember "The Emperors New Clothes"?

[Pauli's neutrinos by the way cannot be the Dark Matter – they are too light and too fast].

(3:17) SWALLOWING THE UNPALATABLE

A juggler can juggle in an aeroplane without having to know its speed. The Laws of Nature – if such Laws there are – must be the same in some distant galaxy as they are on Earth, even if that galaxy is receding away from us at 100,000 kilometres a second due to the 'Expansion of the Universe'. This so called 'Principle of Relativity' – which is scarcely more or less than a definition of what we mean by 'A Universal Law of Nature', implies that such a Law must be independent of your speed, because nearly everybody in our universe is moving at different speeds. The principle goes back at least as far as Galileo and is implicit in the concept of 'universality 'itself.

Then along comes James Clerk Maxwell in 1864 with his collection of abstruse equations claiming to describe the properties of electricity, magnetism and light, claiming in fact that they are all one and the same. To begin with, scarcely anyone could understand Maxwell's arcane mathematics, while not even Maxwell himself could think of a way of generating detectable electromagnetic waves.

Gradually though Maxwell's successors came to grips with and tested out his equations in the lab. They worked! But then it dawned on physicists, and it must have been an uncomfortable awakening, that Maxwell's equations contained a speed – the so called 'speed of light'. How could any 'Law of Nature' contain a speed, which after all must depend on the relative speed of whoever is measuring it? In this respect

Maxwell's Equations appeared to conflict with "The Principle of Relativity" and so could not be entirely correct.

Moreover there were other profound puzzles in the interrelationship between electromagnetism and motion. Oersted had shown that a current leads to a magnetic field. But a current, which is no more than moving electric charge, is surely a relative concept: to an observer moving at the same speed as the charge (and why not?) the charge is not moving at all – so there is no current and therefore no magnetic field. That's OK, it just means that currents and electromagnetic fields are relative concepts which depend, like speed itself, on the relative speed of the observer who measures them. OK that is, so long as their relative measurements lead to no conflict between observer's as to *physical* events. For instance if observer A sees two electrons collide, then so must observer B, even if the two of them cannot agree about their measurements of speed, or current or magnetic field.

Hendrik Antoon Lorentz at the University of Leiden in Holland (1889) set out to find the mathematical transformations which must relate the electromagnetic measurements of one observer to those of another in relative motion to him, so as to ensure that there can be no such conflict between their physical observations. He found these so called "Lorenz Transformations" and very surprising they were. If they were right, and they did seem to work in the laboratory, then they implied that both the distance and the time-interval between two events were no longer absolute but relative quantities which depended on the speed of any observer relative to the events in question.This in turn implied that Newton's hallowed 'Laws of Motion', which relied on the intuitively natural concept of 'Absolute Time', must be slightly wrong.

Now the cat really was amongst the pigeons. Maxwell's Equations, which seemed to predict all the right experimental answers in electromagnetism, only made sense if Absolute Time was abandoned and if Lorentz's mathematical transformations between one observer's measurements and another's prevailed. On the other hand Newton's hallowed Laws of Motion, which had stood the tests of more than two centuries of experiments and observations, certainly did not obey Lorenz's new transformations, for if they did so then fundamental consequences of Newton's laws, such as the Conservation of Momentum, would have to be thrown out. No; Newton's Laws obeyed the old Galilean transformations.

Between 1890 and 1905 Physics almost suffered a collective nervous breakdown. Newton or Maxwell? Lorentz or Galileo? Newton plus Galileo, or Maxwell with Lorentz? Yes and what about the wretched velocity of light, which surely couldn't be an absolute constant, although astonishing measurements by Michelson and Morley seemed to indicate that indeed it was?

Physicists everywhere struggled like beasts caught in a swamp until 1904 when Henri Poincaré – the most distinguished mathematician of his day, suggested a most unpalatable but possibly workable way out. : "Accept Lorenz's ugly looking transformations, and therefore the Maxwell Equations implicit in them, but modify Newton's laws to conform with Lorenz." This was a council of last resort, the very opposite of what physicists had been hoping to hear. Absolute Time would be gone, together with the elegant simplicity of Newton's Laws and Galileo's transformations. Worst of all the speed of light, because it appeared in Maxwell's equations, would have to measure the same to all observers, irrespective of their speed with respect to the source of the light in question. Poincare was suggesting a topsy turvy, repellent program which scattered all history, all sense of physical intuition to the winds.

Poincare published his program in 1904 and set out to look for the necessary modifications to Newton. He found and published it in 1905. Provided the mass of a body could be thought of as relative, that is to say could appear slightly heavier as measured by an observer with respect to which it was moving at higher speed, then all could be reconciled: Maxwell, Newton, Lorenz, Galileo, Michelson and Morley.

This Lorentz-Poincare Theory, the so called 'Special Theory of Relativity' was a daring but to some extent unwelcome step. It worked perfectly, as thousands of subsequent experiments were to prove. But it marked a retreat from 'understanding', as Poincare's predecessors would have understood that word. No one would comprehend the universe in a comfortable sense henceforth. Our baboon-like brains are not equipped to imagine velocities like that of light, invariant as between observers in relative motion with respect to the source and to each other. Nor a Time which is relative. Nor masses which can change with your velocity. Nor spatial distances which are to some extent entangled with time intervals. Nor light waves which can traverse empty space without an aether.

But that is the way the Universe appears to be. Poincare and his contemporaries managed to overcome their baboon-like repugnance and see the truth of the world in spite of themselves. Their powers of thought triumphed over their own intuitions. It wasn't comfortable, any more than it had been comfortable for Kepler to digest the fact that planets move in offset ellipses rather than elegant concentric circles. But it was the truth, or at least as near to it as contemporary man could come.[3]

---

[3] Well what about Einstein? Surely he invented Relativity? Not really. He spent the winter of 1904 discussing Poincare's paper with his friend Besso. Then in 1905 he published his famous and beautiful paper on "The

80

(3:18) ARGUMENTS BASED ON SYMMETTRY

We saw that Oersted was led to the discovery that an electric field produces a magnetic field. Michael Faraday, an imaginative and self-taught physicist, who had been forced by economic circumstances to leave school at age 12, was so excited by this discovery that he set to work on it at once. He found, to his own and to general amazement, that the magnetic field actually rotates around the current in the wire which generates it, a pattern never observed in physics before. For reasons not entirely clear Faraday was seized with the conviction that if electricity could generate magnetism then, by symmetry, the reverse must also be true. Time and again between 1820 and 1831 Faraday devised the most ingenious experiments to determine whether a magnetic field could be made to generate a current. No less than five times he returned to the fray, but each time he failed. By 1831 however he was using an electromagnet rather than a bar magnet to generate his magnetic field. Still it didn't work. Then, almost by accident, he noticed that at the moments when he turned the electro-magnet on and off there were indeed minute transient currents in his electric circuit. That was all the clue he needed. Magnetic fields do indeed generate currents, as his intuition had supposed, but only if they are moving or changing. Faraday had discovered 'Electromagnetic Induction', which subsequently

---

Electrodynamics of Moving Bodies". It was a mathematical working out of Poincare's program, but with no acknowledgement of Poincare whatsoever. At the very least it was ungracious; at the worst plagiarism. Had he redressed this ingratitude later in life when he was incontestably and rightly famous himself, we might excuse Einstein on the grounds of youth (he was only 25 in 1905). However he did not. It might have had something to do with the intense rivalry at the time between the Germans and the French following the Franco-Prussian war. The fact is that both Einstein and Newton were mortal men covetous of glory. The idea of "scientific genius" makes for a good story, particularly when you have wild hair and photogenic eyes like Albert, but not for sound understanding. I used to be an Einstein "groupie" myself – and he remains one of my greatest heroes. But I now believe Oscar Wilde was closer to the truth when he remarked "All of us are lying in the gutter, but some of us are looking at the stars." We return to Einstein's very real contributions to the subject in Chapter 6. He simplified its exposition to the point where it became persuasive to his contemporaries, and later he generalized it in the most beautiful way.

led him and his peers to develop the electric motors, dynamos and transformers which laid the basis for the entire electric power industry.

Although historically vital Faraday's 'Argument by Symmetry' is not perhaps the most instructive one because it is hard to find a sound basis for his argument . It was really no more than a hunch. Today however, and with better reasons, physicists constantly appeal to symmetry arguments of various kinds. When they are looking for the law underlying some particular phenomenon they know that that law, if it is to be universal, must conform to the symmetries of Space and Time. For instance if it is to be true in one spatial direction, it must be as true in another because Space is symmetrical in all directions. Such symmetries, taken together, may impose so many constraints on the mathematical form of the required Law as to rule out all but a handful of possibilities, between which experiment can decide. The beguiling dream of the Theoretical Physicist is to identify so many required symmetries as to render experiment unnecessary. If it were to come off, and it is a very big 'If', we would be able to say that "The world is as it is because it could be no other way. We can excogitate the laws of nature by simply sitting at our desks. " Some few theoretical physicists appear to think we are nearly there already and speak in awe of such a "Theory of Everything." Others think such a notion is no more than quasi-religious Pi-in-the-Sky.

(3:19) BEING SCEPTICAL

Most scientific ideas are wrong, especially one's own, so scepticism rather than the alternative is the healthier scientific stance. Moreover, to quote from Daniel Boorstin, that most readable historian of science: "The greatest obstacle to discovery is not ignorance, but the illusion of knowledge." Certainly, if you know the Earth is flat, you won't set out to circumnavigate it. Whereas silly people tend to have a quick explanation for everything, most good scientists appear to have a naturally sceptical temperament. Before it is widely accepted, even provisionally, a scientific idea should generally have plenty of supporting evidence that has been critically evaluated. Idolatrous acceptance of the work of 'geniuses', or of ideas that have survived for centuries, may be as unwise as gullible adherence to the latest fad or fashion. The more plausible an idea sounds at first, the more rigorously it deserves to be cross-examined. Whereas you can be reasonably sure that an implausible idea that has been allowed into the scientific cannon has first been vigorously interrogated, a plausible one – because of its very plausibility, may have slipped through like a good impostor, largely unquestioned. Take the widely quoted saw: "Exercise is good for you." It sounds plausible but is it really right? Where is the evidence? One can think of counter arguments: if it is true why do farm-workers and other manual labourers who get a great deal of exercise in the course of their occupations have a significantly lower life expectancy than sedentary academics and parsons? And why are fat people, who are forced to take a great deal of exercise every time they climb the stairs, such a poor insurance risk? The more you think about it, the more you realise that "Exercise is good for you" is an assertion that would be very hard to actually prove. One would have to follow a large cohort of people from youth into old age, recording the exercise they take, while controlling for all manner of potentially confusing factors such as diet, occupation, smoking, environment, wealth and available health-care. Lancelot Hogben, the famous medical statistician, who lived to a ripe old age, liked to claim that "The only exercise I take is attending the funerals of my more energetic colleagues. "

This exercise debate nicely illustrates another problem: upon which side of a particular question should the *burden of proof* be laid? One might argue that "Exercise is good for you" is so self-evidently true that it is up to the critics to prove that it is not – something which might take a generation. Equally logically one could take the opposite tack and place the burden of proof on advocates of exercise. The more difficult it is to come up with "proof" the more vital, and consequently the more heated, the burden-of-proof issue becomes. Opposing sides in for instance Environmental debates are more frequently divided by this issue than by any evidence that there may actually be. The current Genetically Modified Food debate for example is actually a religious one because the possibly far-reaching and damaging environmental effects of such crops may not be forthcoming until it is all too late.

Scepticism then is more of a burden-of-proof issue than a scientific tactic in its own right. Nevertheless there are crucial areas of science where the mere assertion that "such and such a thing cannot be done" have led to constructive and absolutely gigantic leaps in progress. Consider Perpetual Motion Machines (PMMs). Many minds, including some great ones, have laboured hard and long to develop contrivances which could generate useable energy while continuing to work indefinitely with no outside intervention or fuel. All failed, although new and ingenious examples are offered to the public every year. Professionals wasted so much of their time testing out new varieties, and finding that they didn't actually work, that they found it easier to declare that "The Perpetuum Mobile is a physical impossibility" and to refuse, as the Paris Acadamy of Science did in 1775, to test any further examples.

This declaration of impotence is of course an asserted scientific principle in itself – which may or may not be right. Today we re-assert it as "The First Law of Thermodynamics" or more commonly as "The Conservation of Energy". In its last guise it asserts that "Energy can neither be created nor destroyed: it can only be converted from one kind into another." The PMM, it claims, could only generate energy perpetually if it were first to steal that same energy from some other source. Whether there is anything more to this resounding statement than an inversion of the sceptical observation that "It is apparently impossible to build PMM's " it is difficult to say. Certainly it has proved extremely useful because it has allowed engineers to design power generating machines, such as windmills or coal-fired stations, in the confident knowledge of how much energy they might in practice extract.

Unperturbed by the official prohibition, inventors came up with "Perpetual Motion Machines of the Second Kind." These more sophisticated designs conformed to The Conservation of Energy but apparently still did not work. Imagine for instance a ship that takes in warmish seawater at the bow, sucks out the heat energy to act as a propellant, and exhausts blocks of ice at the stern. The Conservation of Energy is in no way disobeyed and yet such a PMM 'Of the Second Kind' could still not be made to work in practice. Again, after many such failed attempts, the scientific establishment found it prudent to assert a second principle of impotence, called unsurprisingly "The Second Law of Thermodynamics" which states: "Perpetual Motion Machines of the Second Kind, that is to say ones which extract heat from their surroundings and convert it into mechanical energy(for instance the ship's motion), with no other effects upon their surroundings, can never be made to work." Unlike the First, this Second Law of Thermodynamics became ultimately explicable on a much deeper level. Long after it was asserted men realised the importance of order and disorder in the cosmos. A bull walking into a china shop always causes chaos – never the contrary. You cannot create order out of disorder without doing a great deal of work – which always creates disorder of another kind somewhere else. Our self-propelled ship cannot be made to work because it would be re-arranging disordered energy, in the form of warmish sea water, into more ordered energy in the form of its own motion, plus colder ice. Put in other words it would be separating out a random population of sluggish molecules (making up the sea-water plus the stationary ship) into a fast team – the moving ship, and a slow team – the ice. And you cannot do that without putting in effort from somewhere else. That effort, it turns out, will always be greater, usually much greater, than the effort needed to simply propel the ship.

Engineers continually appeal to this Second Law of Thermodynamics, as they do to the First, to inform their creations. The huge cooling-towers near power stations are there to disperse the disorder necessarily attendant upon the creation of ordered electro-mechanical power from disordered heat. Your airliner flies as high as it does partly in order to conform to the Second Law: its jet engines work much more efficiently when ingesting cold (i.e. more ordered) stratospheric air.

So scepticism is a useful place to start from in science, if only because it saves one from chasing up innumerable unsound leads. And as we have seen it can even be constructive. However one cannot prescribe scepticism as a general scientific nostrum if only because it has sometimes proved spectacularly wrong; it has stood against radio-waves, flight, space-travel, atomic-energy. . . . and so on and so on. In general though history suggests we should remain moderately sceptical of all ideas, whether they be new or old, and that we should continually review the evidence both for and against them. A grain of doubt is always healthy – as we shall prove. After all hypotheses once regarded as almost self-evident – such as The Flat Earth, The Four Elements and The Argument by Design (for the existence of God) have since come under the axe. We return to this important matter in Chapter 8.

(3:20) SO HOW DO SCIENTISTS THINK?

It is time to review our brief foray into scientific history, and to look for guidance in our future search. To that end I have drawn up Table (3.1), with each row corresponding in order to our individual episodes of discovery. The first three columns should remind us of the discovery, while column (4) tabulates, on a scale of 0 to 1, whether the chief mental tactic used is familiar from everyday thinking. If I judge that it definitely is then it receives a mark of 1; if it definitely is not, then zero. Intermediate marks are my rough estimates of where it otherwise lies on the scale in between. The average, given at the bottom, is 0.9, suggesting that the answer is close to 1, and therefore that scientific thinking is close to the everyday variety.

In a like manner column (5) tabulates whether extensive Deduction was needed in making that discovery, again on a scale of 0 to 1. A blank signifies everywhere that the answer is close to zero i.e. little or no deduction employed. Likewise column (6) tells how crucial experiment was to the discovery.

The final three columns are directed at the question "Is there such a thing as scientific genius?" Column (7) records whether the discoverer in question had rivals and/or precursors. If he had then the discovery was not 'out of his own head', but in the air ready to be made, and that if A hadn't made it B or C would probably have done so soon afterwards. The last two columns record whether some recent technical or scientific development had occurred, before which the discovery could hardly have been made. For instance in row 3 Oersted needed a battery to test his idea that a current would generate a magnetic field. Fortunately for him Volta's first battery had just come on the market. Likewise in row 10 both Wallace and Darwin had spent years travelling to many continents comparing their flora and fauna before the idea of Evolution came to them, Wallace as a poor plant collector, Darwin as a wealthy, unpaid gentleman's companion. Much before their time intercontinental travel by scholars was virtually impossible.

Our first inference from the Table must be – see column (4) – that Scientific Thinking can be no more than a refinement of the everyday variety. It follows that if we can find out exactly what those refinements are, and how scientists employ them so effectively, we may learn something of widespread value. That is a chief premise of this book.

Our second inference concerns the primacy of Deductive Thought in science, that is to say following a long chain of reasoning from unarguable premises to unavoidable conclusions. From column (5), and its average value of only 0.1, this primacy would appear to be almost totally incorrect. While the arguments of *pure* mathematicians may be of the kind "... if A then B, if B then C, if C then. ...", there is little sign of this process being vital to our listed discoveries. And for good reason. While mathematicians argue from general rules (premises) to specific instances, the scientist necessarily has to argue in the opposite direction. He aims to *find* the laws of Nature which lie behind the specific instances which come to his attention. He or she is in the business of "Plausible Inference", not Logical Deduction.

Column (6) suggests that the place of experiment in science can be exaggerated, particularly by philosophers of science of the school of Karl Popper, who are overfond of referring to "crucial experiments". No true scientist would neglect to perform, or at least acknowledge such experiments – *if they can be done*. But very often, and for good reasons, they cannot. John Snow could not give contaminated pump-water to healthy citizens to see if they developed cholera, while no meaningful experimental tests of the Theory of Evolution were possible for the first hundred years, simply because Evolution is too slow. Observations generally come first, even in areas where experiments can be done. Kepler, Snow, Oersted, Newton, Maxwell, Wegener, Becquerel, Wallace and Darwin first noticed something which aroused their curiosity. Later, sometimes much later – after their deaths in the case of Maxwell, Newton, Darwin and Wegener – experiments were done. Indeed the greatest, the most synoptic theories are often the hardest to test by direct experiment. Evolution we have remarked on, but what about Heliocentrism, or Universal Gravitation, or Dark Matter, or the Conservation of Energy? And sometimes, don't forget, hypotheses may live on long after so called "crucial" experiments or observations have apparently left them for dead – Heliocentrism comes to mind.

TABLE 3.1. DISCOVERY TACTICS IN SCIENCE

1 TACTIC	2 SCIENTIST	3 DISCOVERY	4 Every-day tactic? 0-1	5 Extensive Deduction? 0-1	6 Experiment crucial? 0-1	7 Rivals Or Precursors? 0-1	8 Rec. Precondition? 0-1	9 The Precondition
1 Recognise Problem	Kepler	Dark Sky	1	0.2			1	Telescope
2 Recognise Coincidence	Snow	Cholera	1	0.2				
3 Connect Different ideas	Oersted	Electro-Magnetism	1		1		1	Battery
4 Numerical Coincidence	Newton	Gravity	0.5	1		1	1	Kepler
5 Recognise Pattern	Wegener	Continental Drift	1			1	1	Geological Maps
6 Pursue Unexpected.	Becquerel	Radio-Activity	1		0.5		1	Photography
7 Careful Experiment	Joule	Conservation Energy	0.5		1	1	1	Electricity
8 Better Tool	Bessel	Astron. Distances	0.5			1	1	Glass Technology
9 Overturn Stones	Giacconi	Hot Universe	1		1	1	1	Space Rocket
10 Argue by Analogy	Darwin Wallace	Evolution	1			1	1	World Travel
11 Mistaken Path	Marconi	Radio	1		1	1	1	Hertz
12 Recognise Similarity	Mendeleev	Periodic Table	1	0.3	0.7	1	1	International Post
13 Eliminate	Grassi	Malaria	1		0.7	1		

Alternatives								
14 Find Correlation	Herzprung	Stellar Structure	0.5			1	1	Precision Telescope
15 Cling to Idea	Pauli	Neutrinos	1	0.2	0.2			Nuclear Physics
16 Swallow Unpalatable	Poincare	Relativity	1	0.4		1	1	Maxwell & Lorentz
17 Suspect Symmetry	Faraday	Electromag. Induction	0.1	0.2	1			Battery
18 Be Sceptical	Several	Sec. Law Thermods.	1			1	1	Steam Engine
19 AVERAGE of column above			0.9	0.1	0.3	0.7	0.9	

Finally we come to the case against 'scientific genius'. They say half of all the scientists who ever lived are alive today. And yet I have never met a scientific genius – nor have any of the other hundreds of scientists I have asked. Geniuses appear not to be alive in science today nor, to judge from columns (7) and (8), were they so common in the past. Most of our discoverers had close and jealous rivals who disputed their primacy in discovery. Newton had Hooke and Leibniz, Darwin had Wallace, and Poincare had Einstein. Moreover most of them – columns (8) and (9) – were the beneficiaries of a recent advance, the lack of which would have crippled any predecessors. Newton stood on the shoulders of Kepler and Galileo, they upon Brahe and Copernicus. Maxwell couldn't have done without Faraday nor Bessel without Fraunhoffer – who finally constructed a telescope, using modern glass technology – precise enough to measure the very tiny stellar parallax. Oersted and Faraday needed the newly invented battery, Giacconi the space rocket, and Mendeleev needed a postal system which kept him in touch with other chemists throughout Europe. In other words they happened to be in the right places at the right times. Yes they had enormous qualities, especially curiosity and doggedness, and yes they got there first. But there is little evidence that transcendental genius was at work. But for quirks of birth and history most would have been forgotten now. And that is well for us. If scientific history can do without genius, so can we. We can ask how it was done without presupposing magic. We can learn from the great men and women without presupposing the supernatural – and that after all is our aim.

It might be objected that our conclusions rest on the analysis of a particular, personal and possibly idiosyncratic list of discoveries. There are two answers to that. First the list was drawn up from memory to provide one example each of the tactics first enumerated in column (1), not at all to colour the outcomes in the remaining columns. Second I have drawn up other lists to meet the same criterion, as could any knowledgeable scientist, and the outcomes are broadly the same.

If nothing else the examples we have chosen illustrate the vast range of scientific experience. No wonder individual scientists are sometimes hard put to explain what unites them with brother or sister scientists in entirely different fields. For instance Alfred Russel Wallace collecting butterflies and orchids in The Spice Islands and James Clerk Maxwell solving differential equations in London both published immortal works in almost the same year (1860). Both leapt to startling conclusions that have benefited mankind. But would either one have understood the other or been able to unpick some golden thread of common culture or thought? That will be a challenge for us.

Ten thousand years ago Man was just another species. In small bands he gathered berries legumes and grubs, hunted when he could but whenever local supplies ran out had to move on, carrying his meagre possessions. Apart from fire and furs nothing marked him out as the future Lord of the Earth.

Then something happened, something that enabled him to build Venice, invent Relativity, vanquish Smallpox and land explorers on the Moon. What was it? In my opinion it could only have been some improvement in his powers of thought – probably something small, something incremental that yet enabled the rapid Ascent of Man. We will be on the lookout for that vital secret and, in Chapter 16, suggest what it is. Once we know that secret we might then be able, in NASA's terms, to 'throttle up' [See our Epilogue].

SUMMARY

Analysing twenty or so scientific discoveries have led us to the following conclusions; scientific thinking is very like the everyday variety; deduction and experiment are far less important than observation; scientific genius seems to be a myth. So now we need to find out just how everyday thinking works.

93

THINKING FOR OURSELVES (Disney)
CHAPTER 4 (14/9/18; 10.7 kw)
NATURAL THINKING AND BAYES' RULE

"....no truth appears to me more evident than that beasts are endow'd with thought and reason as well as men."
David Hume, philosopher. 1737

(4.1) INDUCTION

If we can't use Deduction, that is to say arguing from the general to the particular, to establish the laws of Nature, why don't we use Induction – arguing from the particular to the general? After all that is our stance when we confront Nature: we experience particular instances of it, not abstract generalisations. Unfortunately, as the Scottish philosopher David Hume emphasised, Induction is bound to be an uncertain process: no matter how many white swans you observe you cannot be *certain*, in the deductive sense, that *all* swans are white. But that is exactly what the scientist is hoping to do: to establish general laws of Nature on the basis of observing only a finite number of particular instances. It seems that the best that can be done is to establish the probability, the credibility, the likelihood of some hypothetical general law, on the grounds of our observations. We might be able to say: "On the basis of sighting approximately ten thousand swans over the course of thirty years, all of which were white, I infer that the hypothesis 'All swans are white' has a ninety per cent (or whatever) chance of being right.' Such equivocation was not what philosophers, including 'natural philosophers' – which is what scientists were called until 1831 – wanted to hear. They used, and still do, far more robust language such as "Every body attracts every other body in the Universe, with a gravitational force proportional the product of their masses, and inversely proportional to the square of the distance between them." It's a stretch, it goes beyond the strictly established facts, but it seems to be an extremely useful working hypothesis. It enables Men to calculate tide-tables and navigate spacecraft to Saturn. And perhaps that is the point. As animals we have to *act*, we cannot afford to become paralysed by a lack of absolute certainty as to the way in which our world is going to behave. A sufficiently high probability, a decisive probability, is all we really need. Absolute certainty would be nice, but if it is not available, and it seems that it is not, then we can still fight to survive. What we do need is a means to establish the probabilities of life, including the Laws of Nature, in an optimal manner. If we cannot do that then our rivals, parasites and predators will drive us out of existence. That is the law of Evolution by Natural Selection.

This raises an obvious difficulty. The need to assess probabilities, and act wisely on them, must go far back into pre-human history. Animals have been practicing the art or science of it for hundreds of millions of years. Observe your pets; watch wild birds around a feeding table. The need to assess the relative probabilities of advantage and danger, and then to act, are observable on every side. In the struggle for existence Nature must have hard-wired into our beings an instinctual system for arriving at and acting upon probabilities, and a very successful one at that. The problem for us thinkers about thinking, is therefore not so much to devise a method from scratch, but to try and uncover the instinctual machinery we must already carry inside us. All we can be certain of is that it is there – and it must work pretty well – at least in those contexts in which it evolved. We can hope that by pinning it down, laying it bare, we can exploit it better, perhaps far better, in the artificial world in which we now live, and for which it was never evolved. The obvious difficulty we alluded to is this: we can work out a feasible, perhaps an optimal way for probabilistic thinking – lets call it Plausible Inference (PI) henceforth

– to work, but we can never be certain it is the very one that Nature actually uses. Who would have guessed, before the era of brain-scanning, that skilled chess players seem to exploit their face-recognition abilities to unconsciously memorise positions, and remember the best ways out of them. But perhaps it doesn't matter. Just as we've devised aeroplanes that fly faster than birds so we may devise mechanisms for Plausible Inference that are better than Nature's. Arguing about how we *do* think may be less fruitful than recognising how we *could* think, and think better. In this chapter, and the following two, we unclothe three key mechanisms of PI which scientists seem to use when they are trying to decipher the natural world: Bayes' Rule, The Detective's Equation, and Ockham's Razor.

(4.2) PROBABILITIES AND ODDS

What is 'the Probability of an hypothesis'? It is, for the purposes of this chapter and the next three, *'Our degree of belief' in the hypothesis or idea.* It is a number, conventionally specified in the range between 1 and 0, where 1 connotes complete certainty, 0 complete disbelief. For shorthand purposes it is written P(H) 'the Probability of hypothesis H', *and it is a private quantity, personal to the individual who is expressing it.* Your Probability for the hypothesis that 'It will rain tomorrow' might be 0.5, whereas mine might be stronger at 0.7 [Some people express these Probabilities as '50 per cent' and '70 per cent']. I use the capital P if the probability in question represents a precise number in somebody's mind – not just some vague abstraction.

Another piece of useful shorthand, again a number in the same range, is P(H|E) which simply means 'The Probability of hypothesis H, *given* some piece of Evidence E'. An instance might be P(It will rain tomorrow| the barometer is very low). It can naturally be extended to P(H| $E_1, E_2, ...$)

connoting 'The Probability of hypothesis H, given several pieces of evidence $E_1, E_2, \ldots\ldots$etc' e.g. P(It will rain tomorrow| low barometric pressure, this is Wales in winter, rain is forecast,……). Another convention is to write $\bar{H}$ (pronounced 'H overbar') for 'not-H' or 'the alternative to H' or 'hypothesis H not being true'. Thus, in the above instance, $\bar{H}$ = 'It will *not* rain tomorrow'. Since any hypothesis is either true or not true then it is certain, i.e. equal to 1, that the sum of the two Probabilities P(H) and P($\bar{H}$) must add up to 1, i.e. to certainty. Thus any consistent mind should maintain

$$P(H) + P(\bar{H}) = 1 \qquad (4:1)$$

which, although it is obvious, is sometimes useful.

An alternative form for expressing probabilities is as 'Odds': by O(H) we mean 'The Odds *on* hypothesis H' – again a number in somebody's mind, but not necessarily in the range between 0 and 1. We can have Odds of '20 to 1 on' [O(H)=20], '100-to-1 on' [O(H) =100] or '1 to a 100 on' [O(H)= 1/100 = .01] What we mean by ' The Odds-on H, O(H)' , is 'The Probability that H is true i.e. P(H), divided by the Probability that it is not true i.e. P($\bar{H}$)'. Thus, by definition:

$$O(H) = P(H) / P(\bar{H}) \qquad (4:2)$$

(4:2) makes practical sense , as we can see by example. Suppose our Probability for H is 0.9 then our Probability for not-H i.e. P($\bar{H}$) must be, by equation (4:1), equal to 1- 0.9 = 0.1. Thus our Odds-on H are, by Equation (4:2):

$$O(H) = \frac{0.9}{1-0.9} = \frac{0.9}{0.1} = \frac{9}{1} \quad \text{or '9 to 1 on'.}$$

In other words we would be prepared to bet on H at Odds of '9 to 1 on', but no higher.

Most regular gamblers prefer to use Odds-*against* rather than Odds-on. If the Odds-on are 1 to 100 then the Odds–against are 100 to 1, and in general:

Odds against H = Odds on not-H = $O(\bar{H}) = 1/O(H)$

i.e. $O(\bar{H})$ the Odds-*against* is simply the reciprocal, or upside-down value of the Odds-*on*; and vice versa as is obvious from (4:2).

Thus: $O(H) \times O(\bar{H}) = 1$ (4:3)

[This last equation is for Odds the equivalent of (4:1) for Probabilities. The reason why some equations will turn out to be far more easily expressed in Odds rather than Probabilities (or vice versa) hinges on this comparison. Astonishingly this will turn out to have historical consequences.]

Note that we could extend the notion of P(H|E) to Odds and write O(H|E) as shorthand for "Our Odds on hypothesis H *given* some evidence E." And of course you can reverse both notations in an obvious way. Thus P(E|H) will mean "The Probability of evidence E occurring *given* that hypothesis H is true." which is by no means the same as P(H|E) "The Probability of hypothesis H given evidence E". P(E|H) and P(H|E) are not simply related arithmetically so *be careful to always note which way round the H and E occur.*

Although Odds and Probabilities are so intimately related, we actually need them both, like fractions and decimals. If you are not entirely familiar with both do the following simple;

EXERCISES (4:1) 'ODDS AND PROBABILITIES'
On the website

How could Probabilities and Odds be used in science? When, after my Hidden Galaxies fiasco (later), I began to delve into this subject I became very confused because all the expert authors in the area seemed to be in violent disagreement with one another, starting on page 1. Eventually it occurred to me that the last people I would consult on gambling would be academics – who generally delight in nit-picking. Why not consult a successful gambler instead?

(4:3) THINKING LIKE A GAMBLER.

Gamblers, at least professional gamblers, have to think straight – otherwise they quickly go out of business. My old friend Stella has a very extensive interest and knowledge of horse racing, steeple-chasing in particular. Over the course of a season, so she told me, she places dozens of judicious bets and, more often than not, wins more than she loses – usually enough to pay for an extravagant annual family holiday. And she never misses betting on The Grand National, that Everest of the steeple-chasing calendar. With forty or more horses, thirty fearsome fences and a course over 4 miles long it has everything, not least drama as horses fall, hurling their riders under the hooves of the thundering field.

For months in advance Stella studies the form of the declared runners and places odds in her head (and in her notebook) on likely prospects, with $O(H_j)$ being her odds *on* a particular horse j winning the National; i.e. her hypothesis H_j is 'Horse j will win the National'. She will have several such notional bets in her head long before the race is run.

As the season progresses some prospective runners do well and win steeple-chases whilst others fail too often to appear serious National contenders. If she is to bet successfully, all this evidence from previous races has to be incorporated into Stella's thinking.

For instance Wavehunter, one of the horses she is interested in for the National, is to run in The Gold Cup, another marathon steeplechase held some months beforehand. He wins it ! How does she incorporate this new evidence E into her odds on her hypothesis H_w that Wavehunter will win the National?

What she does, she told me, is go back into her extensive library of record books to find out, historically, what bearing a win in the Gold Cup has had on subsequent success in the National. She discovers that National winners have been 5 times more likely than other National runners to win the previous Gold Cup. In the parlance of Probability, where you will remember P(A| B) means 'the Probability of A given B', she can therefore say:

$$\frac{P(E \mid H_W)}{P(E \mid \bar{H}_W)} = 5$$

where E is the new 'Evidence' i.e. winning the Gold Cup; H_w is the hypothesis that Wavehunter will win the subsequent National, $\bar{H}_W$ that he is *not* going to win it [remember a bar on top of H for Hypothesis means 'not-H'].

So given this new evidence Stella's odds *on* Wavehunter, i.e. her $O(H_w \mid E)$, [her new odds on Wavehunter given the Evidence E that he's won the Gold Cup')] ought to be 5 times greater. But greater than what? Greater than all the other horses in the race? Hardly, because some of the best prospects didn't even run in the Gold Cup. And hardly because some of the prospective runners are such outsiders that surely Wavehunter was a 5 times better bet than they were even before he won the Gold Cup.

What Stella actually does, so she tells me, is multiply by 5 the odds $O(H_w)$ she already had in her head on Wavehunter *before* the Gold Cup was run; i.e. she writes

$$O(H_W \mid E) = \frac{P(E \mid H_W)}{P(E \mid \bar{H}_W)} \times O(H_W) = 5 \times O(H_W)$$

That seems sensible to me because the evidence of Wavehunter's win bears only on him, not on the other prospective runners, many of whom didn't even run in the Gold Cup. Since she had Wavehunter at 25-1 against before the Gold Cup (i.e. 1 to 25 on) she therefore calculates his odds-on afterwards as:

$$O(H_W \mid E) = 5 \times (1 \text{ to } 25) = 5 \times (1/25) = 5 \text{ to } 25, \text{ or } 1 \text{ to }$$

5 on =1/5 [or '5 to 1 against' in the more usual terms].

What she does then is go round looking for bookies who will offer her odds better than 5-1 against. If she can get 10-1 then, she reckons, she is likely to win twice as often as she will lose. And so her experience has proved over the years. Indeed the last time I rang she was about to take her entire extended family, including grandchildren, for a month's holiday in The Seychelles.

After she told me all this, rather diffidently for she is no mathematician, I looked back into all my confusing books on Probability Theory to see if Stella's Equation had some pedigree, and to advise her if she was acting in an optimal way. Eventually I discovered it was called "Bayes' Rule" after a certain Thomas Bayes who died in 1761 and is buried, not far from Daniel Defoe and Robert Blake, in leafy Bunhill Fields in central London, close to Liverpool Street station. More about him anon.

The mathematicians made a great deal of fuss over Bayes' Rule – or 'Bayes' Theorem' as of course they preferred to call it – disagreeing or agreeing with it in almost equal numbers, but violently in both cases. Having had the misfortune to work in a Mathematics department for several years I never take Mathematicians' squabbles too seriously – it's usually hair-splitting. What I could see, and I hope you can see it too, is that Stella's Rule (or 'Bayes' Rule') makes a good deal of common sense. Indeed it is almost self-evident. It's about the apparent association of two things: a piece of evidence E, and an hypothesis H. If E and H are commonly observed in association then it is obvious that in a particular instance, when one or other is observed to be the case, that the other is more likely (more probable) to be the case also. If 5 out of 6 times in the past

E has been observed in association with H then our Odds on H occurring now, after E has definitely been observed, should surely be greater than evens by a factor of 5, or 5 to 1, and that is what Stella has assumed – evidently to good effect.

So for now we shall take Stella's Rule, or Bayes' Rule as we shall call it henceforth, as a generally sensible way to think (If you don't find it self-evident, as I do, we shall look at more grounds for it later). In its full glory it is usually written:

'Posterior Odds'= 'Weight' times 'Prior Odds'

or in shorthand:

$$O(H \mid E) = W(E \mid H) \times O(H) \qquad (4:4)$$

where the "Weight" W(E|H) is shorthand for

$$W(E \mid H) = P(E \mid H) / P(E \mid \bar{H}) \qquad (4:5)$$

i.e. "the Probability of evidence E IF H *is* true, divided by the Probability of E IF H *isn't* true."

Since either P can take any value between 0 and 1 the Weight can take any value between 0 and almost infinity (certainty). W(E|H) has a bland, dispassionate value of 1 when, according to (4:4), the Weight of the evidence in question neither increases nor decreases our Odds on H. In that case the 'Posterior Odds' O(H|E), that is the Odds posterior to, or *after* considering E, remain equal to the 'Prior Odds' O(H), our Odds *before*, or prior to, considering E. In other words that particular evidence is irrelevant to our view of H. We can think of any Weight W(E|H) as the multiplier by which that clue E raises the Odds on H from the neutral value of 1. So all Weights are really odds-on. A Weight of 4 raises O(H) from 1 to 4.

What we are going to do in the remainder of this chapter is to see if and how Bayes' Rule works in a variety of circumstances, forensic, scientific, medical and personal. {3&4}

NB. By the way the 'Mathematics' so far used in this section is as hard as it's going to get. So if you're still with me you should be able to get through to the end of the book. Don't be put off by arithmetical shorthand just because it looks unfamiliar. Yes there are some scary looking formulae later on, but they're only in to demonstrate just why philosophers frightened themselves out of Common Sense.

(4.4) BAYES AND BURGLARS.

A detective is investigating a burglary where she has 7 possible suspects. Lacking other evidence she might suppose that each is a likely as the others to be guilty. To be impartial she ought to suppose that her Odds on any particular suspect, call him X, being guilty, should be 1 to 6 on i.e. we should expect her $O(H_X) = 1/6$ where H_X is 'X is guilty'. In other words her Prior has a value 1/6.

Suppose further that blood was left on the broken window-pane, and that X is found to have a recent and nasty cut on his thumb (this is E the evidence). Some people do get cut by chance but the Detective feels it is 30 times more likely that X cut himself during the burglary. So she believes the Weight of this evidence is

$$W(cut \mid H_X) = \frac{P(cut \mid H_X)}{P(cut \mid \overline{H}_X)} = 30$$

so that Bayes' Rule leads her to alter her Odds to the posterior value

$$O(H_X \mid cut) = W(cut \mid H_X) \times O(X) = 30 \times 1/6 = 5$$

or 5 to 1 *on* the hypothesis that X did it.

This seems obvious and sensible to me, even if in practice she may not think in such a formally quantitative way. After E she has Odds of 5 to 1 on X being the culprit, not conclusive enough to convict perhaps, but sufficient to prompt further digging into X's story – a blood-test perhaps. The advantage of thinking quantitatively, if roughly, is that it can be a useful guide for future action. For instance if she were looking for final Odds of 100 to 1 (perhaps 'beyond any reasonable doubt'- see later) she should be looking for a blood-test of sufficient discriminatory power to boost her Odds by a factor of 20 from 5 to 100. A rough and ready test like Blood-group probably wouldn't do.

(4.5) THE CASE OF THE MISSING GALAXIES.

Somewhere in this book we are going to have to study a piece of real, full-blooded scientific research – by which I mean a story with all the dreams, the failings, the blind alleyways, the disappointments and the odd strokes of luck left in. Since science as a whole is not interested in such personal stories – and for good reasons (chapter 10) – it is almost impossible for the outsider to imagine what the agony and the ecstasy of real research is actually like. In the big picture such personal journeys do not signify; does it matter whether Newton came up with theory of Universal Gravitation lying under an apple tree – or pinched it from his bitter rival Robert Hooke while they were chatting in a coffee house? Neither account should influence our appreciation of the theory's ability to explain the universe – which is why science generally discourages the 'personal account'. But from the point of view of 'Thinking' – which is what we are interested in here – such journeys are absolutely crucial. How did the poor devil stumbling about in the dark find the light-switch – or miss it by a finger's breadth? In particular could he have found the light more directly if he'd been thinking in a different way?

The only person who knows the full messy story is the research worker him or herself. So I am going to have to inflict on you from time to time the story of my personal struggle to understand galaxies. Of course *I* think it is fascinating, but that is not the point; it's in here because it will turn out to illustrate many key issues of significance to thinkers in general. So let us begin.

Galaxies are the biggest, most beautiful and most mysterious objects in the entire universe. And that's just my opinion. If you don't believe me take a look at Fig (4:1) which shows a typical Spiral galaxy not unlike the Milky Way galaxy in which we live. As to size it almost a hundred thousand light-years across, and contains about a hundred thousand million stars. It is obviously spinning and must have rotated about forty times so far, taking 250 million years to make single turn. It is an unimaginably vast distance away – its light having set out towards us long before humans, or even pre-humans existed – and yet it is one of the closest galaxies to us. Even so, using the most ingenious optical, radio and space telescopes, we can tell that it contains stars of many luminosities, colours and ages, a Supermassive Black Hole in its core, vast clouds of both hot and cold gas, and enough smoke to hide most of its internal structure from sight.

FIG (4:1) The Spiral Galaxy NGC 7331 in the constellation of Pegasus, the first one I ever saw or photographed. Down below it are several more distant galaxies in the background. The sharp images with spikes (artefacts) on them are foreground stars in our own galaxy the Milky Way, which is much like 7331. We live in its outermost skirt so we cannot see most of its bright interior which is obscured from us by opaque clouds of smoke (the dark lanes) emitted by stars which have blown up in the past. You can only acquire an image like this by collecting the light from a colossal telescope for an hour or so. Imagine how little you would see looking through the same telescope with the naked eye: NGC 7331 simply becomes an almost imperceptible smudge against the background. And of course there is no indication of depth or distance here: it's actually 50 million light years away. This picture is the result of a telescope with a 3.5 metre diameter mirror collecting light with a highly sensitive electronic camera for over 2 hours! See References for picture credits.

And that's about where our reliable knowledge, painfully gathered over the last two centuries, ends. We don't even know what it is made of because it is spinning far too fast for the stars we can see to hold it together by their mutual gravitation. They can constitute only 5 to 10 per cent of the so called 'Dark Matter' that must glue it together – Dark Matter of whose nature we have not the faintest idea.

And how far out does it stretch? At least twenty times as far as its visible extent; we know that because galaxies are hardly ever isolated – they have companions tens or even hundreds of visible diameters away that must be, in some mysterious way, related to them. For instance the Milky Way has a colossal such neighbour 20 diameters distant in the direction of Andromeda.

Where in heaven did the galaxies come from – after all there are plenty of them out there (we estimate at least a hundred thousand million). Have they always been there? Did they emerge out of the Black Holes so often detected in their cores? Or did they somehow condense out of a primeval gunge? All we can say for sure is that most of them must be at least ten thousand million years old because we can pick out inside them rare variable stars which only begin to pulsate when they reach that age – according to our probably reliable calculations.

The most astonishing thing about galaxies is that they all appear to move away from us (their light is 'red-shifted'), and indeed from one another – as if the entire universe is expanding. For 50 years astronomers couldn't believe their eyes but in the end they were forced to do so by the additional and accidental discovery of huge amounts of cosmic microwave radiation occupying every cubic centimetre of Space. The only explanation we so far have for this colossal reservoir of energy is that it is the remnant radiation from a titanic cosmic explosion – the so called 'Big Bang' – which flung out everything – including the galaxies – about 14 billion years ago. The coincidence here is that galaxies, or at least their oldest pulsating stars, are of the same antiquity. Thus the whole wildly improbable story of cosmic genesis has some basis of evidential support – more at least than any alternative theory suggested so far. Of course that doesn't mean it is actually right. How to judge whether unlikely stories based on circumstantial evidence are right or wrong is what this book is all about. In his time Man has made some spectacularly good guesses – and some spectacularly bad. What could be more fascinating – or more worthwhile – than finding out how to get such guesses right?

Deciphering the secrets of spiral galaxies isn't going to add to the health, wealth or happiness of mankind. It is Pure science. However, learning how we might do such Pure science sets challenges to our powers of thought like nothing quite else in human experience. Learning to meet such challenges might well lead on to wisdom of a more general kind – as it has done so signally in the past. At least that is the hope and the promise.

I recall my first personal encounter with a galaxy as vividly as if it happened last night. I was a novice astronomer out at Steward Observatory in the remote mountains of the Arizona-Sonora desert. I had what seemed like a colossal telescope to myself and when I climbed up all the dark and clanging ladderways and peeked my head tentatively out of the huge dome I found myself lost in a tumbling immensity of stars, a tiny diatom in an ocean of overwhelming night. But where were the spiral galaxies I had read so much about, had so eagerly waited to see? I searched and searched between the giddy constellations and couldn't see a one – not even a glimmer of our supposed giant companion in Andromeda.

Overawed and puzzled I lowered myself gingerly back down the airy ladderways to the control platform, and set the huge howitzer, or so it looked projected against the stars, towards the direction of a well known spiral which I knew ought to be almost directly overhead.

Then I looked through the eyepiece and searched. Nothing, just blackness, and a few tiny pin-prick stars. I moved the beast and the star-pricks moved in the eye-piece. But no galaxy. I moved this way then that. I stared and stared. Nothing – just blackness and pin-pricks of light. I got out my heavily dimmed torch and checked the co-ordinates. Fine. I was pointing right at it: it had to be there; it had to be – unless I'd gone blind. Hither and thither I steered, expecting the monster to swim into the eye-piece at any moment. In vain. Not to the East; not to the West; not to the North; not to the South. It just wasn't there. I steered back again and a glimmer so dim that it vanished when I stared, passed across the field of view. Surely not! Surely not! That couldn't be it – not that wraith, that ghost, that miserable glim of light, lost like a shielded lantern on a mountainside a million miles away?

It took me half a breathless hour to convince myself that I was indeed staring at a galaxy, and another hour to photograph what turned out to be the spectacular monster you can see in Fig (4:1). That image is a total lie – the result of adding together thousands of seconds of light gathered by tens of square feet of mirror. Real galaxies are spread out so thinly against the firmament that they are less visible than fragile cone-jellies lost in the abyssal ocean. I was incredulous, completely incredulous that Man had come to learn anything about them at all.

My life was changed from that moment. I would use it, if I could, to learn some of the secrets of those leviathans of the night, those ghostly wraiths that lay at the very margin of detectability. What could be more exciting, more challenging than that?

The first obvious puzzle, or so it seemed to my ignorant young curiosity, was why we could see them at all. Had the terrestrial night sky been ever so slightly brighter, or spiral galaxies been ever so slightly dimmer, we might have missed them altogether. What a miracle it seemed that Nature had provided them with enough, but only just enough light, for human ingenuity to pick them out!

You can see where such a line of speculation was eventually going to lead. Miracles being rare, and rather frowned upon in science (we'll see why later), perhaps it was true that we were noticing only those few galaxies that were brighter than our local night sky? Maybe the cosmos was stuffed with 'Hidden Galaxies' just too dim for us to detect? It was easy to speculate, but could I find some substantial evidence that such an exciting possibility was true? My experienced colleagues snorted at the notion that a tyro like me was going to find something so spectacular which all our distinguished predecessors had missed. And in my heart I agreed with them. All the same I couldn't prevent the wild idea from drifting into my dreams from time to time whilst I was working insanely hard on other things in order to build an astronomical career. Eventually though, 6 years later, when I was learning radio astronomy at the University of Groningen in Holland, I was rash enough to give a talk on the possibility of Hidden Galaxies. A respected colleague and friend in the audience challenged me to find some skeleton of evidence for such a wild and improbable idea; or otherwise to shut up about it.

He was right of course. I needed to be professional. I went home that night and started calculating. By now I knew exactly how bright the sky was and I asked myself how compact a galaxy *should* be in order to be maximally visible from the Earth. If it was too compact then at any great distance it would look like a star, and not be distinguishable out there as a galaxy. On the other hand if it was too diffuse it would be entirely lost against the sky. It was a pretty tough calculation but I got an answer about four in the morning. I couldn't wait to cycle in to the university library to see if it agreed with the actual observations. To my amazement and delight the agreement was as perfect as the observational uncertainties would allow. It couldn't be a coincidence surely. We WERE seeing only those galaxies optimally spread on the sky. All the rest were simply not showing up! Then I had another idea. Beside Spirals there were very different specimens of galaxy called Ellipticals. Ellipticals don't look like whirlpools but like titanic swarms of bees, with stars instead of insects (Fig 4:2). If my wild idea was right then they too ought to be optimally spread on the sky so as to make themselves maximally visible to us humans here on Earth; but here was the crucial point: they ought to be spread by a *different* amount from spirals because they had a different distribution of light. I calculated and calculated and found what that optimal spread ought to be a couple of nights later. The problem was that nobody appeared to have made suitable observations of Ellipticals against which I could check my calculation. I ransacked the library in vain. But then I remembered a paper I had read years before by a student astronomer named Fish. Fish had alas disappeared from astronomy, as many students do, so I couldn't ring him up. In his PhD thesis Fish had shown that the Masses and Binding-Energies of Elliptical galaxies were related in a fascinating but inexplicable way. Could that relation, I wondered, have a bearing on my problem? I found his 12 year old paper in The Astrophysical Journal and began to reverse his calculations in order to get at his original, but unrecorded, observations on which his calculations must have been based. It took several days and nights but when the answer finally popped out it made me literally dance for joy. Fish's observations actually implied that all his measured Ellipticals were spread out to exactly the same extent on the sky, and that extent was the very one I had earlier calculated would make them maximally visible from the Earth. Either it was another and separate miraculous coincidence, or it meant that we were seeing only that favoured fraction of the true Elliptical population, with all the rest lost to sight. All but one or two of a hundred measured galaxies, of both types now, testified to the likelihood that my wild theory about Hidden Galaxies was *almost certainly* right. I was on my way to fame and fortune – or so I supposed. I gave up everything else to prove that a whole universe of galaxies lay out there waiting to be found, lurking hidden just beneath the night sky. We were like beings living in a brightly lighted room (our own galaxy). And, although we think of it as dark, our night sky *is* brightly illuminated. If, on the darkest moonless night, we look from the very darkest site on Earth- the summit of the Andes- at the darkest patch of sky we can find in between the stars, our naked human eye will actually pick up 50,000 light particles a *second* from an area no larger than the full Moon! The reason that even professional astronomers find this hard to believe is that it takes the human eye several hours to completely adapt to the dark. If you don't believe me, spend a dark-of-the-moon night in some area remote from Man, sleeping out under the stars. Set your alarm, and crawl out of your sleeping bag after several hours asleep. No torch mind; no chink of artificial light. You'll be amazed by how much you can see. And if you do it in the summertime in the Northern Hemisphere you might actually see a great spiral galaxy overhead: and I don't mean the Milky Way. We shouldn't therefore be surprised that when we look out of our lighted window into the night; we will only ever detect other windows as bright as our own, or brighter. The darkened buildings, the trees, the countryside and even the mountains would all be lost to our bedazzled sight.

114

FIG (4:2) Not all galaxies are Spirals. Here are some Ellipticals clustering together in the constellation of Coma Berenices centred around the giant galaxy NGC 4889. Look how its light is sharply condensed in the centre but then fades away imperceptibly into the sky. Where does it 'end'? How much of it is there? What would we see of it if it was more stretched out on the sky? Indeed would we see it at all? And how far is it away? And why are these galaxies so different from Spirals? See what an endless challenge galaxies present us with. This image taken with the Hubble Space Telescope. See References for picture credits.

Convincing myself was one thing. Convincing my rightly sceptical colleagues would be quite another. I was calling for a new continent out there and they would need to see positive proof. A numerical coincidence was all very well; they demanded to see a fair sample of the hypothetical inhabitants.

So, eventually joined by colleagues, assistants, rivals, friends and opponents, I set out to find my hidden continent. I should have realized it wouldn't be easy because staring out of a lighted room into darkness is rarely productive. We would have to find more ingenious alternatives, and over the years we did. Galaxies emit radiation at more than one wavelength. The sky is not equally bright in different colours. The cosmic redshift offers a chance of discriminating between distant galaxy emissions and the local masking emissions of our own Milky Way. And then galaxies ought to absorb as well as emit; thus the light from a very distant Quasar passing through one of my Hidden Galaxies ought to have, etched on its spectrum, very characteristic absorption-lines caused by the atoms in the hidden leviathan intercepting the background Quasar light. And indeed Quasar spectra are heavily contaminated by such Spectral Ghosts, as I shall refer to them. So one way and another dozens of lines of evidence got brought into the discussion over the next forty years, and hundreds of papers. The problem was, and remains, that the evidence was starkly contradictory. Some was conclusively in favour; some was flatly against. There were, understandably, rows. Some thought I was mad to continue; my funding was cut off, and I myself almost gave up the ghost completely at times. In other words I was involved in a typically exciting but messy scientific controversy. And it was in despair that I turned to the philosophical question at the roots of this book: 'How can we, how should we deal with conflicting evidence? It is a question as old as Evolution, and as fundamental as the granite rocks. We must all turn to face it at times, most often in our personal lives. Is there a good way, is there an optimal way to answer it? As we shall see, as a result of struggling with galaxies for nearly fifty years, I think there is. But let's not get ahead of ourselves. Let's see where and why this struggle, this more general quest, really began.

      After twenty years of hard searching, when we had gathered some very promising, but indirect evidence, the International Astronomical Union called together, at my university in Wales, a big conference to discuss the whole question of 'The Invisible Universe'. About a hundred professionals came from every corner of the globe, and almost every branch of astronomy, to discuss The Big Question. Was there a Hidden Universe out there?

Things went very well at first for us enthusiasts. But then, after several days of gathering optimism, there came the dramatic blow. A young Dutch astronomer, let us call him Dr Z, announced that there weren't any Hidden Galaxies, none at all. He and a team of eminent colleagues had pointed the world's largest radio telescope at a narrow strip of sky for several continuous *months*. Independently of their light output they had found fifty radio objects in the strip, and quickly confirmed their existence as radio emitters with another powerful radio telescope. They had at last detected the hydrogen signals that I had predicted would be the signatures of otherwise invisible galaxies. However they had then gone on to look at all of their radio sources with an optical telescope – and in every single case found it to be an easily visible galaxy. Not one of the 50 was invisible. Not one!

Here was the decisive, the crucial test advocated by philosophers such as Karl Popper. And my beloved hypothesis had failed it miserably. I blustered and fumed but there seemed no honest way of denying a test so ingeniously designed, so thoroughly checked and so numerically decisive. Nobody was fooled by my blustering, least of all myself. As a then disciple of Popper it was inconsistent to deny that a theory, no matter how beautiful and beloved, can be murdered by a single ugly fact.

And yet my instinct rebelled. What of all the other positive evidence, admittedly indirect, amassed over the years? Was all that now to be set at nought? Here was a typical messy conflict of evidence: one clearly negative piece set against an agglomeration of separately weaker positive pieces.

It was in the aftermath of this shock that I first came across Bayes' Rule. I wondered if it could be applied to my Missing Galaxies, the Evidence here being Dr Z's failure to find them in his apparently very deep search. In the following table and its footnotes, I have laid out the various factors which enter the discussion of my hypothesis H *"There are numerous Hidden Galaxies out there'* where 'numerous' means 'at least as many as the visible galaxies'. But I have done two calculations: one from my point of view and one from Z's. They might disagree if only because he came to my hypothesis with a sceptical point of view, as he was entitled to do, whereas I came to it as an enthusiast, having been persuaded by previous evidence – of which he might not be aware – or which he discounted. In other words we came to the question with different 'Priors', so that there is no reason why we should agree exactly – should arrive at identical posteriors. And we don't. Study the table and its footnotes carefully.

TABLE (4:1): 'ARE THERE ANY HIDDEN GALAXIES?'
Hypothesis H : "*There are numerous Hidden galaxies*"
Evidence E : "Z's failure to find any in his search."

Person	Prior O(H)	P(E|H)	P(E|	W(E|H)	Posterior O(H|E)
1 Believer (me)	10	0.5 (c)	0.8 (f)	~ 0.6	6
2 Sceptic (Z)	0.2	0.1(d)	0.95	0.1	0.02
3 See Note (g)	0.2	0.5	0.8	0.6	0.1
4 See Note (h)	10	0.1	0.95	0.1	1
5 See Note (i)	0.2	0.9	0.98	1	0.2

FOOTNOTES

(a) My Prior (10 to 1 on) is high because I am persuaded by previous evidence.

(b) Z's Prior is low (0.2) because he is unaware of, or un-persuaded by, my previous evidence.

(c) I feel there is a 50% chance that Z's evidence is flawed.

(d) Z feels there is only a 10% chance that his evidence is flawed.

(e) Z is 95% sure that if H is wrong he would get his observed result.

(f) I am not so sure that if H is wrong he would get his observed result.

(g) What happens if I accept Z's Prior but retain my data assessment $P(E|H)$?

(h) What happens if I keep my Prior but accept X's data assessment?

(i) What Z *would* have had to conclude if he'd factored in his unsuspected low sensitivity (see below).

Because I had strong faith in my hypothesis beforehand [i.e. a high Prior $O(H) = 10$], based on twenty years of earlier work, and a degree of scepticism about the reliability of Z's observations [i.e. my $P(E|H) = 0.5$ only] my posterior Odds remained high [$O(E|H) = 6$] in agreement with my instinctual reaction at the time – despite the damning appearance of Z's result. Z on the other hand came fairly sceptically to my hypothesis, as he was entitled to do, [$O_Z(H) = 0.2$], felt his results were reliable [see his $P_Z(E|H)$ and $P_Z(E|\bar{H})$] and concluded that there was now only a 2 per cent chance of [i.e. the Odds were 50 to 1 against] my hypothesis being correct.

This, my first exercise in using Bayes' Rule, was thoroughly satisfying because it explained so many previously puzzling matters. For instance it explained:

(a) Why Z and I could honestly disagree about the outcome of his observations – because we had very different Priors.

(b) That even so we had both lost some confidence in my hypothesis – which seems sensible. His Odds had dropped from 0.2, i.e. 5 to 1 against, to 50 to 1 against. Mine had dropped from 10 to 1 on to 6 to 1 on. They were still sufficient for me to continue with my project – which was my strong instinctive feeling at the time. (Some of you might feel I was being soft on myself by using a dishonestly optimistic Prior. Fair enough, but then I would have to pay the penalty of wasting my scientific life chasing a dishonest delusion.)

(c) That neither of us was being stupid or perverse and therefore there was no reason to get cross, as in fact, to my shame, I had done.

(d) Why Popper's advocacy of 'The crucial observation' was naïve and made no sense in this context. Real scientific evidence is usually probabilistic not logical, has numerical Odds attached to it, and is not simply Yes or No.

As it turned out in the long run Z's paper proved to be incorrect in two respects. Something had gone wrong in the collecting of his radio data, so that it was vastly less sensitive than he had supposed. That should have been obvious because they had been so quickly confirmed with an apparently far less sensitive radio telescope. More fatally his optical identifications were not reliable. This took me years to even suspect, let alone establish. He and his co-authors had grossly underestimated, as we all did at that time, the tendency of galaxies, be they visible or invisible, to huddle or 'cluster, as we say, close to one another. So any invisible galaxy will very likely have a prominent optical neighbour nearby, which then gets mistakenly identified as the source of the radio signal. So, although we couldn't know it at the time, his interpretation of his data, which was what we both needed, was mostly wrong. Because of his lack of sensitivity he wouldn't have picked up most of my Hidden Galaxies, so he should have used $P(E|H)$ = roughly 0.9, in which case his observations would scarcely have altered his Prior belief because his Weight would have been 0.9/0.95 or close to 1. In other words the evidence was largely irrelevant to the hypothesis under investigation. My pet hypothesis might still be right – or wrong. More or better evidence was needed. Alas most astronomers understandably took Z's evidence at face-value and my campaign to find Hidden Galaxies almost expired. It still hasn't recovered completely, as we shall see over the course of this investigation. Refs {2 to 8} of Chapter 8 deal with Hidden Galaxies.

(4.6) THE CASE OF THE DREADED CANCER TEST. When Mrs. Jones received the letter below she and her family were naturally devastated. They knew, and this was confirmed by reading the family's medical encyclopaedia, that cancer of the woozle is an agonising and invariably fatal disease. And now she appeared to be exhibiting the first signs of it. According to the encyclopaedia she had less than a year to live, and there was little or no hope of a successful operation. The only hope that appeared to remain, and it was a very slim one [apparently a 2% one from the letter] was that her test result was wrong.

                                        Pathological Analysis Laboratory
                                                    North East Glamorgan
Health Authority,
    Our Ref CA/ 41/RSD/15/03/05   Abercwmboy,
                                                        Glamorgan BN23 7EH

       March 15 2014

       Dear Mrs Jones
           A specimen of your blood was forwarded to us by your medical practitioner on Feb 26 last to test for the presence of carcinoma lymostrophia woozeli. We are writing to inform you that the test has now proved positive.
           In accordance with Health Authority guidelines [2001/07/1453c] we are obliged to inform you of the reliability of such a test, as established by the UK Medical Statistics Bureau in 1998.They find:
           Reliability of identifying the disease when it is present :   95%
           Probability of finding a positive result when the disease is in fact absent :   2%

You are advised to get in touch with your medical adviser(s) immediately.

Yours sincerely
R.S.Davies   BSc., Phd., F.R.I.C., M.P.S., D.C.O.G.

For   The Authority.

While she was waiting to see her 'medical adviser', a process, which can take ages in Britain, the poor woman and her entire family, were traumatised. Fortunately however, one of her nieces was studying medical statistics at university. When she heard about the tragedy the niece did a Bayesian calculation of her own. She realised that the true implication of the letter depended on how common cancer of the woozle actually is in the population. She couldn't find this out in a hurry but as no one in the family had heard of anybody who was suffering from, or had died of cancer of the woozle, she guessed that the incidence among the general population was very low, say one in a thousand or less. So her Bayesian calculation looked like this:

Hypothesis H = 'Aunt has disease'.
Evidence E = positive test.
Prior O(H) = 0.001 (general incidence in the population).
And according to the letter P(E|H) =0.95.
Also according to letter P(E|$\bar{H}$) =0.02 (i.e. the test has a 2 percent chance of showing a positive result when in fact the disease is not present). Thus, combining these last two figures the Weight of the evidence is W(E|H)= 0.95/.02=47 and so Bayes' Rule gives:

$$O(H \mid E) = W(E \mid H) \times O(H) = 47 \times .001 = 0.047$$

(i.e. less than 5 per cent!)

Overjoyed she rang up her aunt to explain that the odds-on her actually having the disease was less, and probably much less than 1 to 20. Because cancer of the woozle is so rare in the population at large there will 20 times as many tests which give false positive results as true positives [for that reason the test would not generally be used – at least not on its own.].

What this case illustrates is *the danger of leaving out relevant prior information*, in this example the information that the disease is rare. While prior information may be contentious or uncertain (as it was with the missing galaxies) leaving it out can betray one just as well. Notice that the niece didn't have to know an *exact* Prior: it sufficed to know that the Prior O(H) was very small, i.e. less than 0.001.

In this instance, to be absolutely sure, the family decided to have a second test carried out immediately at a private clinic in London. Unfortunately it (E_2) turned out to be positive too and when the niece did her calculation this time, and with the new Prior arising from the Posterior in the first test i.e. with O(H) =0.045 she found:

$$O(H \mid E_2) = W(E \mid H) \times O(H) = \frac{P(E_2 \mid H)}{P(E_2 \mid \overline{H})} \times O(H)$$

$$= \frac{.95}{.02} \times .045 = 2.1$$

In other words the odds were now roughly 2 to 1 on her aunt being mortally sick.

To be absolutely certain they rushed her to another private clinic in Cardiff where alas too the test (E_3) proved positive, resulting in a Bayesian probability this time of:

$$O(H \mid E_3) = \frac{0.95}{0.02} \times O(H \mid E_2) = 47.5 \times 2.1 \approx 100$$

or 100 to 1 on, which amounted to a virtual sentence of death.

When the desperate woman finally got to see her family doctor he had had time to do his homework. He knew that the first test was far from conclusive while the second and third might be worthless. Worthless because it was known that a woman in menopause, who additionally suffered from low blood sugar, would *always* give a false positive test for cancer of the woozle. He tested her blood sugar on the spot, found that it was indeed very low, and sent her home reassured once again. An alternative type of test carried out later showed that in fact she didn't have cancer.

The problem with the second and third Bayes calculations here was not that they were invalid per se but that they *incorporated an unconscious natural Prior* that was in fact wrong, i.e. that 'false positive tests arise at hazard, in a random manner' when in fact they may be *certain* to arise in particular patients (i.e. those in the menopause with low blood sugar).

So Priors can be extremely significant, easy to neglect, and easier still to slip in unconsciously. One of the great virtues of doing a Bayesian calculation is that it *forces one to be clear about one's Priors*, and to bring them out into the open. Many an argument can be cleared up, if not settled, by recognising exactly why the two sides differ. Unknown to each other, unknown even to themselves, protagonists may have been arguing on the basis of different Priors. If they can reconcile their Priors they might come to agree; or if they still cannot then they can agree to differ – and know exactly why. The argument has shifted ground to a discussion of Priors –which can be very constructive.

(4.7) DID THEY SEE FLYING SAUCERS?

One might have hoped that the accumulation of more and more evidence would, in the end , lead to a convergence of opinions, no matter how far apart the initial Priors, or if you like the initial prejudices, of those in debate. Sometimes it does and sometimes it does not. It turns out that convergence will only occur if both parties are willing to grant the evidence the same degree of belief. If they cannot do that then the same evidence put before them may actually lead to an even greater divergence of opinion. Let us look at an example.

Weird lights are seen in the sky by dozens of inhabitants of the lonely island of Jura. Two 'expert' journalists, a believer B in flying saucers, and a sceptic S, are sent to investigate. B has a prior belief $O_B(H) = 10$ to 1 on the hypothesis H that 'Flying Saucers exist." while for the sceptic S $O_S(H) = 0.1$ or 10 to 1 against. After listening to the transparently sincere evidence (E) of the islanders each must reconsider his position. The following table summarises the outcome from a Bayesian perspective:

WERE THERE ANY FLYING SAUCERS ON JURA?
Hypothesis H : "*Yes there were.*"
Evidence E : "Islanders saw lights in the sky".

Person	O(H)	P(E|H)	P(E|$\bar{H}$)	W(E|H)	O(H|E)
Believer	10	1	0.2	5	50
Sceptic	0.1	1	0.2	5	0.5
Sceptic	0.1	0.5	0.5	1	0.1

Before going to Jura the ratio of their prior beliefs $O_B(H) / O_S(H)$ was $10 / 0.1 = 100$ to 1. Afterwards that same ratio depends on their evaluation of the evidence. Were the sceptic to take the same enthusiastic view of the evidence as the believer [see middle row in Table] – that is to say adopt the same $P(E|H) = 1$ and $P(E|\bar{H}) = 0.2$ so that his Weight was 5 – then his Posterior would increase to 0.5 and the ratio of their Posteriors would still be 100 to 1. In fact the sceptic [last row in table] took the view that the reports could as easily be explained by some other phenomenon as by flying saucers – so that his $P(E|H) = 0.5$ and his $P(E|\bar{H}) = 0.5$ so that his Weight W(E|H) equalled the bland value of 1. In that case his Posterior *cannot* rise above his Prior and he remains as sceptical as ever, i.e. $O_S(H|E)$ =0.1. But because The Believer's Posterior has risen to $O_B(H|E)$ = 50 the ratio $O_B(H|E) / O_S(H|E)$ is 50/0.1= 500, or even higher than before. They have *diverged.*

So the naïve supposition that 'Looking at the evidence' must necessarily lead to agreement, or even to some convergence of views, is by no means guaranteed. It all depends on one's views of the credibility and relevance of that evidence, i.e of its Weight. Whereas the recounting of every new miracle may strengthen a believer in his belief, the disbeliever can legitimately remain unmoved. It is all, according to Bayes, a question of $P(E|H)$ as compared to $P(E|\bar{H})$. If in your view an alternative explanation $\bar{H}$ can as well or better explain the evidence as H, your Weight will assume the bland value 1 and you will not be moved.

(4.8) THE CASE OF THE MIRACULOUS CURE.

Young Mr Smith has developed a most unfortunate condition, Tiptuppititis, which his doctor can do nothing about. It causes acute suffering, painful embarrassment and a general lowering of self-confidence that affects his entire life. Desperate for relief he searches high and low for a remedy. Many are advertised and after discrete inquiries he settles for the Organocalm treatment that, although very expensive, sounds plausible and claims a 90 per cent success rate. Impressed by many ecstatic testimonials from grateful ex-patients to be found on the Organocalm website Smith takes out a loan and starts on a course of treatment.

Sure enough it is entirely successful. A few treatments are enough to demonstrate the efficacy of Organocalm and by the end of the course Smith is restored to perfect health and jubilant self-confidence. He posts his own paean of praise on the Organocalm website and becomes a rabid proselyte for the Organocalm system among his acquaintances. And why not? Organocalm has worked for him and turns out to be just as successful amongst his fellow sufferers. The founder of the Organocalm method eventually becomes rich and famous. His annual books sell in millions; television hosts seek his views on everything from slimming to salvation.

Let us examine Smith's thinking in Bayesian terms. The hypothesis under examination is:

H = '*Organocalm is a reliably successful treatment for Tiptuppititis*'.

While the evidence in favour of it is: E = 'Smith has recovered completely.'

He started with a completely open mind, granting even Odds on both alternatives i.e. that H and its antithesis $\bar{H}$ (that it is not a reliable treatment) have odds O(H) = O($\bar{H}$) =1 because he assumed P(H) = 0.5 and therefore P($\bar{H}$)= 0.5 too – see Equation (4:1).

We can now write Bayes' Theorem as:

$$O(H|E) = W(E|H) \times 1 = \frac{P(E|H)}{P(E|\bar{H})} \times 1$$

If H is true, that is to say if Organocalm is a reliably successful treatment, then the probability that Smith would recover i.e. P(E|H), must be close to 1.Thus,to an excellent approximation:

$$O(H|E) \approx \frac{1}{P(E|\bar{H})} \times 1$$

What is the remaining unquantified P(E|$\bar{H}$)? It is the probability that Smith would have recovered as promptly *even if $\bar{H}$ were true* i.e. even if Organocalm is *not* an effective treatment. The problem for Smith is that he cannot know P(E|$\bar{H}$) because he does not know whether or not he would have recovered spontaneously even without taking Organocalm. For *him* the experiment can be done only once, and it has proved totally successful.

Granted his prompt recovery he feels it most unlikely that the treatment and the recovery are unrelated. If he has to choose between the extremes of setting $P(E|\bar{H}) = 0$ and $P(E|\bar{H}) = 1$ he might reasonably select the former, in which case, according to Bayes:

$$O(H|E) \approx \frac{1}{0} \quad \text{which is infinite!}$$

i.e. the odds on H given E, are so high that H must be almost certain! He has concluded that Organocalm definitely *is* a reliably successful treatment for Tiptuppititis. [Even a $P(E|\bar{H}) = 0.1$, i.e. a 10 percent chance of spontaneous recovery, yields Odds-on the efficacy of Organocalm of 10 to 1.]

We have now encountered one of the most plausible, the most confusing, the most treacherous and the most damaging fallacies in human thought. Starting with an open mind, unquestionable evidence, and the best of intentions Smith has arrived at certainty, or virtual certainty, where in fact none whatever was justified. Thus reputations have been made, fortunes amassed, patients slaughtered, economies wrecked, innocents punished, witches hunted down, parties and religions founded, wars declared, minorities persecuted, uncounted numbers tortured and killed, all on account of what I shall call "The Snake Oil Delusion" or SNOD for short [The official name is "Affirming the Consequent"]. The fallacy in SNOD lies in assigning a plausible value for $P(E|\bar{H})$ when in fact no such value is justified. Smith has no idea whether he would or would not have recovered without Organocalm, or indeed any treatment at all. To assume as he did that $P(E|\bar{H})$ was zero, or was indeed any less than $P(E|H)$, is to argue in a circle. The only such assumption warranted by the evidence, or rather the lack of evidence, is that $P(E|\bar{H})$ is no less than $P(E|H)$, in which case

$$O(H|E) = W(E|H) \times O(H) = \frac{P(E|H)}{P(E|\bar{H})} \times O(H) = 1 \times O(H)$$

In other words the experiment (successful treatment with Organocalm) should not have increased his confidence in Organocalm by one iota above his declared prior belief that the odds on it were evens.

The Snake Oil Delusion [SNOD] is so pernicious precisely because *its' circularity is implicit or concealed*. Nothing wrong is explicitly stated: the fallacy slips in by the back door. The patient is cured, and since no other medicine was given, the medicine must be efficacious. The policy was implemented, the economy crashed, the policy must have been responsible. Children in the village have mysteriously died; no other cause but witchcraft can be imagined; a witch must be found and burned. The fallacy of SNOD lies not so much in lack of rationality as in lack of imagination, an inability to imagine one or more hypotheses alternative to H that will explain the evidence as well. The blind man recovers his sight, no one can explain why: a miracle must have occurred. SNOD underlies conjuring as well as a host of other more pretentious, more profitable and potentially far more harmful forms of mumbo-jumbo and black magic ranging from financial counselling to much of medicine [think of medicinal leeches].

In its most transparent form SNOD was known to the ancients as the fallacy of *"Post hoc ergo propter hoc"* which translates as " After that, therefore because of that" and as such was relatively easy to spot. When it comes in a more elaborate, or apparently more respectable disguise, it can fool all but the most alert of us – including scientists. For instance scientists at The Stanford Research Institute were comprehensively gulled by Yuri Geller the 'spoon bending' magician. Because they couldn't explain rationally what he had shown them, they were willing to entertain an irrational explanation.

At the heart of SNOD lies the often very real need to come to a conclusion about an hypothesis, for instance in order to reach a decision, and the frequent impossibility of assigning a rational value to $P(E|\bar{H})$, i.e. the probability that the evidence could be explained by some hypothesis $\bar{H}$, other than the one under discussion. When decisions cannot be deferred, more or less arbitrary choices must be made which leave any number of doors for SNOD to creep in. For instance it may not have been possible, or in the commercial interest, to carry out the control experiments needed to truly establish the efficacy of Organocalm. Or we may have to trust 'experts' whose expertise is more self-delusional or self-serving (all too frequently the case). Then again we cannot afford to evaluate the potentially infinite number of alternatives to H [e.g. juries] . And if we want a decision at all we may feel we cannot afford to set $P(E|\bar{H})$ equal to $P(E|H)$ [which implies the bland value of exactly 1 for the Weight $W(E|H)$], when perhaps we ought. Choices are necessary, but choices are difficult and potentially treacherous with SNOD waiting in the wings. And SNOD thrives on wish fulfilment. We want to get rich quick so we want to believe in the self-proclaimed financial wizard who urges us to buy those shares. We want to get well again so we want to believe in the wisdom and integrity of Dr X in the white coat.

Regrettably there are no universal nostrums for avoiding SNOD which we return to in Section (13:5) on Poor Thinking. However, thinking out important decisions in terms of Bayes' Rule will always alert one to the component assumptions i.e. the Prior $O(H)$, and the Weight $W(E|H)$ [= $P(E|H)$ /$P(E|\bar{H})$] which must, in one way or another, go to make it up. For instance had Smith known about Bayes it might have prompted him to look into the spontaneous recovery rate $P(E|\bar{H})$, and so he might have saved himself a great deal of money, and worry.

SUMMARY

In this chapter we have recognised that neither Deduction nor Induction can faultlessly make the kind of generalisations that science seeks in its quest for general laws. The best that can be done, it seems, is to work with

Odds, in other words to gamble, hoping to arrive, eventually, at a degree of conviction which, if not absolutely certain, is decisive. We found, through interrogating Stella, that successful gamblers appear to make use of Bayes' Rule:

Posterior Odds = Weight of Evidence E × Prior Odds

where the Posterior Odds have been updated by the inclusion of new evidence E via its Weight, defined as:

$$W(E \mid H) = P(E \mid H) / P(E \mid \bar{H})$$

i.e. the Probability of the evidence E *if H is true*, divided by the Probability of that same evidence E *if H is not true*.

Bayes' Rule appears to give a good account of common-sense thinking, both in a normal ('Burglary') and a scientific ('Hidden Galaxies') context, though it is often criticised for the apparently arbitrary Priors which it allows. But Priors are sometimes essential to rescue sensible thinking ('The Cancer Scare'). However unsound Priors can sometimes slip in unconsciously and undermine one's conclusions ('Tiptuppititis') in ways that benefit rogues and self-proclaimed 'experts'. The Weight one adopts for a new piece of evidence may be, to some extent, a matter of private judgement too. Thus two thinkers provided with the same evidence (see 'Hidden Galaxies') need not come to the same conclusions unless they agree both as to their Priors *and* as to their Weights (i.e. the reliability and relevance they attach to that evidence). Indeed further evidence, available to both parties, can actually lead to a divergence of opinions ('Flying Saucers').

Logicians have heavily criticized Bayes' Rule because of its subjectivity. Its Priors, its Weights, even its definition of Probability ['degree of belief in'] can be subject to cavalier prejudice – therefore, they say, it is 'not scientific'. The question though is 'Have they come up with something better?' Later in the book we shall argue that so far they haven't. Indeed they've often come up with 'Logical' schemes that are far worse. We return to the justification for Bayes' Rule in Chapter 8 on Common Sense and in Appendix 9 on Animal Wisdom. It rests on the association of two ideas (E and H), something even our dog Goch is capable of, though not always reliably.

You should consider using Bayes' Rule when doing your heavy thinking in future. If nothing else it will force you to lay out, in numerical form, the component prejudices (Priors) and clues (Evidence) – as well as the relative Weights you attach to those clues – in reaching your conclusions. That is a clarifying exercise in itself and a useful step towards eventually convincing others. We'll find much stronger arguments for it later.

THINKING FOR OURSELVES (Disney)
CHAPTER 5 (6.9kw)
THE DETECTIVE'S EQUATION
Draft 14/9/18

"Each piece of evidence on its own is not conclusive but put it all together and it's hard to avoid my conclusion." Nicholas Reeves, Egyptologist.

(5:1) DETECTIVES AND ASTRONOMERS

Bayes' Rule (BR) describes how a single piece of evidence can be incorporated into our thinking. It may increase our confidence in some hypothesis H if its Weight $W(E|H)$ is more than 1, or decrease it if $W(E|H)$ is less than 1, but it will seldom suffice, on its own, to bring about decisive conviction. For that several pieces of evidence must be weighed and combined in such a way that the Odds on H can become, one way or the other, convincing.

Evidence however is often conflicting, so a method must be found of combining it nonetheless, a method which is sound, transparent if possible, and potentially decisive. Here we come upon The Detective's Equation, which is the $E = mc^2$ of Common Sense and is the second Rule of Plausible Inference.

Start by considering two examples, one trivial, one momentous:

(a) The body of the young actress was found locked in the library with a stiletto in her heart. One of the other guests at the weekend house-party is the leading lady from the same theatre company. Known to be sexually jealous of the victim she falls under the detective's suspicion. Her alibi is that she was sleeping with the bishop – which he strenuously denies. But he would wouldn't he? And the servants had found a pair of frilly knickers in his bed the morning beforehand. The murderer had obviously escaped through the library window, leaving an 18 centimetre footprint in the flowerbed, whereas the leading lady's foot is only 16 centimetres long. But the footprint expert admits that, in the circumstances, the measurement error is about plus or minus 1 centimetre.

The question is 'How suspicious should the detective be? There is motive, albeit weak; an alibi, albeit fraught, and a footprint which might, or might not, fit: three pieces of evidence, none conclusive, but taken together?

(b) Galaxies are islands of stars in space – the building blocks of the Universe. Early in the 20th. century an astronomer measured the line-of- sight velocities of 6 prominent galaxies. All appeared to be receding from Earth at enormous velocities, which was remarkable in itself, but it further seemed that the more distant galaxies receded systematically faster. Could it be that the whole Universe was expanding? This was the dramatic hypothesis which entered his mind. But was it true? Did he have the evidence to make such an outrageous claim, or would he just be making an almighty fool of himself? Were his results the outcome of mere chance, or was it perhaps the most exciting scientific clue in history? What can be sensibly concluded from the tantalising evidence and what can not?

These are two instances of Plausible Inference (or Common Sense Thinking, CST) – trying to draw rather general conclusions on the basis of limited evidence – and are forms of 'hypothesis testing'. In the first case the hypothesis was '*The leading lady is the murderer*'; in the second '*The Universe is expanding*'. What we want is not absolute certainty – that is surely too much to hope for. We want the odds for or against an hypothesis, given the *combined* evidence. Are the odds 10 to 1 on, or 150 to 1 against? Knowing those odds one might be in a position to act wisely. And if one's method of arriving at those odds is both transparent and plausible, one might induce others to think likewise. We'll come back to these examples later. But first let us look for a sensible way to combine evidence.

(5:2) COMBINING EVIDENCE

Bayes' Rule

$$O(H \mid E_1) = W(E_1 \mid H) \times O(H) \qquad (5:1)$$

describes how a single piece of evidence E, a single clue, can be incorporated into our thinking. It can increase our Odds O(H) on some hypothesis H if the Weight W(E_1 |H) is more than 1, or decrease it if W(E_1|H) is less than 1. It's fine, so far as it goes, but it will seldom suffice, on its own, to bring about decisive conviction. For that several pieces of evidence must be separately weighed and then combined in such a way that the Odds on our H can become, one way or the other, decisive. How could we do that?

Suppose we have a second clue E_2 with Weight (E_2 |H). Using Bayes' Rule again it will further alter our Odds on H by

$$O(H \mid E_2) = W(E_2 \mid H) \times [\text{Odds before considering } E_2]$$

But [Odds before considering E_2] is simply O(H|E_1|) from Equation (5:1). So combining the two equations:

$$O(H \mid E_2) = W(E_2 \mid H) \times W(E_1 \mid H) \times O(H)$$

and so on. Thus we could combine any number N of clues together and write the answer:

$$O(H \mid E_1, E_2 ..... E_N) = W(E_1 \mid H) \times W(E_2 \mid H) \times ..... \times W(E_N \mid H) \times O(\mid H) \qquad (5:2)$$

For instance suppose our Prior Odds were 2 to 1 against i.e. O(H)=1/2, while Clue 1 had Weight W(E_1|H) = 4, Clue 2 had Weight W(E_2 |H) = 8, Clue 3 had Weight = 4 and Clue 4 had Weight = 2 then Equation (5:2) would say:

$$O(H \mid E_1, E_2, E_3, E_4) = 4 \times 8 \times 4 \times 2 \times (1/2) = 256 \times (1/2) = 128$$

i.e. after only 4 clues our Odds *on* H have risen to 128 to 1 – a pretty *decisive* result!

We call (5:2) "THE DETECTIVE'S EQUATION" because it can combine the effect of all the clues from E_1 to E_N that are available. We didn't do anything clever to reach it, we simply considered the clues in sequence, using Bayes' Rule over and over. And yet the Detective's Equation (DE in future) is *far more important than* $E = mc^2$! With a few modifications, which we will come to in later chapters, it is, I contend, the secret which enables living creatures to turn rough and ready clues into wise actions. In other words it is Nature's main survival mechanism. It is the heart, soul, engine and mind of Common Sense Thinking. By multiplication it can reach decisive Odds on any hypothesis from small numbers of weak clues; that is its secret.

This chapter will illustrate the Detective's Equation (DE for short) at work in a wide variety of circumstances which range from solving a murder, to wondering if one's partner has run off, to confronting mammoths.

(5:3) MURDER IN THE LIBRARY

Let us return to the murder described in Section (5:1) (a) above and the problem faced by the detective who has to try and solve it. The question is 'How suspicious should she be? There is motive, albeit weak; an alibi, albeit fraught, and a footprint which might, or might not, fit: three pieces of evidence, none conclusive, but taken together?

After our detective has interviewed everybody relevant, collected all the clues she has, including the footprint in the flowerbed outside the library, she sits down to consider the hypothesis H that '*The leading lady is the murderer*'. She proceeded to use the DE as follows:

(a) E_1 is the motive, here sexual jealousy of the murdered young actress. But sexual jealousy is commonplace while murder is very rare. The detective concedes that the leading lady's jealousy might double the odds against her – but no more than that. So she writes $W(E_1 \mid H) = 2$ to 1, or 2. Note that she has gone straight for a Weight without bothering to estimate it from the intermediate Probabilities. She's quite free to do that, and indeed it may make more sense when the Weight $W(E_1 \mid H)$ is going to be as crude as it is bound to be here. Her gut feeling, bred of 20 years experience as a detective, which is not to be ignored, comes in.

(b) E_2 is the actress's claimed alibi. The fact that the bishop challenges it must count against her – but not by much because he is a proven liar. The detective crudely estimates that $W(E_2 \mid H)$ is 4 to 1 on, or 4.

(c) Finally comes E_3, the slightly too large footprint. What is the Weight of that? The forensic expert who measured it has claimed that, more often than not, say two-thirds of the time, foot and print will lie within 1 centimetre of each other. So the 2 centimetre difference in the leading lady's case is rather a lot. The detective needs to know the likelihood that the measurement could be, due to measurement errors, as much as 2 centimetres discrepant from 16 centimetres, the actual size of the suspect's foot. She herself is in no position to gauge this so she rings the forensic lab to ask the expert. His reply, after some minutes for reflection, is rather technical : "If the measurement errors have a normal distribution, with a standard deviation of plus or minus one centimetre, then the likelihood of finding 18 centimetres or greater, is about 40 to 1 against". After some discussion the detective takes this to mean that $W(E_3 \mid H)$ is 1/40. It would then be sensible for her to write in her notebook:

$O(H \mid E_1, E_2, E_3) = 2 \times 4 \times (1/40) \times O(H)$

But what is O(H)? It is her Odds on H (that the leading lady is guilty) *before* considering *any* evidence. It is therefore her 'Prior'. How should this be quantified? There are 7 surviving guests in the house party and if one, but only one of them is guilty, the odds on any particular one, say the leading lady, are, being impartial, 6 to 1 against her being guilty. i.e. O(H) = 1/6 leading to:

$O(H \mid E_1, E_2, E_3) = 2 \times 4 \times (1/40) \times (1/6) = 1/30$

or 30 to 1 *against*.

She decided to write down her thinking in the form of a table as follows where H, her hypothesis in question, is *'The leading lady is guilty'*:

INFERENCE TABLE (5:1): MURDER IN THE LIBRARY

Clue (E)	O(H)	W(E|H)	O(H|E)
Prior	1/6		
Motive	1/6	2	2× 1/6 = 1/3
Opportunity	1/3	4	4/3
Footprint	4/3	1/40	1/30

So the leading lady is, in the detective's mind, clear *for now*, though she will be interested to reconsider the case when similar odds have been calculated for all the other suspects. The significance of Odds like these can only be judged in the context in which they might be used. 30 to 1 against might be quite adequate to deflect the detective's immediate suspicions. But would 30 to 1 *on* suffice for a jury to convict? What do you think?

This example brings three important matters to light:

(a) Firstly the sequential order in which the various clues are considered cannot affect the final conclusion because 1/40 times 4 times 2 is the same as 2 times 4 times 1/40 – indeed one doesn't have to think sequentially at all. Instead one can, if one chooses, consider all the evidence *together* and use it at once – and that is what the detective has done here.

(b) The Detective's Equation can handle *conflicting* evidence, as it has done here: here two clues with Weights 2 and 4 favour the hypothesis, while one with Weight 1/40 is strongly against. This is an extremely powerful and useful property which makes the DE as useful in everyday life as it is in the courtroom or laboratory.

(c) Writing out one's thinking in a Table like (5:1), which I call an 'Inference Table', is a very useful trick to learn. It summarises very clearly the way you've reached a decision; it's a great reminder for later consideration and it's often the best way to persuade others of your conclusion. We shall use Inference Tables all the time from now on, and I hope you will too.

(5.4) THE CASE OF THE MISSING WIFE.

Most of our thinking doesn't involve murders or Cosmology. So let's consider a much more mundane if painful domestic situation where the Detective's Equation could be equally useful.

A married couple were having breakfast when a post-card from Australia dropped through the letterbox. An old flame of the wife, returning to Britain for a few days, was inviting her out to dinner that very evening. An almighty row ensued with both parties banging out of the door vowing never to speak to one another again.

At 7pm the husband returns from work to find an unusually darkened house. By 9.30, with still no sign of his wife, the jealous husband begins to ponder whether she really has left him. The idea, the hypothesis H that '*She has run away with* X' begins to form in his mind, but with a low Prior odds of O(H) =1/100 because they have been reasonably happily married for nearly a year.[O(H) can only be a very rough guess – as it is. He might feel O(H) is not as high as 1 to 10 on, nor as low as 1 to a thousand, so an O(H) of 1:100 on, or 1/100, is an intermediate compromise.] All the same it is something, though not much of a coincidence, that she should be out tonight because 3 nights out of 4 she will be home before him – the other 1 in 4 being either delayed at work, or visiting family or friends. How much should the evidence (= E_1) that she is not home tonight worry him?

Sometimes it is easier to assign Weights directly, as the detective did in the last example, but sometimes it is better or easier to calculate them from the definition of a Weight which you will recall from Chapter 4 was:

$$W(E|H) = \frac{P(E|H)}{P(E|\bar{H})} \qquad (5:3)$$

where $P(E|\bar{H})$ means "the Probability of E if H isn't true". Here it turns out to be easier to find the two probabilities above and below in the equation above and to calculate the Weight therefrom.

He knows that $P(E_1|\bar{H}) = 1/4$ whereas $P(E_1|H) = 1$ i.e. she would certainly be absent if she had actually run off with X. That implies the Weight of E_1, i.e. $W(E_1|H) = 1/$

(1/4) = 4. If he cared to, the husband could update his odds on H (his worry) using Bayes' Rule:

$$O(H \mid E_1) = \frac{1}{(1/4)} \times \frac{1}{100} = 4 \times \frac{1}{100} = \frac{1}{25}$$

or 1 to 25 on. In other words the odds are now 25 to 1 against her having run away. We've laid out his thinking in the 'thinking table' which I call an 'Inference Table' (5:2) below. His thinking so far is contained in the top line of that table. The figure at the right of that line is 4/100 or 1/25 as we calculated above.

When she is still not back by 10.30 he rings round to her friends and relations because he would have expected her, with 90 per cent Probability, to be with one of them if she hadn't run away. But alas she is not! The evidence that she is not i.e. E_2, thus has a Probability in his mind of P($E_2 \mid \bar{H}$) =1/10, whereas if she had run away, E_2 would be certain i.e. P(E_2 |H) =1. That enables him to calculate the Weight W(E_2 |H) as 1/(1/10) = 10 and hence to use BR again to find his updated Odds O(H| E_2) = 4/10 which appears at the end of the second line in the table. Now he's worried; he believes there is a forty per chance she's left him. He embarks on a vigorous search for new evidence. He finds her credit cards are missing (E_3) but her suitcase, passport and jewels are not. He calculates, after each clue his new Odds on her being missing and his thinking is laid out line by line in the Inference Table below. (Take your time to go through it carefully; indeed re-constitute it for yourself). It's only different from the Inference Table (5:1) in being more complete, i.e. including the Probabilities P(E|H) and P(E|$\bar{H}$) as well as the Weight W(E|H) calculated from them.

INFERENCE TABLE (5:2) THE CASE OF THE MISSING WIFE
Hypothesis H: "She has run off with X"

Clue	O(H)	P(E|H)	P(E|$\bar{H}$)	W(E|H)	O(H|E)
Wife Out	1/100 Prior	1	1/4	4	4/100
Friends?	4/100	1	1/10	10	40/100
C. cards?	4/10	1	1/10	10	4
Suitcase?	4	1/5	1	1/5	4/5
Passport?	4/5	1/10	1	1/10	4/50
Jewels?	4/50	1/10	8/10	1/8	1/100
REORDER					
Jewels?	4	1/10	8/10	1/8	½
Passport?	½	1/10	1	1/10	1/20
Suitcase?	1/20	1/5	1	1/5	1/100

This calculation seems to track the kind of Common-Sense reasoning that might naturally pass through our minds when we are faced with such an unpleasant situation. The first three pieces of evidence, or clues, all had Weights in the husbands mind in excess of 1 and so they drove his suspicion, his Odds, from 100 to 1 against the hypothesis '*wife has left me*', to 4 to 1 *on*. This roused him to strenuous efforts to find new clues, and E_4 to E_6, finding her suitcase, passport and jewels still at home, all turned out to have fractional Weights, driving his suspicion down to 100 to 1 against once again so that he could go to bed and sleep soundly.

Notice, as we mentioned above, that the ordering of the evidence is immaterial to the final conclusion. To demonstrate the truth of that the clues E_4 to E_6 are also factored into the assessment in reverse order – see the last 3 rows in the Inference Table. While the intermediate conclusions naturally differ from those above in the table, the final result is identical i.e.

$$O(H \mid E_1, E_2....E_6) = W(E_1 \mid H) \times W(E_2 \mid H)....\times O(H) = 1/100$$

The Detective's Equation utilizes the Weight $W(E_a \mid H)$ of each separate clue. Recall that the Weight is a measure of the degree of association of the clue E_a to H in the thinkers mind. If H was indifferent to E_a, i.e. E_a was as likely to occur whether H was true or untrue [i.e. $P(E_a \mid H) = P(E_a \mid \overline{H})$], then the Weight is clearly equal to 1, see Equation (5:3). This is only common sense. But if E_a is far more likely to occur when H is true than when it is not (i.e. $\overline{H}$) then the Weight is much greater than 1, increasing our Odds on H significantly. This too is common sense. Finally, when E_a is less likely to occur when H is true, than when it is not Eqn (5:3) yields a fractional Weight, decreasing our Odds on H. So the Detective's Equation is common-sense from beginning to end, and the whole burden of the various pieces of evidence is summarized in their individual 'Weights'.

The Weight, being the ratio of two Probabilities [see Eqn (5:3)] – which are themselves a matter of private conviction, must be a personal number too. It isn't universal or objective – it belongs to somebody, in this case the husband. And in many instances it must be very rough and ready. How could the Weight $W(E_1 \mid H)$ of coming home to find the house empty be anything else? For a start it depends on whose house. Nevertheless, using a collection of rough and ready, not to say personal and conflicting Weights, the husband has arrived at decisive conviction. How has he done it – or rather how has The Detective's Equation done it for him ? Through *Multiplication*! That is the essence of our new equation. Multiplying small numbers together that are more than one quickly leads to a large number – decisive Odds (e.g. $3 \times 3 \times 3 \times 3 \times 3 = 243$) , whilst multiplying fractional numbers together quickly leads to a very small number – decisive Odds in the other direction (multiply 1/5 together 4 times,: answer 1/625).
Multiplication alone can turn crude evidence into decisive conviction. More than any other factor that is why Common Sense works – even though it can't always be right. Even if the husband doesn't get all of his Weights numerically exact (whatever that

means) , but they are all generally in the same sense (i.e. all more than or all less than 1), he will move in a consistent direction. Some may be too high, some too low, but he can hope that, generally speaking, such ill-weightings will, in the end, roughly balance out.

Not always however. A jealous husband will generally assign higher Weights to evidence supportive of H ('*she has run away*') [i.e. W(E|H) more than 1], and lower Weights to evidence inimical to H ('*no she hasn't*'). But that's his affair. He might be wrong, or he might be right. Either way he's the one who will have to live with his conclusions, as we all have to do with our own.

When they are discussing Bayes' Rule logical types have fits over the Prior O(H) – because it seems private, or 'unscientific' as they say. But in the context of the Detective's Equation we can see that it may be no more private than *some* of the Weights. Some Weights can be objective (e.g. the Weight that a fair coin will come up with so many more heads than tails.) but that doesn't mean (I submit) that many other useful clues E_x, E_y, E_z .... should be discarded simply because their Weights must necessarily be a matter of rough judgement (e.g. motives).

In summary the Detective's Equation is a tool for combining all sorts of evidence together, including conflicting evidence, to produce a decisive conclusion. The soundness of the conclusion must depend entirely on the soundness of the various Weights, and the Prior, that are fed into it by the tool user. Multiplication is the key factor which can lead to decisiveness. That multiplication arose from considering the clues in sequence. But the particular sequence is immaterial. Thus the multiplicative nature in which evidence is combined must be quite general, and that evidence could, if preferred, be combined together in a single operation. This is absolutely fundamental. Sophisticated minds capable of holding several pieces of evidence, and their Weights, in memory, can tackle far more sophisticated problems, *and* reach far wiser decisions. But if you can write you don't need a stupendous memory to achieve the same supercharged capability. You can write your thinking out in an Inference Table. [Could this be the secret of civilization? See later.]

(5:5) ANIMAL THINKING

At a time (1737) when Edinburgh was known as 'The Athens of the North' David Hume, the great philosopher of the Scottish Enlightenment, maintained that you can't justify Inductive Thinking on logical grounds. He argued that you can't use Deductive Logic to justify the Inductive kind, and you can't use Inductive Logic – because that would be to argue in a circle. There being no other kind of logic, we are stymied. This is called 'Hume's Problem' and it has never been resolved. Nevertheless we have to use Induction and we, that is to say humans and their fellow creatures, do so every day. Hume further remarked : "…no truth appears to me more evident than that beasts are endow'd with thought and reason as well as men…" As one who grew up amidst a virtual menagerie of animals including dogs, cats, pigs, ducks, hamsters, bantams, rabbits, mice, guinea-pigs and a horse I totally concur. One of our temporary house-guests was a jackdaw nestling who fell down the chimney, joined the family and taught himself to count to fourteen before he flew off to join his peers in the wild [See essay 'Our Jack' on our website]. Whether or not they worried about The Meaning of Life, they all had to conclude and decide, otherwise they wouldn't have survived. To attribute all animal behaviour to 'instinct' is to deny that animals can adapt to modern conditions, which many of them very plainly do. Take Herring Gulls for instance; once foragers of the tidal shore, they now make their living off modern man: picking over his garbage dumps, trailing his fishing vessels and following his plough. As a glider pilot I was fascinated to shadow small groups of such gulls as they thermalled from cloud to cloud over long distances across country. No observer who went with them could doubt that they knew exactly what they were doing and where they were going. Somehow they had formulated a communal plan and were sticking to a compass bearing, on their way to an objective far beyond our mutual horizon. It is a great conceit to imagine that other animals cannot think or plan – just because we do not know how they do it.

Unfortunately, very unfortunately, we here run into the philosophical, indeed religious legacy of the past. The philosopher Descartes (who was educated as a Jesuit) preached Duality – the complete separation of body and mind. Our Minds, bequeathed by God as he thought, were an incorporeal presence capable of wonder, intellect, imagination, empathy, forethought....It followed that animals, not so favoured by God, were incapable of such feats. Descartes' follower the philosopher Nicolas Mallebranche wrote: "Animals eat without pleasure, cry without pain, grow without knowing it; they desire nothing, fear nothing, know nothing."

Have you ever read such rubbish? How could anyone who had ridden a horse, as presumably he had in the 16th. century, believe it didn't experience pain when he carried instruments such as whips and spurs to deliberately torment it? I suppose such men were influenced by the Bible which taught; ' Then God said, "Let us make man in our image, after our likeness. And let them have dominion over the fish of the sea and over the birds of the heavens and over the livestock and over all the earth and over every creeping thing that creeps on the Earth." Three hundred years later Charles Darwin, who had beloved pets of his own, and more importantly had closely observed many animals all over the globe, argued that animals and humans differed in their mental powers only in degree, and his follower Georges Romanes wrote the influential 'Animal Intelligence' (1886) partly based on Darwin's notes.

They were however heavily criticised for their 'anecdotal' approach and it became the fashion across science (and Psychology – which is *not* a science) to regard animals as incapable of intention, no more than automata responding to 'instincts'. There was a certain intellectual snobbery in this "Behaviourist" approach: it demonstrated that you were more 'scientific' than the other chap. If you wanted to say any more about animals you had to "prove it" through the medium of their behaviour in carefully controlled laboratory experiments, something it was very often impossible or very difficult to do. Thus when Jane Goodall (arguably the most influential scientist alive today) came to observe chimpanzees making and using tools in the wild (1964) she had to pretend they did so without intention, and even today 'Behaviourism' is still respectable in both Ethology (Animal Behaviour) and Psychology.

Modern research, beginning with Jane Goodall's famous study of the Gombe chimps, described in her book 'In the Shadow of Man', is turning this old-fashioned nonsense about animal mentality on its head. If you aren't convinced then I have, in the list of references{1 to 9}, suggested several fascinating books you might like to read. A good place to start is Virginia Morrel's 'Animal Wise' which gives a fair summary of the modern scientific work on many animals from birds to fish to wolves and dolphins. Did you know that many animals make and use tools, that some birds can cache 10,000 seeds over a wide area (miles) and then find nearly all of them again when they are buried under deep snow, that crows can solve very complex puzzles, that magpies hold funerals, that baboons are much nicer than academics………? (Which actually isn't saying much.)

It is worth watching an animal trying to decide. Take our dog Goch. He evidently makes associations between observations and hypotheses; for example between 'master is putting on boots' and 'we are going walkies'. Goch knows from past experience that 4 times out of 5, when I've put on my boots, that I've taken him for a walk. He becomes alert, but not overly excited because he's had his disappointments too. He waits for more evidence. If I pick up my brief-case he goes back to his corner, evidently downhearted. But if I pick up my walking stick he wags his tail and runs for his lead. In the first instance he has obviously associated evidence (boots) and weighed it in connection with his hypothesis (walkies). What else is that but Bayes' Rule? In the second instance he's combining a new clue (stick) with the first (boots) to reach a virtually certain conclusion. Neither clue is convincing in itself, but combined together they appear conclusive enough to spur him into action. That looks very much like multiplicative combination to me, very much like The Detective's Equation in action. Neither walking-stick nor boots excite him overmuch but, in combination, they make him practically certain. What is that if not multiplication? Other of my experiments with him suggest, but do not prove, that Goch is capable of combining three weak clues multiplicatively.

Bayes' Rule makes explicit the degree of association between two concepts – a clue and an hypothesis. The Weight $W(E|H)$ that you give that clue is some measure of how close you think that association is. If past experience has taught you that that clue has invariably been associated with that hypothesis you will sensibly grant it a high Weight. Some (rare) times there will be a logical basis for that Weight's numerical value (e.g. when playing cards); more often it will be little more than an intuitive 'high', 'medium' or 'low'. To me then Bayes Rule is self-evident, an obvious mental tactic available to any creature capable of memory. It assumes that the future will be much like the past. No you can't prove that that is logically true, but on what other working model could one base a desire to survive? To deny that the future will probably be much like the past would be deprive oneself of most of the basis for rational behaviour.

On its own Bayes' Rule is a feeble tool. It considers only one clue and employs a single Weight to transform Prior Odds into Posterior Odds by means of multiplication. It will seldom be decisive.

However there is nothing to stop a creature from considering several such clues sequentially, multiplying by a new Weight each time. Such a combination of Weights, if they are mostly in the same sense, can quickly lead to decisive Odds. I am suggesting that The Detective's Equation is the recipe (*algorithm* in posh terms) which we animals continually use to reach the decisions, from the mundane to the existential, we all need to make in life. Most philosophers of Thinking didn't care what animals thought; that was their fatal mistake

(5:6) THINKING LIKE STONE-AGERS

Now that we know The Theory of Evolution, we can see how ridiculous it was to believe, as our great great grandfathers did, that we couldn't think until either the Ancient Greeks taught us, or Saint Paul was struck dumb on the road to Tarsus. If animals can think, as they certainly can, then stone-age hunter-gatherers must have been capable of taking decisions based on complex data. To see how it *might* have been done let us imagine ourselves back 100,000 years in a small Palaeolithic band. Having inadvertently trespassed and hunted on some other and larger band's territory we are now being pursued by them with murderous intent. How might Bayes' Rule and the Detective's Equation, which we must have been using even then – though we didn't know it of course, have helped us to escape?

We reach a valley which we desperately need to cross. Unfortunately it is being grazed by mammoths, enormous beasts we have never encountered before. Overawed by their size and their tusks we must estimate whether they are dangerous or not. Since they are obviously herbivorous, and apparently cumbersome, the majority opinion among the 4 seniors of the band is 3 to 1 on our working hypothesis H that '*Mammoths are harmless*'. That is to say our Prior $O(H) = 3$ as we cautiously set off down the hillside to cross the valley. At first the mammoths are oblivious, but then the giant herd-bull, previously invisible behind some trees, charges out trumpeting with rage. Targeting one unfortunate old lady he chases her so fast through the long grass that she only narrowly makes the safety of the rocks.

The tribe assembles to reconsider the situation. We still desperately need to cross the valley for there are enemies in close pursuit, so that making a detour could itself prove fatal. In reconsidering our options the tribe now have some evidence E to go on. One mammoth at least looks vicious, even if the rest appear docile. It is time to re-evaluate our hypothesis H that "*Mammoths are Harmless*" in the light of the experience E that one of them has charged. We certainly would not have made an explicit "Bayes' Calculation" but a rational discussion among ourselves would have amounted to much the same thing.

In order to arrive at a rational assessment of the *new* $O(H|E)$ we would need to arrive at estimates for $P(E|H)$ and $P(E|\bar{H})$ in order to calculate our Weight $W(E|H)$. Now $P(E|H)$ is the probability that mammoths will sometimes charge (= E), even though they are harmless (= H). The tribe have encountered such behaviour before among other large beasts such as oxen, which make a tentative charge but then turn tail when challenged boldly enough. Perhaps if the old lady hadn't run away the bull mammoth would have behaved more politely. Perhaps. It may seem unlikely, but it has to be considered because the tribe simply must make its escape from pursuit. We decide, on the basis of previous experiences with large herbivores, that P(E|H), that is to say the probability of mammoths making mock charges like oxen, is a pessimistic 1 in 20 or 1/20.

Finally $P(E \mid \bar{H})$ is the probability that mammoths will charge, given that they *are* dangerous (i.e. not harmless, therefore $\bar{H}$). This must be high but is not necessarily certain [i.e. =1]. Mammoths might have poor vision, or be docile when well fed, or ignore you if you are down wind. A rough guess at $P(E \mid \bar{H})$ might therefore be 9/10, which amounts to saying that mammoths, if indeed they are dangerous, will nearly always charge if approached, but one time in 10 = 1/10 = 0.10 will leave you alone.

We would now be in a position to re-evaluate our hypothesis, initially thought likely, that *mammoths are harmless*, because after the experiment:

P(E|H) =1/20; $P(E \mid \bar{H})$ =9/10 so the Weight:

$$W(E \mid H) = (1/20) \div (9/10) = (1/20) \times (10/9) = 1/18$$

while our Prior O(H) was 3 to 1 on, or 3. Were we to use BR we would then find:

$$O(H \mid E) = W(E \mid H) \times O(H) = (1/18) \times 3 = 1/6$$

i.e. 1 to 6 on, or 6 to 1 *against* mammoths being harmless. This calculation appears to give the kind of answer we might have expected intuitively – after the event E. It seems to embody learning from experience in a sensible way.

Unfortunately odds of 6 to 1 against are not decisive. With our children exhausted, and murderous pursuers closing in behind, we cannot afford to ignore the 1 in 7 chance of making our escape across the valley unscathed.

A second attempt (experiment?) is therefore made, and a second time the bull charges malevolently. Does that mean that the odds are now twice as great, or 12 to 1 against us crossing safely – in which case it might be worth making a third try – or what?

With this second piece of evidence (E_2) our Bayes' calculation could be updated in the light of it as follows. Nothing has changed except that now O(H) must be amended from 3 to 1 on to 6 to 1 against – as we had previously calculated. In that case the new BR calculation becomes:

$$O(H \mid E_2) = W(E_2 \mid H) \times O(H)$$
$$= W(E_2 \mid H) \times O(H \mid E_1)$$
$$= \frac{1}{18} \times \frac{1}{6} = \frac{1}{108}$$

so the odds *on* H have dropped to roughly 1 to 108, or roughly 100 to 1 *against*.

Because our Weight was 1/18 and doesn't change, then each time we made the experiment and a mammoth charged it reduced our Odds on "*Mammoths are harmless*" by a factor of (1/18). Two charges have thus reduced our faith by (1/18) times (1/18), from 3 to 1 on to 108 to 1 against. That's pretty decisive. We decided to take the long way round – and obviously survived.

The above argument used Bayes' Rule twice in succession, which amounts to going through an Inference Table line by line twice, as follows:

INFERENCE TABLE (5:3) CONFRONTING MAMMOTHS

HYPOTHESIS 'Mammoths are harmless'

Clue	O(H)	P(E|H)	P(E|$\bar{H}$)	W(E|H)	O(H|E)
	3/1 Prior				
1st charge	3	1/20	9/10	1/18	3 × 1/18=1/6
2nd charge	1/6	1/20	9/10	1/18	1/18 × 1/6=1/108

We can also see we have arrived at the Detective's Equation, i.e. successive applications of Bayes' Rule, in an utterly natural way. Serious Thinking is very old indeed. If it wasn't we wouldn't be here.

(5:7) WHEN SHOULD ONE USE THE DETECTIVE'S EQUATION?

The short answer is 'Almost always'. If you have at least a few minutes before having to make a decision it is always wise to compose an Inference Table, fill in the Probabilities (Or go directly to the Weights, whichever you prefer) then multiply up to reach your conclusion, and perhaps your decision. Some people are reluctant to proceed so, not knowing how to fill in their Weights with precision. So here I will let you into an amazing secret whose good sense will only become apparent later: USE ONLY ROUGH AND MODEST WEIGHTS either in favour or against any hypothesis as follows:

'Strongly for' use W= 4   'Strongly against' use W = 1/4
'Weakly for' use W= 2   'Weakly against' use W= ½
'Neutral' use W = 1

It is unnecessary, and indeed imprudent, to be either more forthright or more precise. When I first discovered this prohibition it came as a great shock to me, but, as the pages will show, it is as wise as it is convenient. And we don't have to be geniuses to pick the optimum Weight for any clue when there are only 5 categories to choose from.

EXERCISES(5:1) DETECTIVE'S EQUATION

SUMMARY

Different pieces of Evidence $E_a, E_b, \ldots$ can be combined using The Detective's Equation:

$$O(H \mid E_a, E_b \ldots E_z) = W(E_a \mid H) \times W(E_b \mid H) \ldots W(E_z \mid H) \times O(H)$$

where the Left Hand Side is our Odds on hypothesis H – taking all of the clues into account. O(H) on the RHS is our Odds on H *before* we considered the evidence and is known as 'The Prior', just as the Odds on the left are known as 'The Posterior Odds'.

The Weight W(E |H) of each piece of evidence is a personal estimate of its relevance to H and its reliability and is the weight it should carry in the argument. Values of W greater than 1 increase our Odds on H by W, values less than 1 decrease it. Your Weights W(E|H) can be guessed directly or calculated from the Probabilities P(E|H) and $P(E|\bar{H})$ [where $\bar{H}$ means 'not-H' i.e. all the other hypotheses bar H which might give rise to E] using the definition W(E |H) = P(E|H) / $P(E|\bar{H})$. The ordering of the evidence is immaterial to the conclusion. The steps in the argument, clue by clue, can be laid down in the succeeding rows of an Inference Table as in Table (5:1). Such a table is useful for clarifying your own ideas, reminding you not to leave important factors such as $P(E|\bar{H})$ out, and for later explaining your thinking to others.

The crucial aspect of The Detectives Equation is its *multiplicative* nature. Thus a few pieces of consistent evidence, each separately of modest Weight, but mostly in the same sense, can nevertheless combine to yield Odds on H of a decisive nature – Odds on which we can act.

Strong (high Weight) evidence is more decisive than weak (low Weight) and should be sought by experiment, observation, research, analysis etc.

Weak evidence though should not be dismissed because, in combination, several pieces of weak evidence can over-rule a single strong piece. And weak evidence may be much 'cheaper' to find, and often more reliable.

Quantitative evidence *can* be very strong and should be sought whenever possible. However its strength (Weight) depends entirely on understanding the uncertainties and errors involved in its estimation – which is often extremely tricky because *we have no inbuilt instinct for weighting precise numbers*. Precision and Weight are by no means the same thing! Being precisely wrong is much worse than being roughly wrong. It is wise to be doubly cautious about *any* Weighty clue. We devote Chapters 9, 10 and 11 to discussing Errors.

The Detective's Equation is the Second Law of Plausible Inference, of Common-Sense Thinking (CST). Like Bayes' Rule it too almost certainly has pre-human roots, so that we use it unconsciously.

The Detective's Equation very rarely leads to absolute certainty in the Logical sense. It is provisional and can be overturned by new evidence – that is the nature of Plausible Inference. However it can lead to *Inductive Conviction*, that is to say to Odds on H so high that there seems little point in challenging H any further. *We would then be wise to act as if H was certain.* Such *Inductive Conviction* allows most of us to live quite happily without the Deductive Certainty which some philosophers have insisted upon.

Take your philosopher, your Karl Popper for instance, who claims that Induction is impossible {13/1} and therefore that there is little hope of establishing the laws of aerodynamics on a sound footing. He flies confidently none the less because sub-consciously he is applying The Detective's Equation all the time. Every successful flight that he completes, indeed every flight that takes off and lands successfully somewhere on the globe, carries a positive Weight in favour of the aerodynamic laws. Multiplied together tens of thousands of times a day they lead to a confidence so decisive that no one needs to bother about the matter any further. And note that such Inductive Conviction is far more robust than Deductive Certainty. The former can only be overturned by a longish series of aircraft crashes whereas the latter can be shattered by a single trivial occurrence inconsistent with the premises upon which any deduction must rest. Thus the discovery of a single triangle containing 181 degrees would send the whole of Euclidean Geometry into a wobble. Thank goodness then for Inductive Conviction; it is the basis for Common Sense Thinking.

THINKING FOR OURSELVES (Disney)
CHAPTER 6 (4.4 kW)
NUMBERS AND THINKING (draft 14/9/18)

" .....*better to be roughly right than precisely wrong*....."
John Maynard Keynes, Economist.

(6:1) INTRODUCTION

Animals have been 'Thinking', that is to say making decisions about what to do next, for hundreds of millions of years. And if we can think then it is because we inherited most of that vital survival mechanism – call it Common Sense if you like – from our animal forbears. But animals do not appear to count – at least not much – whereas homo has been counting for at least forty thousand years – to judge from the oldest tally marks found scratched on Palaeolithic bones. What they were counting we do not know – the numbers of their flocks perhaps? Inevitably numbers would be dragged into Thinking, so much of which relies on reasoning of the "more than – less than" kind. We have already incorporated numbers in the form of Weights – to which we have assigned numerical values – in the form of Odds. But that transcription of Weights into numbers is artificial – very convenient, but artificial none the less. If innumerate animals can think using Common Sense they have to weight clues into categories such as 'very strong', 'weak', 'neutral' and so on without assigning actual numbers to them. And presumably they have a mechanism for compounding two weak clues to make one strong one. But they don't need numbers to do so, any more than they need symbols like $O(H|E)$ for Odds. We have introduced numbers and multiplication as a means for explicating and exploiting Bayes' Rule – which we feel underlies the entire process for suspecting that some hypotheses is either strengthened, or weakened, by a particular piece of evidence. Once we admit that artificiality – and I don't see how we can avoid doing so – then we are bound to be sceptical about *precise* numerical Weights henceforth. Anyway it is not the precision of Weights which matters, but their combined value after multiplication. '125 to 1 on' is no better than '100 to 1 on', but is vastly more convincing than '4 to 1 on'.

Passing on from Weights, what about clues themselves being numerical? – £430; 25,000 voters; 1.73516 millimetres; 0.089 micrograms; 18 deaths among 73 treated patients; 16 Million light years; 4.1 % growth per annum……. Numerical evidence appears to open whole new continents for analytical thinking, and on those continents new and flourishing populations have multiplied: physicists, statisticians, economists, pollsters, management consultants, operational research scientists, accountants, finance experts, forensic scientists, Quants, informatics experts…… In one or way or another they all use esoteric techniques to analyse numerical data and arrive at conclusions which they eventually try to sell on to the rest of us.

Take two simple examples we have encountered already. If my Theory of the Heavens predicts that sunrise will occur at 05:27:31, and it actually occurs within 2 minutes of prediction, you will grant that my theory has strong evidential support – far stronger than if it was 89 minutes out. Then there was the forensic evidence in the case of The Murder in the Library. The discrepancy in measured size between the leading lady's foot (16 plus or minus 1 centimetres) and the footprint in the flower bed (18 plus or minus 1 centimetres), was sufficient to supply a very strong Weight (40) against her guilt, a Weight at least 10 times stronger than those accorded to either Motive (2) or Opportunity (4). You can see the attraction of numerical evidence. It is often easy to obtain, precise, and therefore decisive. No wonder whole new professions have mushroomed around the business of getting it and then exploiting it.

There are however several vitally important caveats. All CST requires Weights. No number or measurement can be incorporated into any calculation of Odds without first having a Weight assigned to it (Recall that it is the Weight, and not the number itself, which contributes toward a decision). And that, as we shall see, can be very tricky. Secondly, numerical evidence is, in Evolutionary terms, a very new and purely human device. Thus we have not had time to develop reliable instincts for dealing with it, as we have for older kinds of evidence. Indeed we have to be taught how to analyse it – always a problematic undertaking which must assume that the teacher knows what he-she is talking about. Thirdly the techniques for dealing with numerical data can be highly technical – and so opaque. This can certainly fool outsiders but, as we shall see, it can fool the insiders too – and often has. Advanced Mathematics is then appealed to and horrors like The Central Limit Theorem (of which more anon) turn up – which scarcely anybody understands. The opportunities for self-deception on the one hand, and deliberate bamboozle on the other, multiply exponentially. Suffice to say that much numerical evidence is not at all what it seems, not anywhere as decisive as its proponents claim. This is a hard lesson to learn, and often very painful. But we must all beware of numerical evidence – and those who preach it. Even if they are not rogues, they may be naive. Thus Long Term Capital Management was a hedge fund set up by two Nobel Prize winning mathematical economists which lost 4 billion $ in a few weeks in 1998 and had to be bailed out by a consortium of big banks before it brought the entire financial system down.

(6:2) LOOKING AT NUMERICAL DATA.

The Detective's Equation (DE) is far more decisive than Bayes' Rule because it compounds several pieces of evidence – not just one. Likewise with numerical data. One good fit between hypothesis and measurement should never be convincing; several rough fits are nearly always more convincing than a single precise one. It helps to look at the matter pictorially, which is what scientists nearly always do (See Fig 6:1)

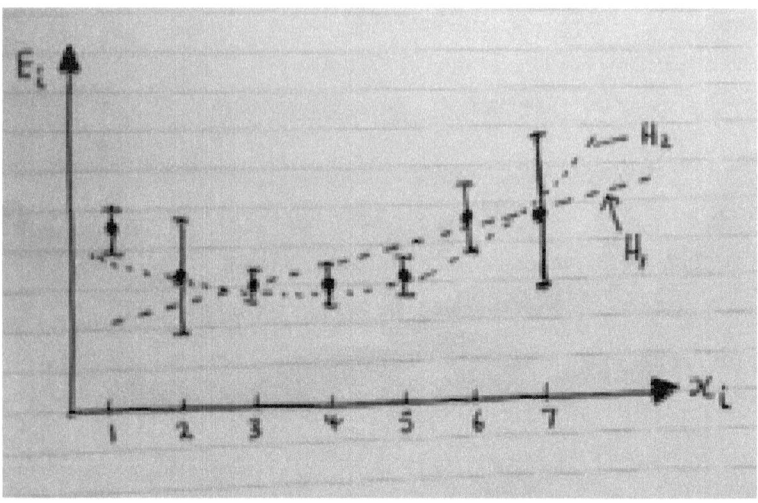

Fig (6:1) A typical graph of numerical data: they might represent temperatures measured against time, profits against investment, ... Here there are 7 data points each with an estimated error indicated by the error bars. Those estimates might have been arrived at through making repeated measurements of the same quantity. The bars are such that the datum is found to lie within the bars in two thirds of such measurements. Here two theories, or hypotheses H_1 and H_2, have been proposed to account for the data, with each generating an expected curve of E against x. The aim is to find what the Odds are on either of the hypotheses, and then perhaps to choose one.

Fig (6:1) shows a typical situation in which numerical measurements are being used to decide whether some hypothesis H_2 is likely to be right, or at least whether H_2 is better than some alternative hypothesis H_1. There is a limited number of data, and none of those measurements are perfect; they have estimated errors associated with them whose size is indicated by the error-bars. What we are looking for is a good overall fit between the data and a curve generated by our hypothesis. Where the data is precise (small error bars) we look for a close fit between hypothesis and measurement, but where the data is looser (larger error-bars) a larger discrepancy would be tolerable. Thus the Weight of each datum, as it bears on hypothesis H, should be related to the discrepancy, and to the size of the corresponding error-bar [call that σ_1, 'sigma-1' for point 1]. It is customary to write the discrepancy $t_m \equiv (E_m - E_H)/\sigma_m$, where E_m is the measured value at point 1, and E_H is the hypothetical value there. Then if we could estimate a Weight W(t_m|H) for each measurement E_m the DE tells us:

$$O(H \mid t_1, t_2, \ldots t_N) = W(t_1 \mid H) \times W(t_2 \mid H) \times \ldots t_N) \times O(H)$$

which makes eminent Common Sense. It is always helpful to have a picture in one's mind of how any abstract concept works and Fig (6:1) is really a drawing of how the DE operates. You will scarcely find a scientist's office without a sketch much like Fig (6:1) on the blackboard. The smaller a particular discrepancy t_m is the higher its Weight W(t_m|H) will presumably be, and the picture tells us that the overall better fit a hypothetical curve makes with the whole set of numerical data, the more confidence (i.e. higher Odds) we can have on H. Furthermore we have a numerical way for discriminating between the two hypotheses H_1 and H_2 because one will make a better fit, and have numerically higher corresponding Odds. Yes it all makes Common Sense.

But wait a moment. Where did the various Weights W(t_m|H) come from? Somebody has to estimate them, and have some rationale for doing so. You can see there's a potential problem there. For the moment we are going to brush that problem aside, in order to see how the whole scheme could work in practice; see

159

how a set of numerical data points $E_1, E_2, ... E_N$ with their estimated errors $\sigma_1, \sigma_2, ... \sigma_N$ can be used to find the Odds that some, or indeed any hypothesis H, could have generated them.

As a starting point we have to introduce a specific Table of Weights. In the next section we are going to do a realistic problem in extragalactic astronomy. Our astronomer is going to use the Table (6:1) of Weights below which he extracted with much difficulty from a very thick tome of Statistics which he found in his Observatory's library. They look at least plausible in the sense that the smaller the Discrepancy t the higher the tabulated Weight. Whether they are the best Weights he could have used I leave out of the discussion for now because in this chapter and the next it is not the particular Weights we are interested in so much as the way hypotheses can be manipulated to try and exploit them. Later on, and in Appendix (A1) we go into the tricky business of finding appropriate Weights in considerable detail.

TABLE (6:1) THE ASTRONOMER'S WEIGHTS

Discrepancy t	Weight W(t |H)
.1	12
.2	5
.4	2
.7	1
.8	1
1	½

(6:3) IS THE UNIVERSE EXPANDING?

Now let's do a real example. Galaxies are islands of stars in space – the building blocks of the Universe. Early in the 20th. century an astronomer measured the line-of-sight velocities of 6 prominent galaxies, with results as portrayed in Fig (6:2). All appeared to be receding from Earth at enormous velocities (hundreds or even thousands of kilometres a second), which was remarkable in itself, but it further seemed that the more distant galaxies receded systematically faster. Could it be that the whole Universe was expanding? This was the dramatic hypothesis which entered his mind. But was it true? Did he have the evidence to make such an outrageous claim, or would he just be making an almighty fool of himself? Could Fig (6:2) be the outcome of mere chance, or was it perhaps the most exciting scientific clue in history? It was difficult to estimate the velocities, and the error-bars portray his uncertainties in those estimates. If the Universe *was* expanding smoothly then the measurements ought to lie on a straight line like the one shown, sloping up to the right. They sort of fit, but then again they sort of don't. What can be sensibly concluded from the tantalising evidence and what can not? Here is a dramatic example of using numerical clues to try and reach a sound conclusion.

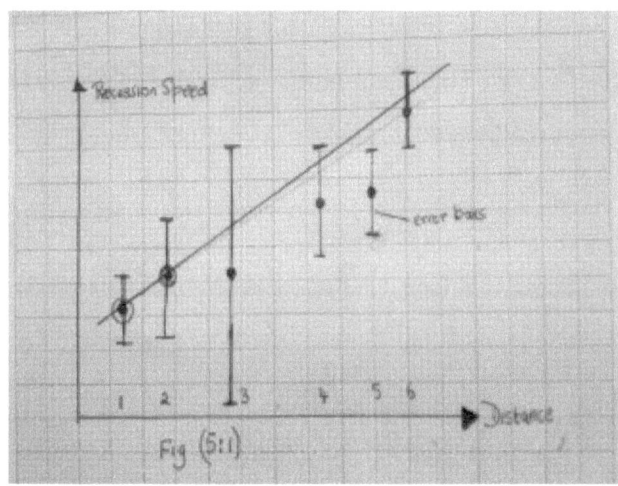

Fig (6:2) Galaxy recession speeds plotted as a function of their distances away (horizontal). Note the apparent trend of increasing speed with distance which *might* imply that the Universe is expanding. But there are large errors denoted by the error-bars. They mean that there's a 2/3 chance that the true value lies between the bars of each measurement. Some measurements, like 3, evidently have much larger errors than others, e.g. 5.

Should he dare to claim this astounding discovery? The practical problem for him was that it was extremely difficult to come by more and better data. It had taken him years to accumulate what he had. The spectrum of a galaxy was so faint that sometimes it took him a week of long clear freezing nights up on a mountain peak to measure the recession speed of a single object, and then only crudely. And as for the distances they were plainly so enormous that it was difficult to see anything recognizable within his galaxies, and so he had to hazard a guess. But he did his best and the results of a decade of work, together with his estimates of his errors, his plus and minus probable scatter on the crude velocities, are all in the figure. Each velocity is given a different error-bar corresponding to the difficulty he found in estimating it. (He didn't dare to estimate the errors on his distances; they were just too problematic).

The question facing him was: "Is there a positive slope (trend) in my data, and if so how credible, how significant is it? Will it convince my colleagues or merely label me as a buffoon? " This was plainly a case for The Detective's Equation and its use in an Inference Table. There were 6 independent measurements, each with its associated error, and a simple but astounding hypothesis (*'The universe is expanding'*) to test.

"If it is expanding smoothly (Hypothesis H)" he argued " All my points ought, within the errors, to fit on a straight line because if a galaxy is twice as far away it ought to be receding twice as fast. But which straight line? Suppose I fit one through the leftmost 2 points, on the grounds that their estimated errors are probably the lowest because they are the nearest. Then the 4 remaining points can be used to test or 'validate' my hypothesis.[See Fig (6:2)]. For instance point 3 is discrepant from that line by 0.2 error bars (he wrote $t_3 = 0.2$ in his Inference Table), point 4 by 0.4 ($t_4 = 0.4$), point 5 by 1.0 and point 6 by 0.1($t_6 = 0.1$). Clearly the further a point is away from that line, relative to the size of the appropriate error-bar (i.e. the higher its t-value is), the less support, the less Weight it should give to the hypothesis on which it is supposed to fit. Conversely the closer a point is to that line (i.e. the lower its t-value) the more support, the more Weight it should give to my hypothesis."

Now he couldn't use the Detective's Equation without first defining his Prior. What was he to choose for his O(H)? After some thought he argued that the Universe must either be expanding, contracting, or static. Lacking further information, he decided all three hypotheses should be granted equal status, in which case the impartial Odds *against* any one of them are 2 to 1, or 1 to 2 on, i.e. ½.

Having filled in his Prior and his discrepancies our astronomer now needed values for his Weights. He had a hazy recollection of his undergraduate lectures where the Stats. lecturer had talked about 'The Normal Distribution of Errors and its wide application to science'. So our astronomer went to his observatory library and extracted the figures in the above Table (6:1) from the back of an authoritative looking textbook on Statistics.

THE ASTRONOMER'S INFERENCE TABLE (6:2) : THE EXPANDING UNIVERSE

HYPOTHESIS: *'The Universe is expanding'*

Clue	O(H)	$E = t_x$ Discrepancy	Weight $W(t_x\|H)$ from Table (6:1)	O(H|E)
Prior	1/2			
Point 3	1/2	0.2	5	5/2
Point 4	5/2	0.4	2	5
Point 5	5	1.0	½	5/2
Point 6	5/2	0.1	12	30

With growing excitement the astronomer worked his way down the Inference Table clue by clue until his last galaxy, galaxy number 6, his last point on his graph, yielded a Weight of no less than 12 and raised the Odds on his amazing hypothesis that the whole Universe is expanding to 30 to 1.

Poor Devil! Try to imagine yourself in his shoes. Either he has made the most exciting scientific discovery of all time, and would be hailed as a colossus alongside Galileo and Newton – or he would become a laughing stock throughout his profession, and probably far further afield. He couldn't sleep for a fortnight. He couldn't tell his wife because, like most astronomers wives, she thought he was mad already. As for his children – well they had far more important matters to worry about. And the one colleague he felt he could confide in warned him not to breathe a word about his ridiculous idea. He could lose his job, or even worse, finish in the asylum.

What do you think? What would you have done in his place? You know all the facts available, and you have seen his argument laid out. Would you have made a bid for glory at Odds of 30 to 1 on, or would you have slunk off into obscurity with the prospect of having to later watch some other bolder man seize your one opportunity for immortal fame?

LONG PAUSE FOR THOUGHT

I should tell you that this story is not entirely imaginary although I have made the numbers up. An astronomer called Vesto Slipher working at the Lowell Observatory at Flagstaff in Arizona was left in very much this tantalising position back in 1917 . What went through his head and his heart we do not know. Certainly he didn't use the Detective's Equation because it hadn't been discovered then but probabilistic thinking of a like kind must have kept him awake at night. For now we'll leave the poor devil to his agony. But we'll return to him in the next chapter, a discussion of muddy thinking in general. There it will lead us to the third vital rule of Plausible Inference – Ockham's Razor.

(6:4) PROBLEMS WITH FINDING WEIGHTS

Weights are everything when it comes to including clues, be they numerical or otherwise, in any calculation of Odds. But reliable Weights can be hard, very hard to come by. In the last section we glossed over that difficulty; in this section we discuss the problem of estimating Weights in a little more depth.

Recall that to fit the sense of Bayes' Rule we defined the Weight of clue E as:

$$W(E \mid H) \equiv \frac{P(E \mid H)}{P(E \mid \bar{H})} \qquad (1)$$

Where $\bar{H}$ means 'not-H', i.e. any other hypothesis but H for explaining E. Eqn. (1) makes sense because if there is little chance that any hypothesis but H could explain E then $P(E \mid \bar{H})$ must be small while $P(E \mid H)$ must be close to 1 and so, according to (1) the Weight will be high [because dividing 1 by a small quantity yields a large quantity].

Equation (1) tells us that to find a reliable Weight for E we need a clear understanding of $P(E \mid \bar{H})$ − the possibility, *indeed the numerical probability that E could have some other cause but our chosen hypothesis H*. Now that is a tall order when you think about it. How can we ever be certain that some other phenomenon but the one we have in mind (i.e. H) is not responsible for E? To see the challenge let us look at two examples:

In our mediaeval village some of the children develop a strangling throat from which half of them die. Terrified that our own children will follow them into the grave we desperately seek for a remedy. Witches are known to cause diphtheria so a witch-finder from the next town is sent for. Old ladies lacking powerful relatives are rounded up and he commences his grisly investigation. Eventually Mrs Jenkins, the widow who bakes the village bread, is found to be possessed of the devil. How clever the witch-finder was not to be deceived by her outwardly kindly behaviour and her fondness for children. Thank goodness we brought in an expert. The whole village turns out to watch her dragged to the stake and burnt alive, still protesting her innocence.

Mrs Jenkins was a victim of Equation (1). The microscope hadn't been invented and the whole concept of bacteria ($\bar{H}$) lay centuries in the future. With $P(E \mid \bar{H})$ considered vanishingly small the Weight of evidence W(dead children| Mrs J. killed them with witchcraft) was very high. As a result thousands of poor witches were hunted down by experts and burned alive − while children continued to die.

Thus ignorance, or more general lack of imagination, can lead to Weights on certain hypotheses becoming far too high. Beware of $\bar{H}$ – some alternative hypothesis.

Even if there were no $\bar{H}$ there is still the probability of Chance, call it $\bar{H} = C$. Then from (1)

$$W(E \mid H) = \frac{P(E \mid H)}{P(E \mid C)}$$

from which it is clear that before we assign a high Weight W(E|H) we must be certain that P(E|C) – the Probability of E arising by chance – is low, or indeed very low before we take drastic action. Suppose there is even a 25 % [P(E|C) = 0.25)] Probability of chance being responsible we cannot assign a Weight to H, according to the above equation of more than

$$W(E \mid H) = \frac{1}{0.25} = 4$$

Now here is the rub: unless one has a reliable theory of Chance – as it applies to *the specific context in which we are interested* – one cannot estimate P(E|C) – and hence W(E|H) – with any confidence. In that case, in that context, one ought to be cautious.

The fact is that, outside trivial games of chance, we seldom have reliable estimates for P(E|C). For instance if I am measuring star positions with a sextant I may mis-read the scale, or mis-record the correct reading into my notebook – especially when I am seasick. No theory of Errors, no Theory of Chance, can take seasickness into account.

We don't burn witches any more, but occasionally we do send people to prison for life for crimes they did not commit. In Chapter 11 we take up the very sad case of Sally Clark who was so condemned in 1997 for murdering her two babies. All I will say for now is that it came about not through fallacious evidence, nor poor police work, but through unsound thinking on the part of the Court – and in particular the judge. But don't blame the judge. All around us intelligent people,

Statisticians in particular, are reaching unsound decisions because they insufficiently understand Equation (3). So take a very good look at it.

PAUSE FOR THOUGHT

And what about the forensic scientist in the Murder in the Library case? If he had understood Equation (3) he might have hesitated before employing "Normal Weights" as he did. He should have measured his errors, not assumed some theoretical distribution for them. Alas he had a muddled notion of his undergraduate notes on Statistics, and condemned an innocent parlour maid to 20 years hard labour.

SUMMARY

Counting and measuring are purely human skills of very recent vintage. Numerical evidence is so attractive because it is often easy to obtain, precise and therefore apparently *decisive*. Thus whole professions such as Physics, Statistics, and Management Science have grown up around the gathering and analysis of numerical data.

Precise though it may be, *numerical data cannot be incorporated into thinking until after it has been Weighted!* The thinking process then goes on to calculate the Odds on some hypothesis using the Detective's Equation, a process that can fruitfully be pictured graphically as trying to fit data-points, with their estimated errors, to some curve generated by the hypothesis in question. The better the fit, the higher the Odds on H, given the data values $E_1, E_2, ...E_N$ (See Fig(6:1)) [Different hypotheses will give rise to different curves].

But how do we estimate the Weights W(E|H) etc? That remains to be discussed later, and will turn out to be tricky. It cannot be emphasised too strongly that precise measurements do not necessarily lead to precise or reliable Weights. We argued that:

$$W(E \mid H) = \frac{P(E \mid H)}{P(E \mid \bar{H})}$$

which means that our Weights are only as good as our knowledge of hypotheses $\bar{H}$ *other than the one we are considering*. And how can we be certain that there is no better hypothesis ($\bar{H}$ = bacteria) than H (= witchcraft)? And even if there is no other such but Chance ($\bar{H}$ = C) we still need a reliable knowledge of Chance, or 'Error', as it occurs in the very particular context in which we are interested. For instance Measurement Errors generally have to be *measured* (e.g. by repetition), and not assigned from the same bell-shaped distribution applying to Counting Errors. Although that distribution is called 'The Normal Distribution of Errors' it is not normal at all in the sense of 'usual' – so beware of using it, and especially beware of some one using it against you illegitimately.

Because they dismiss Common Sense, scholars generally have a very poor understanding of the distinction between precise evidence and reliable Weight. Statisticians in particular use numerical tables accurate to 4 figures (1 part in ten thousand) when it is often true that their Weights can be known to no better than a factor of two (i.e. 100% or worse). Such folly is hard to credit but most often comes about because they try to use Mathematics – where Common Sense is required instead.

We have so far only touched on what is a very sensitive matter – but one which is absolutely vital to Thinking. Chapter 9 is however devoted to Error Analysis; Chapter 10 to the Principle of Animal Wisdom (PAW) – which is all about being cautious with Weights, while Chapter 11 is on 'Terror Analysis' which is what I call Statistics. I will argue there that Statistics is a folly that has grown out of mistaking Mathematics for Common Sense, precision for reliability. If I am right, that is excellent news because we will be able to analyse numerical data using

Common Sense without ever having to resort to, or pay attention to, complex Statistics.

At this point you might take a first look at Appendix A8 *"CertaintyFalsifiability and Common Sense"*.

THINKING FOR OURSELVES (Disney)
CHAPTER 7 (7.9 kw)
WOOLLY THINKING AND OCKHAM'S RAZOR

(draft 14/9/18)

" ...*the scientific enterprise has never produced, and never will produce, a single conclusion without invoking parsimony. Parsimony is absolutely essential and pervasive.*" Hugh Gauch Jr, "Scientific Method in Practice" 2003, CUP

(7:1) A REALLY WISE MOVE

All hypotheses should be tested against evidence. *The best reason to believe any hypothesis is that it fits the existing evidence better than it has any right to do by chance* . A simple hypothesis has little chance of fitting more complex data unless it is actually right. Complex hypotheses on the other hand are much harder to test because they have more flexibility to adjust to potentially threatening evidence. An infinitely flexible hypothesis is invulnerable to test, and therefore there are no grounds whatsoever for believing it to be right. A principle called Ockham's Razor (or sometimes 'Parsimony'), urges thinkers in general (not just scientists) to always prefer simpler hypotheses to elaborate ones for two reasons. First because, if both fit the existing evidence, the Odds on the simpler one being right are higher, perhaps much higher. Second because if a simple hypothesis is wrong it is much easier to prove it so, and move on. In other words simpler hypotheses are more practical and more progressive than elaborate ones which can clog up the works for ages, even millennia, as we shall see.

Suppose I have lost my specs. I could formulate a number of hypotheses as to what has happened to them:

I have left them up in my bedroom.

They have fallen down behind some furniture.
I left them in a shop while I was out.
One of my flatmates took them to work by accident.
That house gremlin has been up to his evil ways again.
Interstellar travellers have broken into the house and stolen them.

Disregarding which of these hypotheses is most probable, which may be a matter of personal judgement, there is no doubt they are ordered in a sequence of increasing complexity. Therefore it makes good sense to pick the simplest first and go up to my bedroom to test it. Testing the second will mean a good deal of work because there's all too much furniture. As for shops I must have been to a dozen earlier today. And so on. As for the last two hypotheses don't go anywhere near them because they are invulnerable. Whether they are plausible or not there's no way of testing them out so that entertaining them, however tempting, will get us exactly nowhere.

In such a simplistic example so much is obvious, not least because the difference in complexity between the various hypotheses is so glaring that appealing to Ockham's Razor is superfluous. In many practical instances however things are not so simple. Then we need to appeal to the Detectives Equation to see exactly why, despite the temptations, simpler hypotheses are always better. We shall later see how time and again Ockhams Razor has played a key role not only in scientific history but in military intelligence (Chapt 14). I would go so far as to contend that anybody who doesn't use Ockham's Razor on a regular basis, and doesn't understand exactly why it works, cannot become wise. They are fated to remain woolly thinkers for ever.

(7:2) THE EXPANDING UNIVERSE AND OCKHAM'S RAZOR.

The best way to introduce Ockham's Razor is through an example – indeed it is the only way. And it is an example from real life – as we shall see later.

Let us go back to the poor astronomer in the last chapter who, you will remember, was faced with tantalising evidence that the whole universe *might* be expanding. Unfortunately, when he applied the Detective's Equation to his observational evidence the Odds were only 30 to 1 on this dramatic hypothesis. The poor devil couldn't sleep. He was tantalisingly within reach of immortal fame – but not quite near enough to snatch it. The Odds were not quite high enough, in his personal opinion, to justify such an outrageous claim. The obvious remedy of getting more data simply wasn't practical. He'd taken a decade to get his meagre data and he didn't have two or three more decades in front of him.

He wondered if he could instead adjust his hypothesis slightly and get a more decisive result.

It occurred to him that if the Universe *was* expanding, the force of gravitation between the parts would slow the expansion down. In that case a decelerating Universe, his new hypothesis H_2, should yield a curve (dashed, see Fig (7:1) defined by 3 'parameters' (height, shape and curvature) and which he fitted exactly through his best data points 1,5,and 6. Why does it look that shape? Well the steeper the slope the faster the expansion. And, because light has a finite speed, when you look into the distance you are looking back in time – to when the expansion was supposedly faster and the slope greater – and that is what the curve labelled Model II does in the figure – it is steeper in the distance than it is nearby (i.e. today).

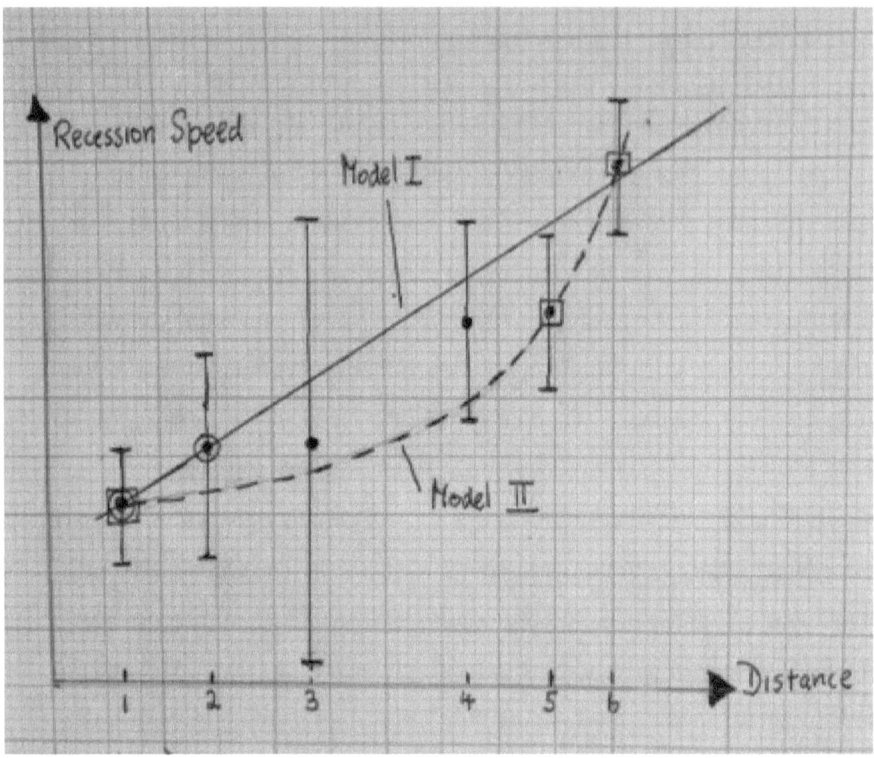

Fig (7:1) The two cosmological hypotheses I (no deceleration) and II (decelerating) compared with the data, including the estimated error-bars.(a) Hypothesis or 'Model' I (solid line) was deliberately fitted through the two data points 1 and 2 (circles) – see Chapter 6. The deviations of the Model from the *other* 4 data points (3,4,5 and 6) can then be read off as a fraction of their individual error-bars. Thus at point 3 the hypothetical solid line is about 30 per cent of the local error-bar-width above the measurement point so we would say its $t_3 = 0.3$. (b) Hypothesis or 'Model' II (dashed line), the decelerating model, was steeper in the past and was deliberately fitted through the 3 data points 1,5 and 6 (squares). The deviations of Model II (the dashed curve) from the *other* data points (2, 3 and 4) can likewise be read off as a fraction of the local error-bar width. Thus at point 4 the Model II is about 80 per cent of a bar-width below the measurement there, i.e. its $t_4 = 0.8$

He was now in a position to evaluate the Odds on his new hypothesis H_2 by measuring the discrepancies t from the *new* dashed curve, and using Table (6:1) for the Weights to fill in an Inference Table (i.e. employ the Detective's Equation) once again. His results were as follows:

TABLE (7:1)

DISCREPANCY ODDS FOR DECELERATING UNIVERSE

Point	Discrepancy t	Weight $W(t| H_2)$
2	0.4	2 to 1 on (2/1)
3	0.1	12 to 1 on (12/1)
4	0.8	1 to 1 on (evens)

His Prior Odds O(H) also had to be adjusted, for now he had 4 hypotheses to choose from: static, collapsing, expanding steadily or expanding but decelerating; thus the impartial odds against deceleration are 3 to 1, or $O(H_2) = 1/3$. So the Detective's Equation now yielded:

$$O(H_2 | t_2, t_3, t_5) = W(t_2 | H_2) \times W(t_3 | H_2) \times W(t_4 | H_2) \times O(H)$$
$$= (2/1) \times (12/1) \times (1/1) \times (1/3) = 8 \text{ to } 1 \text{ only}$$

The situation now was *less* decisive than before. How odd. How surprising. How damned frustrating! After all he was using an improved model containing more physics (gravitation) which, by design, fits perfectly through 3 data points (1,5 and 6) instead of only 2 (1 and 2) hitherto. And yet he's finished up with a *less* decisive result. His dreams of glory have melted in the morning sun.

You might like to think about this. Can you come up with an explanation for what has happened? Good for you if you can because it is a deep business.

LONG PAUSE FOR THOUGHT.

The answer is to be found partly in comparing this table (7:1) with the corresponding table (6:2) in Chapt 6 . Now there are only three pieces of evidence, three data points (2,3 and 4) which can be used to calculate the Odds on the hypothesis H_2 instead of 4 (points 3,4,5 and 6) before i.e. for H; one of them has been swallowed up in specifying the new hypothesis H_2 (i.e. had been fitted to the 3 points on curve in advance.) And partly it is because the Prior Odds on the new hypothesis (1/3) are lower than they were on the old (1/2) because now there are 4 alternatives to choose between instead of only 3. And you can see which way the trend is going. If he were to choose a yet more elaborate hypothesis H_3 with 4 free parameters instead of 3 there would be only 2 data points left from which to calculate the Odds on it while the Prior would fall to $(1/4)^4$. Eventually a hypothesis H_5 with as many free parameters (i.e. 6) as data points could be formulated which would have no evidence whatever left to estimate any Odds on it at all. It would be a Just So story and nothing more.

The best reason to believe any hypothesis is that it fits the existing evidence better than it has any right to do by chance. The more such evidence it fits the more likely it is to be true or, as scientists say, the more 'significant' it is because the Odds against it fitting *by chance* are higher. On the other hand if we deliberately mould the hypothesis to fit most of the existing evidence, the Odds that it can fit the shrinking remainder by chance increase and so the less significant the hypothesis becomes. That is the trouble with elaborate hypotheses; they are by definition, very flexible and so they can be deliberately moulded to fit much of the existing evidence, leaving too little evidence left over to judge if they are 'significant'. That is Ockham's Razor, Parsimony, stated in modern words. There's nothing philosophical about it, no

---

[4] Worse still he would have naturally used up his four best points, the ones with the lowest error bars, to specify his model, leaving only the two worst to do the testing with. They cannot give a strong result – either way.

assumption about the simplicity of the natural world. It is a straightforward consequence of the Odds approach to reaching conclusions – embodied in The Detective's Equation. And indeed, without appealing to the DE it is damned hard to either explain or justify Ockham's Razor (the explanations given in most text-books wander off into stratospheric Mathematics and are, I find, almost wholly unpersuasive).

In truth between 1911 and 1917 Vesto Slipher, an astronomer at The Lowell Observatory in Flagstaff Arizona, managed to measure the recession speeds of a couple of dozen galaxies using a small wooden telescope. All but one or two had enormous recession velocities, and the fainter looking ones seemed to recede the fastest. His raw results excited storms of speculation but no one, not even Slipher himself, mentioned the 'Expansion' word, though I dare say they thought about it.

Then George Ellery Hale completed a far larger telescope, 'The One Hundred Inch', on Mount Wilson in California., and one of his assistants Edwin Hubble took up Slipher's project. A decade later he produced Fig (7:2) where a trend seems to be obvious though error bars are lacking.

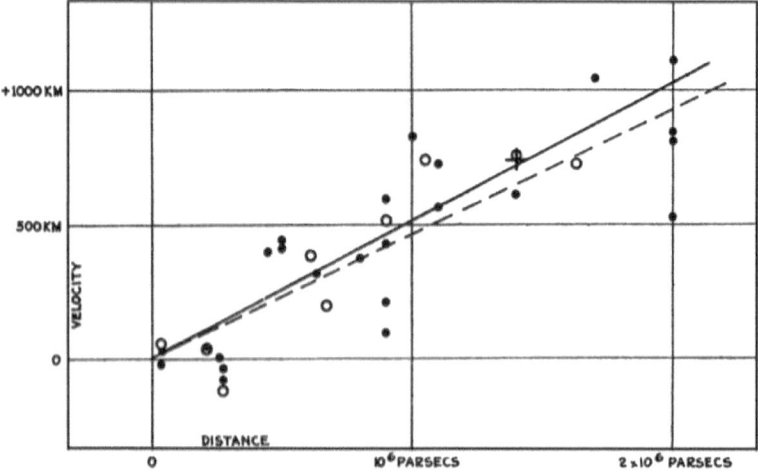

Fig(7:2) Edwin Hubble's original data on the relation between the measured recession speeds of bright galaxies (vertical: '500 KM' means 500 kilometres a second) plotted against their estimated distances (horizontal); 1 parsec is about 3 light-years. Note they are plotted 'logarithmically', that is to say the scales increase in multiples, not by constant amounts (i.e 'linearly'). This data, obtained in the 1920's with the 100-inch telescope in California, was used, though not by Hubble, to argue that the entire universe is expanding. Note that there were no error-bars, which would not be acceptable today. Would you have been persuaded? If you think about it the slope can be used to estimate the expansion age of the universe, i.e. the time since it was all squashed together

Again no mention of 'expansion' appears in the paper though (1927) a Belgian priest and trained mathematical astronomer, the Abbé Lemaître, had claimed in a paper that hardly anybody read that the cosmos must be expanding out of what he called 'The Primeval Atom' – a claim that seems to have disappeared from his subsequent papers. Nevertheless the idea was now 'in the air' and theoreticians fiddled with their equations so as to at least allow for the possibility of Expansion. Hubble died in 1956 still sceptical of expansion. It was only in 1965, when microwave radiation from all over the sky was accidently picked up with a satellite-antenna, that nearly everyone accepted that there must have been a Big Bang – a very hot beginning. By one of those perverse ironies of history cosmologists then awarded Hubble the palm for discovering expansion, although he'd never dared to believe in it himself (personally I think that is ridiculous). Had poor old Slipher known about the Detective's Equation he might have won the glory himself – and some 30 years before-hand. [A complicating factor was that Flagstaff gets clouded out for half of every year so half of Slipher's sky was unavailable for observation. Hubble suffered no such handicap in California.]

What would you have done in Slipher's place ? Published and be damned? Shut up and slunk away like Slipher to become embittered later in life? You need character in science, the moral courage to withstand a storm of criticism if you make a claim which your contemporaries will not accept. In my experience very few scientists possess that kind of fortitude. We prefer to pass by on the other side than be accused of hubris. One wonders how many big discoveries have thus been lost along the way.

The third alternative would be to make the claim alongside your workings of The Detective's Equation and let the others argue over the assigned values of your Weights and your Prior. If they don't like yours let them defend their own instead. Thus the Detective's Equation can have a role in helping to persuade as well as to decide. Had Slipher known about the DE he might have done just that, and gained the posthumous glory awarded to Hubble.{1}

(7:3) THE COMPLEXITY OF HYPOTHESES

It's all very well saying 'Avoid Complex hypotheses' but how do you assess complexity in this context? It's not always obvious. The astronomer's second hypothesis was more complex than his first because it allowed for the rate of expansion to change with time. It required an extra number or 'parameter' to describe it – the acceleration of the expansion – which could be positive, negative or zero. His first hypothesis had no such freedom; the rate of expansion was assumed fixed once and for all. It only required two "free parameters" to describe the Universe: its present size and its rate of expansion. The second hypothesis required an additional free parameter – the rate of deceleration of that expansion. We say it is more complex because it requires 3 "free parameters" to describe (or define) it as opposed to the first which required only 2. Each such Free Parameter (FP) gobbles up one more piece of the available data to define it, leaving one less piece to test the hypothesis in question. And if you go on complexifying your hypothesis you will eventually run out of data to test it with – which means you will have no grounds to believe in it at all. That's more or less Ockham's Razor. You could put the matter the other way round. If you have a fixed amount of data (evidence) you have only to manufacture an hypothesis with an equivalent number of free parameters to be certain of making a perfect fit to the data (evidence). That being so there are no grounds for believing that the hypothesis is 'right' i.e. will apply in other circumstances. It can't be *generalised*. And that's the point isn't it. We are looking for generalisations that *will* apply in a wider range of circumstances than those already experienced. A generalisation that is widely applicable is a source of wisdom and power. It is an inference that goes beyond the immediate evidence into the wider world of Space and Time; to another continent and a future year. The grounds we have for believing in it are that it fits the existing evidence *better than it has any right to do by chance.* It doesn't have enough free parameters to fit that evidence as a matter of course. It is 'significant' – more than any old Just-So story dreamed up merely to fit the available facts. And if there are many less free parameters than pieces of evidence that fit the hypothesis then there are very strong grounds for believing it to be right. It is a generalisation we can believe in with some conviction, and act upon. Such generalisations can only be made with confidence where Ockham's Razor applies; where the generalisation (hypothesis) is simpler (contains less free parameters) than the original evidence which gave it birth. Science would be impossible without Ockham; hence Hugh Gauch's strong statement which heads this chapter.

There's another way to look at complexity – in graphical terms. Fig (7:3) shows a series of graphs of increasing complexity as you go down the page. It turns out that there's a one-to-one correspondence between a graph's obvious complexity and the number of free parameters required to define it. The upper (a) has only one – its height. The second requires 2: height at the origin where x= 0, and slope. (c) requires a third – the position of the minimum.(d) requires one for each maximum and minimum plus 2 more – and so on. The one at the bottom – which certainly looks complex, has no less than 10 free parameters.

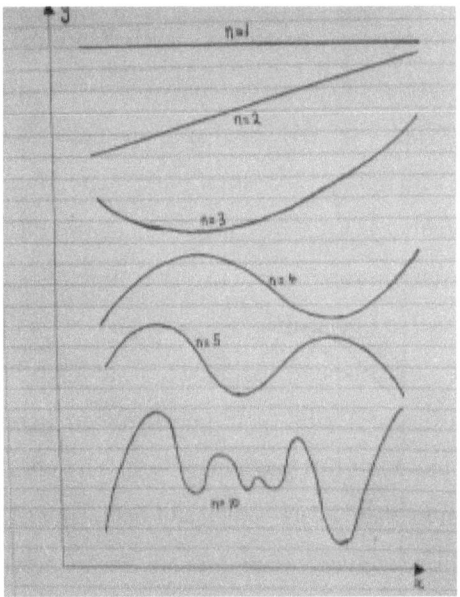

Fig (7:3) Graphs of increasing complexity as you move downward. The number beside each denotes the number of free parameters required to define it. Thus the top one requires only one, its height. The second requires height and slope. The third requires another one to describe its curvature, and so on. The obvious complexity (wriggliness) rises with the number of free parameters.

Now very often hypotheses amount to hypothetical relationships between factors, between say the exam-performances and class-sizes of school children. Thus hypotheses can be translated into mathematical relationships, which is to say into graphs.

Fig(7:4) shows several graphs of equal complexity – each controlled by 4 FPs. Note that by changing the values of the 4 FPs you can make the graph wriggle all over the place. It has a degree of flexibility. Also on that figure I've drawn 6 measured data points (evidence) with their accompanying error-bars. By choosing the FPs of my curve judiciously I can always force my curve to fit almost perfectly through 4 of the points. There is thus no significance to a fit between this particular curve and 4 data points because such a fit can always be arranged by a suitable choice of FPs. However such a curve cannot in general be forced to fit through 5 or more pieces of data because it lacks sufficient flexibility. If it does fit then there are some grounds for suspecting that the hypothesis that generated this graph in the first place may be right. And if it fits a lot more than 4 data points, and fits them well (by comparison with the error bars) there are very strong grounds for believing in that hypothesis.

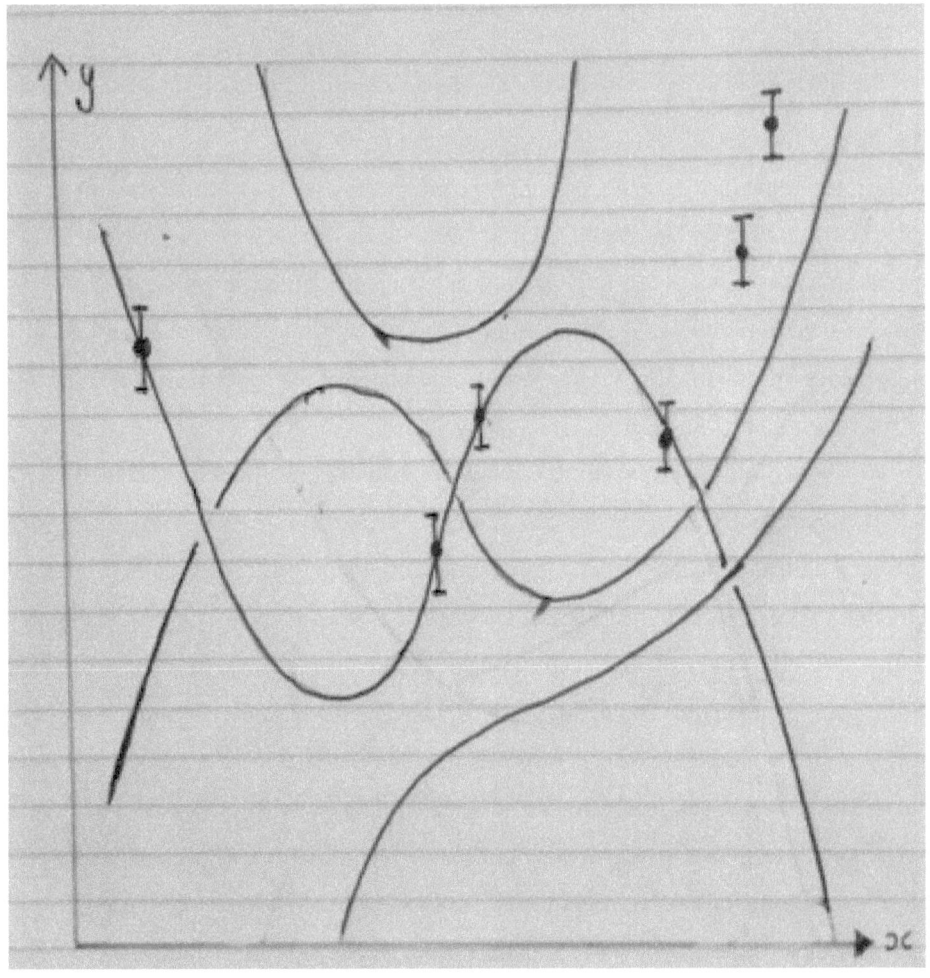

Fig(7:4) Four different graphs all defined by 4 free parameters. The shape of each graph is defined by the actual values selected for each of the 4. One of the curves (bottom right) has values selected to fit exactly through 4 of the depicted data points as shown. But it cannot also fit the other 2 points because it's not complex enough; it doesn't have enough free parameters. Each extra one makes it more wriggly, more flexible. If it had two more it could fit all six points perfectly.

In graphical terms then Ockham's Razor can be restated as; "A good fit between a curve and some data points is only significant if the curve is controlled by less FPs than the number of such points. If there are notably less FPs than points there are strong grounds for believing in the hypothesis which generated the curve." Note that this explanation highlights the importance of the errors. If the estimated error-bars are large a wide variety of curves (hypotheses) could be made to fit through them, and so such a fit would be less discriminating (significant).

In our daily lives we are constantly evaluating hypotheses without counting free parameters or measuring graphical wriggliness. To what extent we possess an informal mechanism for estimating complexity it is difficult to say but I certainly feel suspicious when I hear a story that's 'too good to be true'. An elaborate lie often sounds – like a lie. It exactly excuses the 'crime' in question without offering any other significant evidence of its veracity. An absolutely minimum criterion for a credible hypothesis is that it should be falsifiable. Without some vulnerability to test, and preferably some testing evidence, what grounds could one have for believing in it? The philosopher Karl Popper in particular has emphasised such falsifiability with regard to political theories such as Marxism which used to claim for themselves a degree of 'scientific' respectability. But since they weren't falsifiable, being endlessly elastic, such claims were spurious. Always ask yourself if your own hypotheses are falsifiable, and if not then what other grounds do you have for believing in them.

Ockham's Razor is not something you're likely to meet in the school playground. Scientists and philosophers have got themselves into tangles over it (see later) – and yet it is one of the wellsprings of wisdom. It says *"Of the hypotheses which fit the data (evidence) the simplest is most likely to be right; i.e. the Odds on it are likely to be higher than on the alternatives."*

(7:4) OCKHAM'S RAZOR IN ACTION.

We are so used to the notion that the Earth goes round the Sun, and not vice versa, that it is hard to imagine that for millennia learned men thought otherwise. They didn't do so lightly either. They weren't driven by blind prejudice. There were strong arguments against the Heliocentric hypothesis; for instance if the Earth orbited the Sun why didn't we see the foreground stars nodding back and forth annually against those in the background? This lack of measurable 'parallax' defeated Aristarchus's Heliocentric hypothesis when it was first mooted around 300 BC.

The idea was resurrected by Copernicus again in 1543 AD in his 'De Revolutionibus'. Copernicus didn't have any brilliant new insights or any new data of his own but he did point out how unwieldy the alternative Earth-centred hypothesis had become in the meantime. Observations of the planets' motions accumulated over the centuries didn't quite square with the simple form of that hypothesis. It had had to be adjusted time and again by adding new free parameters in the form of new 'epicycles'. Never mind the details. Each such new FP made the Earth-centred hypothesis more complex and thus less likely to be right. Eventually a point was reached when the Heliocentric hypothesis came to seem more likely than the alternative. True one had to adjust the Heliocentric hypothesis as well in order to remove the unobserved parallaxes. But that could be done in a single bold stroke by assuming (As Aristarchus himself had suggested) that the stars must lie at such enormous distances from us that the parallaxes, though still present, would be too small to be detectable with contemporary instruments – basically the eye-ball.

Thanks largely to Parsimony, i.e. to Ockham's Razor, Heliocentrism gradually prevailed. But it was to be 1727 before the Earth's orbital motion could be detected directly by James Bradley (see Chapt 3) and 1848 before telescopes of sufficient precision to measure the parallax were built by Joseph Frauenhofer in Germany.

Continuing in the astronomical vein consider Newton's idea of Universal Gravitation. Even Newton himself thought it preposterous: he wrote:

"That Gravity should be innate, inherent and essential to Matter, so that one body may act upon another at a distance thro' a Vacuum, without the

Mediation of any thing else, by and through which their Action and Force may be conveyed from one to another, is to me so great an Absurdity that I believe no Man who has in philosophical Matters a competent Faculty of thinking can ever fall into it."

Nevertheless it was quickly and universally accepted? Why so? Because of Ockham's Razor, because of Parsimony. That one simply stated law with only two FPs could be wrong when it could explain the complex motions of all the planets and their moons, the comets, the tides, the shape of the Earth, the eclipses..... seemed so unlikely that it was accepted as having to be right. Einstein removed Newton's philosophical objections to it in 1915 by reinterpreting the law not as force at a distance but as a modest warping in local Space-Time which permeates the whole neighbourhood of a massive body. Each portion acts on the piece next to it like the fabric of a balloon. It has become a 'field-theory' as we say rather than the objectionable 'action at a distance'.

Einstein's own reputation was largely built on a piece of outrageous parsimony (see Sect 3:17) In his 1905 paper 'On the Electrodynamics of Moving Bodies' he finessed all the profound difficulties identified by his illustrious predecessors Lorenz and Poincare by making a single grotesque assumption. He assumed, with hardly any proof, defence or justification, that the velocity of light must be the same for all observers – whatever their state of motion in relation to the source of that light. From that drastic assumption he was then able to derive key formulae such as the already known Lorenz Transformations, embodying weird relationships between Space and Time measurements made by observers in motion relative to one another. These are what later became known as 'The Laws of Special Relativity'.

It's not clear what was in Einstein's mind when he made his indefensible assumption – and later, like his famous predecessor Newton, he became 'economical with the truth'. To this day nobody has a satisfactory explanation for why the speed of light (but not of sound) should be constant. But if you are prepared to swallow that one bitter pill all the other complications of Space and Time – which have been confirmed by innumerable experiments – can be derived from it. It is justified by its sheer simplicity. It absolves one from any further hard thinking. Follow this one drastic rule and you can always work out the many other necessary modifications required of Newtonian Physics. The truth is that Special Relativity is a marvellous conjuring trick. Einstein found a way to pull the rabbit out of the hat – without worrying how it got there.

When the idea of Evolution by Natural Selection first came (independently) to the two world-travelling naturalists Charles Darwin and Alfred Russel Wallace there were insuperable objections to it. Insuperable. No one expressed those objections better than Darwin himself in the quite wonderful final chapter of his "Origin of Species" entitled "Recapitulation and Conclusions" :

"That many and grave objections may be admitted against the theory of descent with modification through natural selection, I do not deny. I endeavour to give them full force." For instance how could large animals, evidently related to one another, be found on continents unconnected by any land bridges? Where were the missing-link fossils say between birds and reptiles, or between monkeys and men? How could complex organs like the eye evolve by gradual change? What natural force could possibly drive the exotic plumage of a peacock? And indeed how could anything evolve much during the short estimated age of the Earth? (Then 20 million years).

And yet. Against his naturally religious instincts (and his wife's) Darwin was eventually forced into believing in his theory almost entirely on the grounds of Parsimony. Natural Selection was so much simpler an explanation of what he had seen in Nature than Creationism. And in persuading himself Darwin gradually and gracefully persuaded the majority of his readers. Consider some passages:

"The framework of bones being the same in the hand of a man, wing of a bat, fin of a porpoise, and leg of the horse – the same number of vertebrae forming the neck of the giraffe and of the elephant – and innumerable other such facts, at once explain themselves on the theory of descent with slow and successive modifications."

"Such facts as the presence of peculiar species of bats, and the absence of other mammals on oceanic islands, are utterly inexplicable on the theory of independent acts of creation."

"These are strange relations on the view of each species being independently created, but are intelligible if all species first existed as varieties."

"We can plainly see why nature is prodigal in variety but niggardly in innovation. But why this should be a law of nature if each species has been independently created, no man can explain."

"Nevertheless all living things have much in common, in their chemical composition, their germinal vesicles, their cellular structure, and in their laws of growth and reproduction. We see this even in so trifling a circumstance as that the same poison often similarly affects plants and animals or that the poison secreted by the gall-fly produces monstrous growths on the wild rose or oak-tree. Therefore I should infer from analogy that probably all organic things which have ever lived on this earth have descended from some primordial form, into which life was first breathed."

Darwin's last resounding chapter is entirely an appeal to Parsimony, to the notion that simpler hypotheses are far more likely to be right than complex ones. And of course Darwin's and Wallace's theory has stood the test of time even though fascinating questions remain. For instance at what level (or levels) does natural selection work; at the species level, the gene level or the level of an entire ecosystem? {1/ 2&3}

(7:5) PARSIMONY, NOISE AND PREDICTION

A simpler hypothesis is also easier to remember, easier to use and easier to falsify. And a simpler hypothesis, even if it is not right, is more likely to be a reliable predictor outside its sphere of immediate applicability. Why so? Because of the important phenomenon of 'Noise' illustrated in Fig (7:5). In the real world phenomena have to be detected and measured against a background of 'noise', a clutter of smaller extraneous signals arising from a variety of other background and foreground sources. Because of that very variety the noise jiggles all over the place, it is far more complex than the signal which may be describable by a curve with a smaller number of free parameters than the noise Thus fitting a simpler hypothesis to all of the data selects out the signal you want from the noise, which you do not. A slavish attempt to fit too many details may confuse one as to the salient facts.

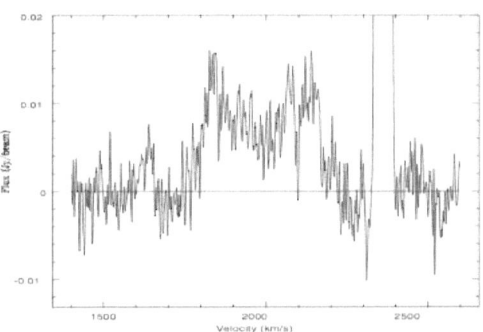

Fig (7:5) Signal and Noise. This shows the output from the big radio dish at Jodrell Bank when we were pointing it in the direction of the Virgo Cluster. It shows incoming signal plotted vertically as a function of recession speed in the expanding universe plotted horizontally. There's a lot of jiggling about or 'Noise' arising from the receiver hiss, signals from faint sources in the cosmos, interference from local security systems (the big spike at 2,300), and goodness knows what else. But more or less in the middle is a wide double hump which is less complex, doesn't look like noise, and is in fact the signal from a galaxy in the Virgo Cluster. We can only detect it because the ratio of signal to noise is high enough. Much of science is concerned with digging signal out of noise. How likely is it that the double hump is a chance superposition of noise? It's all about Odds again.

Suppose you were trying to predict the rate of exchange for your pounds against the Euro in 3 months time when you propose to take a holiday. The rate changes from day to day for all sorts of unknowable and uninteresting reasons. It's 'noisy' as scientists would say. If you smooth out the noise by fitting a smoother (simpler) curve through the day-to-day variations you're more likely to see the underlying trend, if there is one, and to find that trend a more reliable guide to the value of your pound in 3 months time than the trend from yesterday to today projected 90 days into the future. In other words a simpler model is more reliable because it takes account of, averages over, more of the data – even if in detail it's wrong.

(7:6) EXAMPLES OF WOOLLY THINKING

Like the other components of CST Ockham's Razor is constantly in use without being widely understood. If asked most scientists think it's got something to do with Philosophy, of which they generally have a low opinion. And those that do understand it are divided into those who think it has got to do with the universe itself being simple and those who think of it as merely wise advice on how to think productively. Wise advice it certainly is, and of everyday application. Whether the Universe really is simple seems to me a question impossible to answer definitively. Certainly it is far simpler than it might have been. There are only 100 different types of atom (Element) when there might have been myriads, and even they are organised into families with similar properties [Mendeleev's Periodic Table] . Fundamental particles are made up of only 3 Quarks and 3 Leptons (and their anti-particles). Proteins are made up out of only 20 amino-acids. The same genetic Code applies to all organisms on Earth. All electromagnetic processes are described by only 4 laws while stars, for all their immense size and complexity, are controlled by only 4 simple equations. And so on. Maybe its because we preferentially come to understand the simple bits while never even recognizing profounder issues. What is certain is that without it we're likely to lapse into woolly thinking, the very hardest kind from which to escape. Consider 3 infamous examples.

(A) The Four Elements was an idea promulgated by Aristotle around 300 BC in which all substances were supposed to be composed of a mixture of Earth, Water, Air and Fire. By adjusting this hypothetical mixture, and a deal of plausible sophistry, the old thinkers could explain everything – and nothing. So long as it was widely believed serious chemistry was unnecessary and therefore unpursued. This monster was a many headed gorgon bristling with free parameters. Whenever something didn't fit you ascribed further properties (parameters) to your 4 imaginary elements and lo everything could be made to fit once again. Thus it couldn't be overturned.

The Universe of order or disorder; one marching grandly towards a predestined end, or one in chaos, the temporary affect of innumerable atoms and organisms in tumultuous conflict? Both visions go back to Classical Greece and each has had its turns at governing scientific debate. Chaos is abhorrent yet order illegitimately imposed may be worse still. Each branch of science, each age has tried to strike some balance between the two extremes. What we would call chemistry was ruled for over 2000 years by Aristotle's hypothesis of the four Elements: Earth, Water, Air and Fire naturally finding their appointed places in an ordered cosmos. Almost no material evidence supported his scheme but it appealed to religions such as Christianity and Islam looking for a comforting order to life. It wasn't overturned so much as left behind by crisis, the firewood crisis which struck Britain in the 17th century after it had cut down most of its forests to build houses and ships. A new source of power had to be found and the mining of coal led to a desperate search for new materials and new contrivances such as pumps to stop the mines flooding. Experiments were necessary, and from experiments came evidence that made no sense in Aristotle's floppy scheme. For instance burned in air some substances became heavier not lighter. And careful balance measurements initiated by Joseph Black in Glasgow University (1750, the English universities didn't teach science then!) showed that substances combined in precise ratios to form compounds. Out of such observations the notions of atoms and molecules grew. These in turn gave rise to materials of great strength like steel, and to new compounds of great value such as artificial dyes. Crisis, experiment, discovery, understanding, wealth: a virtuous circle had begun, epitomised by The Lunar Society formed in the Midlands around 1750. Its dozen or so members met in one or another's houses every Full Moon to discuss their various scientific and industrial interests ranging from engines to geology, from medicine to metal-ware, from ballooning to botany, from electricity to pottery. Unencumbered by a classical education, which was dominated by philosophers like Aristotle, such men looked for experimental evidence because they knew from experience that good ideas could make a mint of money. It's probably no accident that most of the pioneers of the Industrial Revolution were Non-Conformists, educated outside the official system with its emphasis on the classics. Observation and experiment were a long time in coming but when they did progressive chemistry quickly followed. The infinitely flexible, therefore unprogressive chemistry of Aristotle was simply left behind as unprofitable.

(B) The Four Humours was another hypothesis of ancient origin, modelled on the Four Elements, in which it was imagined that health was controlled by a balance of four liquids: choler, melancholer, phlegm and blood. Physicians who were learned in such jiggery-pokery dosed us, leeched us and charged us, shortening our lives as they impoverished our purses. Again the hypothesis was immune to criticism partly because it was impossible to experiment much on the human body but mainly because it was infinitely adaptable. In place of bones it had an infinitude of free paramaters – and what was more could earn good money. Again it couldn't be displaced by evidence, being infinitely flexible. It was gradually superseded by more useful notions about physiology such as the germ theory of disease, a direct result of the invention of the microscope.

(C) The Argument by Design opined that all things wise and wonderful, all creatures great and small, were instances of The Creator's wonderful powers of invention. After all no other cause could be imagined for the intricacies of Nature's architecture, from the perfect spiral of a sea shell to the extravagance of a Rainbow Lorikeet's plumage. Science at the ancient universities, even to the end of the nineteenth century, was aimed at uncovering such wonderful manifestations of the Almighty. Since nothing was outside his (oh yes he was masculine) powers everything could be explained. There was no possibility of bringing Him down, since nothing was beyond Him, even burying fossils of inexplicable design merely to challenge our faith in Him. This Argument by Design could have been rejected by nothing else but Parsimony, by the discovery of an alternative theory which was simpler, far simpler than a Great Designer in the Sky. The trouble with Him was that he had an infinite number of free parameters.

Ockham's Razor underscores the importance of challenging hypothesis with evidence . Simple hypotheses can be thoroughly challenged while complex ones are much harder to test. Really vague hypotheses of the kinds sketched above, are virtually invulnerable – *which makes them not strong but weak*. Scientists are a pretty irreverent lot and I suspect that the invulnerability of religious and political hypotheses make them unattractive to 'the scientific mind'. The dangerous thing about an invulnerable hypothesis is that it may become immortal, obstructing all progress for generations, centuries, even millennia. The historian Daniel Boorstin opined that "The greatest obstacle to discovery has been the illusion that we already know." That is reason enough to be sceptical, however plausible certain theories seem.

Sometimes hypotheses that start off vulnerable may eventually become virtually immortal through the gradual accretion of more and more free parameters. A good example of such is Big Bang Cosmology, the current paradigm for the subject. It started in a modest way back in 1917 with only 2 free parameters but has grown into a thriving industry today with, by my count, no less than 17 and an annual budget in the hundreds of millions (mainly for specialised spacecraft). In the Table (7:2) below I have compared the number of pieces of independent evidence (observations) bearing on it (column 4) compared to the number of free parameters that have been added from time to time to square the theory with those observations (column 3). The details don't matter only comparison of the numbers in the two columns. As you can see (column 5) at no time have the observations exceeded the free parameters in number. By my reckoning that makes the Big bang Theory 'insignificant' i.e. implausible. And matters have not improved with time. For instance in 1998 when Supernovae observations revealed the astonishing and entirely unpredicted discovery that the expansion was apparently accelerating, cosmologists barely halted in their stride. They shamelessly devised a new free parameter, gave it the sexy but meaningless title 'Dark Energy' and carried on regardless. Whether that's scientific I leave it for the reader to decide. We'll return to the matter again.

The 'Standard Model of Particle Physics' involving Quarks, Gluons and the Higgs Boson is another theory involving no less than 17 free parameters which is why it is widely regarded as a makeshift construction which will have to be superseded one day by a much simpler theory.

TABLE (7:2) FREE PARAMETERS IN BIG BANG COSMOLOGY

	(1) DATE	(2) NEW STEP	(3) NEW FREE PARAMS	(4) NEW MEASUREMENTS	(5) CURRENT SIGNIFICANCE.
1	1917	Einstein's model	H_0, k_0, Ω_0	One equation between them.	-2
2	1921	Cosmological constant	Ω_Λ		-3
3	1929	Galaxy Redshifts		H_0	-2
4	1965	Cosmic Background Radiation(CBR)	η	η	-2
5	1970's	Big Bang Nucleosynthesis	Ω_b	Ω_b	-2
6	1974	Cosmochronology		$(\sim 1/H_0)$	-2
7	1978	Dark Matter	Ω_M		-3
8	1970,s	Initial Seeds	A, n_s		-5
9	1978	Gravitational Waves	r		-6
10	1981	Inflation	N		-7
11	1980's	Large Scale Structure	b, σ_8, ξ		-10
12	1990	COBE		A	-9
13	1998	Supernovae	w	Ω_Λ	-9
14	1998	Clustering		σ_8	-8
15	2000	Galaxy Infall		ξ	-7
16	2000	BOOMERANG		n_s, Ω_M, Ω_0 (k_0 inferred from equation in row 1)	-4
17	2003	WMAP	$dn_s/d\log k, \tau, \tau_0$	$\tau, dn_s/d\log k, b$	-4

The 17 free parameters which define Big Bang Cosmology (Col 3) compared with obsrvations available (Col 4) as a function of epoch. Never mind the details but note that the Significance in the last column, that is to say the number of pieces of independent relevant data (evidence) *minus* the number of free parameters, has always been and remains negative. That's very worrying, or it should be, because it means the evidence is 'insignificant' i.e. insufficient to support the theory. To me it is nothing more than a 'Just So' story

(7:7) OCKHAM'S RAZOR AND OUR BOOK

Acting prudently on the basis of incomplete and imperfect information is the basis for survival in both the human and animal worlds. Humans call it Common Sense Thinking (CST).

Can animals think in the same sense? Some would argue 'Yes' because to suppose otherwise requires that some new 'spirit' has entered the human brain and imbued it with the quality of 'Mind' – a requirement that is unnecessary, unnatural and not parsimonious'. Others would argue, to the contrary, that the achievements of humans obviously go so much beyond the achievements of other animals that we must be imbued with thought processes that the others clearly lack. Thus to attribute to animals thoughts, feelings and intentions akin to ours is unnecessary, unnatural and not parsimonious. It is 'anthropomorphism' and as such a hindrance to our understanding of animal behaviour – which should be based purely on observation and experiment.

Here Ockham's Razor is faced with two very different, almost antagonistically opposed candidates, both claiming to be maximally parsimonious, to be simpler than the other. The former supposes there to be only one type of thinking, the latter supposes two. If the former is true then human thinking can be no more than an evolutionary extension of the animal variety, different perhaps in degree, but not in kind. It then makes sense to ask the question 'How do animals think?' in order to answer the question 'How do we think ourselves?' But if it is untrue we will be wasting our time and avoiding the really interesting question of why we are so obviously different, the question that has obsessed philosophers, scholars and divines for at least three millennia.

It has not always been obvious which side Friar Ockham would urge us to choose. But everything changed in 1859 with the publication of Charles Darwin's 'Origin of Species'. We now know that superficially different organisms share much the same underlying anatomy, physiology, chemistry and molecular biology, so why not that fundamental survival mechanism – CST? It would have been hard for scholars, invested in 3000 years of culture, to make that choice, and by and large they did not, though even before

Darwin the great Scottish philosopher David Hume, who grew up on a farm, wrote (1737) *"....no truth appears to me more evident than that beasts are endow'd with thought and reason as well as men."*

Our investigation very definitely begins from the presupposition that if we can think it is because our animal forbears, going back many millions of years, could think too. It means identifying a mechanism, or mechanisms, which all could have shared, but which could have evolved gradually by natural selection to the point where, together, they could compose operas and launch the Hubble Space Telescope. That is the real challenge for us. How do we explain the dramatic Ascent of Man in a mere few millennia without invoking miracles? We give our answer in (16:8), close to the end of the book. We never expected to find it, but with the unmasking of CST it simply fell out of the sky.

P.S. There is a more general point here. It is not always clear which of two competing hypotheses is the most parsimonious. If you choose one, be prepared to defend your choice.

SUMMARY

All hypotheses should be tested against evidence. *The best reason to believe any hypothesis is that it fits the existing evidence better than it has any right to do by chance* . A simple hypothesis has little chance of fitting more complex data unless it is actually right. Complex hypotheses are much harder to test because they have more flexibility to adjust to potentially threatening evidence. An infinitely flexible hypothesis is invulnerable to test, and therefore there are no grounds whatsoever for believing it to be right. Ockham's Razor, otherwise known as Parsimony, urges thinkers in general (not just scientists) to always prefer simpler hypotheses to elaborate ones because, if both fit the existing evidence, the Odds on the simpler one being right are higher, perhaps much higher. A simpler hypothesis is also likely to be a source of more reliable predictions because it is less the plaything of Noise.

197

Time and again Ockham's Razor has played an indispensible role in scientific history by rejecting long held but invulnerable theories in favour of simpler models that can be tested. It is therefore progressive and provides the logical underpinning to a generally sceptical point of view, reminding the thinker to always ask "Has this hypothesis been tested and if so how well?"

The complexity of an hypothesis can be judged either by the number of free parameters (FPs) required to define it, or by the wriggliness of the graphs it can generate. Always ask if an hypothesis is vulnerable to test. If not then what other rational grounds are there to believe in it?

Despite its' absolutely crucial importance and ancient lineage Ockham's Razor has often proved a slippery concept to grasp. For instance, although they both relied on it extensively, neither Newton nor Einstein appear to have grasped why it works – or if they did they left spurious explanations – Einstein for instance waffled about God. So it is worth dwelling on this chapter, particularly the Expanding Universe example in Section (7:2).

THINKING FOR OURSELVES (Disney)
CHAPTER 8 (11.0 kw)
COMMON SENSE
(draft 14/9/18)

"Good sense is the most evenly shared thing in the world, for each of us thinks he is so well endowed with it that even those who are hardest to please in all other respects are not in the habit of wanting more than they have."

Rene Descartes, French philosopher, 1637.

(8:1) REACHING THE WATERSHED.

This is the watershed chapter. From here we can look back and see the zig-zag path that has led us up to Common Sense Thinking (CST), and we can overlook the main features that remain to be explored. Here we tackle more or less realistic problems head on, using the tools we have found. In particular we train ourselves to use the Inference Table so that it becomes a powerful but natural extension of our everyday thinking. Thus we find ourselves exploiting Bayes' Rule and the Detective's Equation without having to recall them in detail. We learn how to use crude Weights, how and why to set Priors, and how and why conviction can emerge from confusion. We learn about the dangers of bias, of bully clues, and of omitting information. We discuss 'reasonable doubt', reaching a verdict, and 'Blackstone's Ratio.' In turn we find out how tricky it is to sit on a jury, how messy the world of real scientific research can be, and how easy it is to fool oneself, or be fooled by fashion. In a thicket of confusion sound thinking is seldom easy but common sense, expertly wielded, can usually cut through a deal of undergrowth.

(8:2) INFERENCE TABLES

Stripped of all inessentials Common Sense Thinking (CST) can be guided and summarized by an Inference Table (IT). It starts with an hypothesis H and some preliminary ('Prior') Odds O(H) you would put on it. You then search for evidence bearing on that hypothesis and assign to each clue E_a, E_b..... ...a numerical Weight W(E_a|H), W(E_b|H)......., more than 1 if it increases your Odds on H, less than 1 if it is unfavourable and decreases those Odds. You then multiply the existing O(H) by W(E_a|H) to obtain your updated Odds O(H|E_a). Each line of the table is then an update on the running Odds on H, taking into account a new clue on each line. Thus for the murder in the library (5:1) the detective produced an IT on her hypothesis that *'The leading lady is the murderer'* as follows:

INFERENCE TABLE (8:1): MURDER IN THE LIBRARY

Clue (E)	O(H)	W(E|H)	O(H|E)
Prior	1/6		
Motive	1/6	2	1/3
Opportunity	1/3	4	4/3
Footprint	4/3	1/40	1/30

What could be simpler? Anyone can follow this prescription for CST, and in my opinion they should. All references to Bayes' Rule, and The Detective's Equation have disappeared – and yet one is using them throughout – but in a natural informal way. The 1/6 is called 'The Prior Odds' or just 'The Prior' because it is the Odds you assign to H *prior* to considering any of the evidence. The detective sensibly gave this a value of 1/6 here because she thought there were 7 equally plausible suspects and she wanted to be unbiased. She then assembled 3 relevant clues E_x and assigned to each of them a Weight W(E_x|H). Each Weight, by multiplication, then updated her Odds on H, taking that particular clue into account. Two Weights were rather vague: W(motive|H) = 2 and W(opportunity|H) = 4 – but what else could they be? She could only use her judgement ('intuition', 'experience' – call it what you like) in the matter. It would be more foolish to leave them out just because she cannot assign them a precise value. At least her Weights reflect the 'sense' of that evidence i.e. in which direction it points. Then there's the quantitative evidence, the foot size. On the face of it that looks decisive. But she had to rely on an outside 'expert' – the forensic scientist – to supply the decisive Weight W(E_x|H) = 1/40. In so doing she has surrendered her own judgement. She's probably been trained to do so. But was her training sound? At all events she's arrived at an O(H) of 30 to 1 against H – the best she can do in the circumstances. Could any one do better, or argue with her? Whatever she decides to do next on the basis of those Odds, she could defend her decision to an outsider, or to her boss, by going through her Inference Table. At least it is methodical and transparent, even if it may not be 'optimal' – that can only depend on the soundness of her various Weights.

The Weights of course are crucial. What can we say about them *in general*? Not much. They depend so much on personal experience and training. One hopes she is better qualified than we are to judge those Weights. Sherlock Holmes might have disagreed. The charm of the

Conan Doyle stories{1} lay in Holmes' ability to observe and weigh trifling clues overlooked by his more methodical colleagues from Scotland Yard (e.g. The Dog in the Night). I would suspect (hope) that in real life training and experience would most of the time trump brilliance and opium. But we can remind ourselves that the Weight $W(E_1|H)$ is the ratio of two Probabilities:

$$W(E_1 | H) = \frac{P(E_1 | H)}{P(E_1 | \bar{H})}$$

each Probability lying in the range between zero (impossible) to one (certain). It's easy to forget this sometimes because it may be easier to draw a Weight directly out of ones judgement. Thus the Weight W(motive| H) wouldn't necessarily be improved by assessing P(motive| H) and P(motive| $\bar{H}$) separately. But sometimes mistakes, even fatal mistakes, can be made by omitting to consider P(E| $\bar{H}$) i.e. the Probability of the evidence even if H is *not* true [We recall from Chapter 4 that in the Cancer Scare there was a high, but neglected, Probability of getting a positive test-result among certain kinds of healthy women. And the Tiptuppititis sufferer Smith misattributed his 'cure' to Organocalm because he didn't realize how high the Probability P(E| $\bar{H}$) was that he would recover anyway]. Thus it is often wiser to write out your IT in full, including an explicit column for each of P(E|H) and P(E| $\bar{H}$) as we did in the case of The Missing Wife:

INFERENCE TABLE (8:2): THE MISSING WIFE

Hypothesis: *'She has run off with X'*

Clue	O(H)	P(E|H)	P(E|$\bar{H}$)	W(E|H)	O(H|E)
Wife out	Prior 1/100	1	¼	4	4/100
Friends?	4/100	1	1/10	10	4/10
C. cards?	4/10	1	1/10	10	4
Suitcase?	4	1/5	1	1/5	4/5
Passport?	4/5	1/10	1	1/10	4/50
Jewels?	4/50	1/10	8/10	1/8	1/100

Written out in full such an Inference Table will remind you, both before and afterwards, not to omit vital components. If you want to go directly to W(E|H), omitting the intermediate steps P(E|H) and P(E|$\bar{H}$), you can always do so; but then it is a deliberate choice, not an oversight you might regret afterwards.

Ockham's Razor (or Parsimony) enters the process of CST at the very outset, in choosing your hypothesis. Always go for the simple hypothesis first because you have a better chance of inferring whether it is decisively right – or decisively wrong. It is thus more progressive.

So there we have most of the recipe for CST: observe the situation; gather clues; formulate an hypothesis to explain what you see – as simple a one as possible. Select Prior Odds on H. Write out an Inference Table in 6 columns, and work your way down through the table updating your Odds in the light of each fresh piece of evidence, the ordering of the evidence being immaterial. The Odds which you reach at the bottom of the right hand column are provisional. Only you can decide whether they are decisive enough for you to act upon. Indeed they may do no more than urge you to go out and look for new clues bearing on H. On the other hand because the process is inherently multiplicative a relatively small number of clues can lead to very high Odds either on or against your hypothesis, Odds on which you may feel safe to act. For the same reason a large number of clues, even weak ones, acting in the same sense, can lead to Inductive Conviction [or conversely Inductive Dismissal], which may be more secure than a Deductive conclusion (because that necessarily relies on perfect information). For example every safe airline flight, although he's not personally aboard it, should move even a philosopher towards Inductive Conviction in airline safety.

So the Inference Table is the skeleton of Common Sense Thinking. If it seems childlike then so it should because that is how animals presumably learned to think. But it is neither superficial nor superfluous. By writing out our thinking in a formal IT we can reap several benefits: (a) We finish with Odds, with something more than a vague idea of how certain or uncertain we are about the situation we are worrying about. It may for instance turn out that we already have the evidence to act upon, but we didn't put it together multiplicatively as we should have done. That has often time happened to me, and most gratifying it is. (b) On the other hand we can be prevented from making a hasty decision that was over-weighted by one particular clue to which we overreacted. I'm a rather emotional person and I find the reflection that writing out an IT enjoins is often calming and helpful. (c) An IT folds in individually trifling clues which might have been separately left out but which, in multiplicative combination, may be worth listening to. (d) It reminds us to include and weigh key factors which might otherwise be left out – to sometimes serious effect (e.g. the rarity of cancer of the woozle in the example of Section (4:6). (e) An IT throws the spotlight on those clues carrying most weight. We are then forced to examine them most carefully – as we should. Has a suitably thorough error-analysis been carried out on them? A high Weight based on an inadequate appreciation of the real errors is a frequent cause of disasters in science e.g.: Cosmology (Sect (5:5), Law, Cot deaths (Appendix A6), Economics (Long Term Capital Management: Sect 12:3) and Politics (Sect. 16:4);(f) Because the IT is never complete it is a salutary reminder that Induction is always provisional. It is work in progress, always open to the addition of a new line if and when new evidence comes to hand. It is thus very valuable for assessing the significance of a *new* clue. Does it change the combined Odds by enough to call for a complete re-appraisal of the hypothesis / decision? Or is it merely marginal?

(g) An IT is a compact but very necessary record of one's thinking. Memory can be treacherous while distraction can sweep valuable insights out of mind. Important matters or big projects may take years or decades to resolve so it is vital to have a faithful record. (next chapter)

(h) Finally, when you later need to convince others of your decision, an IT is an excellent place from which to start (Chapt 12). After all it embodies your own thinking. All that is missing from it is a consideration of the error involved in multiplying lots of numbers together (That we tackle in the next chapter).

In summary then Inference Tables systematize the process of Common Sense Thinking in such a way as to help one avoid mistakes and omissions, and hence to think better. Try using them. I think you'll find them invaluable. [Worked examples for you to do later.]

(8:3) THE LOGIC OF COMMON SENSE THINKING

Nature doesn't have to be logical. Her creations simply have to evolve, reproduce and survive – sometimes in the oddest niches. So it may be with CST. It simply has to work, and have no rivals that are competitively superior. Thus providing a strictly logical basis for it is not absolutely necessary. I have tried to persuade you, without being absolutely positive, that we scientists, even when we are not aware of it, use Bayes' Rule, the Detective's Equation and Ockham's Razor to arrive at both trivial and far reaching conclusions. We're all using CST because we haven't come up with a better alternative. If you are satisfied with that reassurance you can ignore the rest of this section. If you are not then I try to provide a little more logical comfort.

In my view The Detective's Equation is the crucial tool of CST. BR is merely a sub-case of it when there is only one clue, and as such not so convincing. The DE is so powerful because it is capable of dealing with all of the evidence, including conflicting evidence, *at once*. It highlights the fundamentally *multiplicative* nature of Inference and it can be decisive. Furthermore, on its own, it provides the underlying explanation for Ockham's Razor [Chapter 7].

I would argue that the simple reasoning we used to move from Bayes Rule to the Detective's Equation in Section (5:1) is elementary, fundamental, trivial: after all we were simply using BR over and over again. If that is so then any doubt about the DE must hinge on the validity of BR itself. However Richard Cox {*} has proved (using lengthy but not hard algebra) that to deny BR is to argue inconsistently with the information you have i.e. in a flawed way. But Evolution would surely drive such inconsistent minds towards extinction. Thus BR, and hence the

DE, provides a sound basis for CST; a nice instance of Evolution nudging Logic down the correct path.

What is there to argue with? The Weights? Well they are undefined in any detail; they have to be. After all they depend on the particular clue, the circumstances and the individual thinker involved . They may be rough and ready, for example 'high', medium', or 'low' or they may be more precise where observation and Error-Analysis permit of it. There is room for judgement, intuition, experience, logic and folly within every Weight (and the Prior) on the right hand side of the Detective's Equation. This allows two individuals to wildly disagree about the same evidence. Logicians understandably don't like this; it is not so with their preferred Deductive Logic. Alas Deductive Logic is not much use in the real world of imperfect information where moreover there are few rigidly defined rules. And Nature doesn't care if you reach daft conclusions or make foolish decisions. She will allow Natural Selection to replace you with one more wise.

Implementing the Detective's Equation in the animal mind requires mechanisms for classifying clues, memorizing experiences, associating ideas, assessing Weights, and combining them together. What those mechanisms are on a physiological level we may never know. And anyway the precise mechanisms don't matter because they might not be unique— e.g. you can build an electronic computer out of either valves or transistors. Clearly the more memory there is the more clues that can be considered *together* before taking action. And that is obviously desirable from the point of view of survival. A simple mind presumably reacts to one clue at a time; a complex mind can weigh and combine several, even clues which conflict with one another.

The great value of writing the DE out (or using an Inference Table – which amounts to exactly the same thing) is that it allows the thinker to escape the limits of memory; any number of perceptible clues can be consulted at once, their Weights considered and re-considered at leisure. Surely that is invaluable.

We have emphasised the multiplicative nature of the DE and one might wonder how animals –which apparently don't use numbers to any great extent – would evolve the ability to multiply. There are two answers to that. Firstly mindless electronic circuits (e.g. valves) can be constructed to multiply, they are called amplifiers, so surely physiology could do likewise. Secondly, and more profoundly, it turns out that multiplication is only a convenient allegory, or model if you like, for a much deeper and simpler process for combining Weights which is described in Appendix A9 on 'Categorical Inference'. If you are unhappy with the notion of animal minds multiplying, read that – but I would defer doing so until after Chapter 10, when it will make much more sense. Successful minds have to combine the Weights they put upon clues in such a way as to reach wise decisions – and it turns out that that is more than one way to do that – but for now multiplication is the easiest one to describe.[Older readers will recall that multiplication used to be done without multiplying, using logarithms or slide-rules.]

If we haven't provided an unavoidably logical basis for CST we have teased out a procedure, an 'Algorithm'(a posh name for a recipe) by which an individual mind can reach generally sensible decisions by combining several clues together. It supposes that the future will be much like the past; it associates ideas together, it weighs the degree of their association and finally it multiplies those Weights together to update the Odds on the hypothesis in question. It provides no guarantee that two individuals faced with the same evidence will agree. However, foolish Odds will, sooner or later, lead to extinction. That is some kind of loose assurance that we survivors should be able to agree more often than not, where there is enough evidence. [Appendix A9 on Categorical Inference gives, I believe, an almost incontestable account of CS 'logic' but it cannot be tackled until after Chapter 10.]

(8:4) TIPPING WEIGHTS

The Weights are everything in assessing evidence and I have said they are almost entirely a private matter for the thinker in question. They must come from his or her personal judgement, training, experience, intuition, calculation or whatever, which ought to be far more relevant than the general strictures of an author like myself. However there are a few useful tips that are worth handing on.

At the outset it's worth asking yourself at what Odds will you be prepared to close the discussion and act. Presumably that will depend upon the circumstances. You might desire higher Odds on success when deciding on whether to undergo a life-threatening operation than when choosing a holiday destination. Of course they might not be available.

The size of those decisive Odds reflects back on the kind of clues you will wish to consider. If your decisive Odds are 100 to 1 on then as $2^7 = 128$ you would need *at least* 7 clues with Weights of 2 to reach a decision in favour (or 7 clues with Weights of ½ to reach a decision against). For my part I don't generally bother with clues with Weights in favour of less than 2 (or clues with Weights against of more than ½) – my justification being that I cannot usually judge Weights to better precision than a factor of 2.

When you are stuck for a Weight it's not a bad idea to estimate a maximum likely value and a minimum likely value then choose a mean, the right kind of mean, somewhere in between. Consider an important example in Law. As a jury person, at what Odds on guilt would you be prepared to convict? What do you mean by 'beyond reasonable doubt'? When I ask myself that question I find that Odds of 5 to 1 are too low, Odds of 20 to 1 too high. A reasonable compromise lies between the two. But where is that compromise? Remember we are in a *multiplicative* not an additive situation. In that case the right thing to do is to multiply the two extreme numbers together and then take the square root of their product; i.e. take the square root of 5 times 20 = 100 which is 10 (In other words take 10 which is twice as high as 5 and half as high as 20 – which seems 'not unreasonable'). So I choose the Odds on which I will convict as 10 to 1 on guilt, or better. That's arbitrary I will admit but it happens to be the very Odds enjoined by long-standing practice in British and American Common Law. That law, which in its very nature is Inductive (because it erects generalizations on a number of past case-verdicts, rather than deductive – reaching judgements from codified principles, as on the Continent) was most famously expounded by Sir William Blackstone in his 'Commentaries on the Laws of England' (1765-69). In it he said "It is better that ten guilty men escape than one innocent suffer." Thenceforth 'Blackstone's Ratio', 10 to 1, became a common yardstick for 'reasonable doubt' (Lawyers don't like to admit that it does imply that up to 9 per cent of the convicts in gaol will be innocent).

In more general terms the Square Root is the appropriate mean value of two Weights – its sometimes called their 'geometrical mean' (as opposed to the 'arithmetical mean' or average). Square rooting is a useful trick when you are dealing with a doubtful Weight for a clue. Personally, had I been the detective in the Library-murder case I would have suspected that the forensic scientist's foot-print Odds (40 to 1 against you will recall) were too 'strong'. I would quietly have taken the square root and as $\sqrt{40} \sim 7$ and amended them to 7 to 1 against. Square rooting always retains the sense of a clue while reducing its leverage (or Weight) one way or the other (the square root of 1/40 is roughly 1/7). I will say more about this in Chapter 10 when we come upon 'The Principle of Animal Wisdom' or PAW.

You might remark 'How arbitrary, how unscientific' and you might be right. But how certain are you that the 40 to 1 was more 'scientific' ? Don't tell me it's because the forensic chap was 'an expert'. Experts can very often be wrong too. Look at Arthur Stanley Eddington (later Professor Sir of that name) massaging the eclipse measurements back in 1919 (Sect 2:3). Unlike the detective, I know enough Statistics to suspect the expert's use of the word 'Normal' in describing his errors to her over the phone. I very much doubt that they could in fact have been 'Normal' in the technical sense of that word. And if not then his Discrepancy-Odds of 40 to 1 against are more or less meaningless [Actually the leading lady *was* the murderer but she got off because of the forensic evidence. The maidservant was convicted instead, mainly on the grounds that she was working class. She served 20 years hard labour before the murderess left a posthumous confession in her will].

Square rooting Weights may seem a very blunt tool but it's not out of place when it is so difficult to choose Weights in the first place. One notorious factor is bias. It's so easy to fall in love with one's hypothesis – then fail to see the blemishes in its soul. Scientists need above all to be dogged, and our pantheon is full of dogged heroes, but doggedness commonly follows upon what the French call a 'coupe de foudre' – falling instantly and hopelessly in love. I've done so myself and remained wedded to my hypothesis on Hidden Galaxies (Sect (4:8)) half a lifetime after she drifted so beguilingly into my youthful gaze. Francis Bacon warned us when he said ; "The human understanding when it has once adopted an opinion……. draws all things to support and agree with it. And though there be a greater number and weight of instances to be found on the other side, yet these it either neglects or despises." More recently the astronomer Saul Perlmutter has remarked: " Your job as a scientist is to figure out how you are fooling yourself." I agree wholeheartedly .

The best remedy I can think of when you are romantically challenged is to get out the old Square Root and wield it ruthlessly on the Weights of those clues which best support your beloved hypothesis. If it still hangs in there, good for you. If not it's probably time to look for either new clues or new hypotheses. Which leads to a second prescription: if possible never weigh an hypothesis in isolation. If you court two or three at once you'll be less likely to fall for one. Alas that is often a council of perfection. It may be hard enough to come up with a single hypothesis consistent with the available facts – let alone two or three. But when you can…..

Never underestimate the pernicious effects of bias in thinking, *especially in your own thinking*. It can take so many forms: selecting and/or over-weighting favourable clues, ignoring or under-weighting unfavourable ones; likewise people, notably listening to friends and dependent juniors in contrast to enemies and peers. It is a form of infantilism which most of us never completely grow out of, not even scientists. I'm sorry if I sound cynical but science is performed by flesh and blood people with ambitions, egos, impatience and families to feed. Everyone worthwhile in it wants to make discoveries, or win promotion, get funding and of course go down in history. Why not? But the temptations……As some one said of science: " To be first is paramount. To be second is to be forgotten." {*}

To try and avoid bias those of a mathematical turn of mind come up with elaborate theories for 'calculating' Weights rather than plucking them out of the air as they could claim we laymen often do. Statisticians in particular produce numerical tables accurate to 4 places of decimals. Although their tables may be an accurate representation of their theories, their theories are *seldom* an accurate representation of the real world. They are fooling themselves, and alas all too often in their foolish certainty, fooling others who are less numerate. In Chapt 11 we take on the pretensions of Statistics. Suffice it for now to ask a simple question of our Statistical friends: 'What on Earth is the point of calculating some clues to high precision when you are going to have to combine them with others of very low precision, accurate to a factor of no better than 2?' There is no sensible answer. Precision is no substitute for Common Sense, no substitute at all.

Given the difficulty of choosing Weights with any exactitude I often resort to the following crude factor-of-two scale (as I suspect does Goch):

TABLE (8:3): WEIGHTS AND WORDS
'Weakly in favour': $W(E|H) = 2$; 'Weakly against' $= 1/2$
'Strongly in favour'': $= 4$;   'Strongly against' $= 1/4$

When in doubt go for *more* evidence ; one more moderate clue in favour can tip the argument into Inductive Certainty. The great thing about science is that there usually *is* more evidence out there to be found; that may not be true in law, history, economics, psychology……..

One final warning. Don't go back over the calculation to alter Weights just because you don't like the final outcome. That defeats the whole purpose! A way to avoid this cardinal sin is not to carry out any of the multiplications until the very end, when all the clues and Weights have been assembled. Reassure yourself they are the best you can manage – then go forth and multiply. [NB The whole logic of Thinking is laid out in Appx. A13 *A Sketch map of Thinking*]

(8:5) SETTING PRIORS

The most problematic factor in Common Sense Thinking is the Prior O(H) – the Odds you put on your hypothesis *before* you consider any evidence. Some thinkers, particularly those of a mathematical persuasion, throw up their hands in horror at the very idea of a Prior. Fine – but what is their alternative? Invariably something equally objectionable – but cunningly concealed from sight – their own in particular. We shall argue that the situation is far from hopeless.

We can begin by looking back over past chapters to see how values for the Prior were set up in particular cases:

(a) The detective in the Library-murder case convinced herself that one of the 7 house guests must be responsible. Thus the unbiased, dispassionate Odds O(H), her Prior in the case of the leading lady, were 6 to 1 against her guilt. Don't you think that was reasonable?

(b) Dr Z the young astronomer in the Hidden Galaxy controversy found against my beloved theory because he set his Prior Odds on it at 1/5. If he knew little of the subject beforehand I don't find that unreasonable myself. If dozens of astronomers had spent dozens of years looking for dim galaxies – without finding a single specimen, the hypothesis that they exist does seem improbable, though, given the practical difficulties, not impossible.

(c) In the case of the expanding universe the astronomer considered that there could only be 3 possible alternatives: expanding, collapsing or static. Thus his Prior of 2 to 1 against expansion is at least dispassionate.

(d) In the case of the mammoths the band elicited a Prior by asking its 4 wisest old heads. They came down 3 to 1 in favour of mammoths being harmless [i.e. O(H) = 3]. What better could they do? Their Prior was weak and quickly overruled by new evidence.

(e) In the case of the missing wife the quality of the couple's marriage is plainly important, even paramount. Any spouse will have some sort of idea about whether their partner is likely to stray. That deserves to be included in any calculation – indeed it demands to be included. That is one advantage of the Prior: it does enable one to incorporate information which might otherwise be left out. Take the hypothesis 'The Moon is made of green cheese'. Prior to Apollo 11 it might have been difficult to comment upon. I would nevertheless have assigned it Prior Odds of less than 1 in a million (10^{-6}) on the grounds that I cannot imagine cow's milk finding its way into lunar orbit.

(f) In the case of the cancer scare a very vague Prior (Cancer of the woozle is rarer than one in a thousand women) sufficed to transform the argument in a constructive way. Priors don't have to be exact.

(g) In the case of the flying saucer neither party had much to go on apart from blind prejudice. Good luck to them if that's how they feel. But both parties need to understand that a Prior set too low, or too high, could close their minds to all future conversion. If that's what they wanted then why did they bother to go to Jura?

(h) The Tiptuppititis sufferer set up what he considered to be a dispassionate Prior on Organocalm of O(H) =1. But think of the implication; he allowed that a virtually unknown substance had an even chance of curing his disease. Applied to a whole pharmacy he would be allowing that half of all drugs on its shelves would be effective against Tiptuppititis. In other words his Prior Odds on Organocalm were, unconsciously, way too high. So you always need to think hard about your Prior.

The mere fact that we are willing to evaluate some hypothesis against the evidence tells us that we are unconsciously setting Odds *against* it that are not too high. Imagine that the typical favourable clue we will consider has a Weight between 2 (low) and 8(high) i.e. has a geometric mean of 4 (moderate). If we were fortunate enough to find a *net* excess of 5 such clues with the same sense we would have combined Odds of $4 \times 4 \times 4 \times 4 \times 4 = 2^{10} = 1024$ or more than a thousand to one in its favour to set against our Prior Odds. That crudely suggests that no hypothesis worth considering in detail should have unconscious Prior Odds against it of worse than 1000 to 1. (or in its favour of less than 1/1000, which is the same thing). It amounts to an unconscious estimate of how much accessible information (evidence) there is out there to change our minds. If it seems arbitrary it is not foolish, at least not for natural science. For if there were more relevant information out there the problem (hypothesis) would probably have been solved already. And if there's much less there's no point in considering it at all. In other words hypotheses only become viable (worth considering) at Prior Odds on of more than 1/1000 ($1/10^3$ or 10^{-3}). When all other arguments fail that is the rule of thumb I use. To what extent you want to borrow it is up to you. Nature, that is to say Natural Selection, has probably hard-wired some such 'Null Prior' into our minds. If it was too low we'd never make up our minds, never act. And if it was too high we'd be too incautious and pay the penalty in blood.

I don't pretend that this discussion of Priors has been logically watertight. It will be easy to criticise. But there's no point in that unless you're prepared to offer a better alternative. Some have tried but with (to me) unconvincing and in some cases ridiculous results (see Chap 13 for the emotive discussion). I like Francis Bacon's famous remark: " If a man will begin with certainties, he shall end in doubts; but if he will be content to begin with doubts he shall end in certainties."

In summary then try to find a Prior as best you can. But think about it very carefully and try to set it neither too high nor too low for the conceivable evidence to change your mind. Otherwise there's no point in trying to use Common Sense in the matter. If all else fails a 'Null' Prior Odds O(H) of 2^{-5} or 32 to 1 against, isn't a bad rule-of-thumb.

(8:6) JURY DUTY   You find yourself called to stand on the jury for a murder case involving a young student who was raped before her strangled corpse was dumped in a pond outside her university town. You will be presented with the evidence, and the judge's summing up, and be asked to try and reach a verdict using the various tools for CST rehearsed in this chapter

Vicky and her flatmate Georgina live amongst 10,000 other students and assorted young people in a warren of old buildings euphemistically called 'the student village'. On the night she disappeared they went out at 9.30 pm to pick up a noodle take-away. Leaving Vicky in a busy queue Georgina went round the corner to get a bottle of wine before returning to their flat to wait for Vicky. When Vicky hadn't returned by 10 Georgina went down to look for her. At 1 am. the next morning she informed the police that her friend had gone missing.

Several days later, as the result of an extensive search, the body was found half sunk in the pond. She'd been raped and strangled but the only useful forensic evidence was the discovery of blood under her finger-nails belonging to a rare group present in the population with a frequency of about 1 in a thousand. Tyre marks in the surrounding wood revealed that the body had been transported there in a vehicle with nearly bald tyres.

In the course of a big man-hunt blood samples were taken from most men living in the student village. Vicky's friends, student colleagues and neighbours were interviewed and eventually the defendant, an illegal immigrant from Lithuania, one Robertus Vilnius was charged with Vicky's murder. The evidence against him included:

He was Vicky's ex boyfriend. She had broken it off, but several letters found in her bedroom showed that he passionately wanted her back. Under questioning Georgina revealed that Vicky had found Vilnius to be an habitual liar and 'not very interested in sex'.

Originally Vilnius had had a good alibi – at the time he was at the other side of town with his English teacher Pavinder Singh and hadn't returned home to the village, and block of flats where he and both girls lived, until after midnight. But later when he found out about the timing of the murder Singh went to the police and confessed that he'd unwittingly supplied a false alibi at Vilnius's request.

Vilnius maintained that he'd only lied because he was innocent but knew he was bound to fall under suspicion in the circumstances.

'Anyway', he said, ' I couldn't have committed the murder because I don't have a car and I can't drive.' This was confirmed by a local driving school who had terminated his course after only two lessons after his failure to pay their charges promptly.

In his summing up the prosecuting barrister made much of Vilnius' dishonesty. He'd lied to the immigration authority, lied to social workers, lied to the housing authority, lied to the police, lived well on entirely undeserved and illegally obtained benefits, and never lifted a hand to work during his entire five years in Britain. And of course he knew Victoria, which was significant because most murder victims are killed by people who know them well. When searched his flat was full of filthy pictures and he had a motive – revenge for being cast aside as unworthy – which he was. He had had the opportunity, a suspicious false alibi, and most tellingly a very rare blood group present in only one in a thousand, the very same as found under the poor girls fingernails. So what if he couldn't drive? The body hadn't been found for several days – leaving ample opportunity to organise a lift from one of his unsavoury illegal associates.

The judge in his summing up agreed that Vilnius was a nasty young man, a liar, a fraud and a benefit cheat on a massive scale. He had sneaked into Britain and abused our compassion and hospitality to the tune of nearly one hundred thousand pounds. But what had all that got to do with murder? There were no doubt adulterers, tax fraudsters and even wife-beaters in his very court room, but it didn't follow that they'd murdered Vicky Parker. He directed the jury to focus its attention on the relevant clues only and in particular on the unlikely coincidence of the defendant's blood group with that in the victim's finger-nails. But they should also note his lack of a vehicle and his inability to drive. And as the accused and the victim had lived high up in the same block of busy flats with no lift it was difficult to see how he'd managed to get her into his room and her body out again, without being detected.

So what do you think? I'm asking you to assemble a 6-column Inference Table for the whole case, from Prior to final combined Odds and then use it to deliver your verdict; guilty or not? Fill in your Prior, the clues you think relevant and their attendant Probabilities (where appropriate) and Weights. Then multiply everything together. It might prove trickier than you imagine. Then, when you've made up your mind, and only then, turn to the discussion in Appendix 7A 'Murder in a university town.' I hope you'll find this an illuminating exercise in formal Common Sense Thinking.

(8:7) COMMON SENSE THINKING IN THE REAL WORLD

The weakness of the Jury problem, as of most problems in this book so far, is that it is artificial. Such problems have been deliberately constructed or distilled to illustrate this or that distinct point. Quite intentionally they miss the noise and confusion of real life – which makes it so much more interesting and difficult to live. It is time to put that reality back in to see what further lessons we can learn about CST. I'm going to do so by looking at a program of real scientific research with all the mess, the noise, the confusions, the conflicts and the uncertainties left in. I hope you'll forgive me for using my own research program on 'Hidden Galaxies'. I do so because I am familiar with all the mess in this case —which I am not in any other. It is practically impossible to dig out other scientists' research histories because we are not encouraged to publish all the vicissitudes and misunderstandings that really occurred.

Quite understandably the community only wants to see your final conclusion, with only your best evidence for believing in it. That is what makes it so hard to reconstruct scientific history in a credible way. Only occasionally do you get an account like Jim Watson's 'Double Helix' with some of the unvarnished and discreditable truths left in. Scientist are no better than other people when opportunities come to burnish their public reputations. For instance Sir Isaac Newton, the greatest of our kind, appears to have been a serial liar who may have pinched the notion of Universal Gravitation, for which he is most famous, from his bitter rival Robert Hooke – who made the huge mistake of dying first.

The danger of quoting my own research is of course that I'll be biased. But that doesn't matter here because I'm not trying to convince you of my hypothesis. You probably won't care a damn whether Hidden Galaxies (HGs) do or do not exist – why should you – so my prejudices in the matter will be of no account to you. What matters in this context is the mess – and I think you'll find there's certainly plenty of that. But first of all a little necessary background.

Galaxies you will recall (Sect.4:5), are immense islands of stars in Space and the building blocks of the universe on the cosmic scale. They contain most of the stars, give off almost all the light and contain a significant fraction of all the *detectable* mass in the cosmos . We live in one such spiral galaxy The Milky Way which contains no less than 10^{11} stars within its 10^5 light-year diameter. We have a nearest equivalent neighbour at a distance of only 20 diameters away (see Fig. 4:2) and yet Andromeda, despite its huge size, is virtually impossible to see because it is scarcely brighter than the night sky. Because the individual stars within galaxies are so spaced out from one another galaxies are dim and thus very hard to observe, which explains why they are still largely mysterious. That's why I find them so fascinating and why they have taken up forty years of my working life.

Back in 1975 I came up with 'My Beautiful Theory' (MBT) suggesting that most galaxies are hidden completely from our sight by their elusive properties, and ever since then I and my associates have been trying to prove, amidst general scepticism, that this unlikely theory is right. We're still not certain whether it is right or wrong even now. The Inference Table below is a history and a summary of the project's successes and failures over the course of 38 years. In important respects it is not atypical of other ambitious pieces of scientific research. It started as the result of a casual observation; became an obsession; sucked others in and generated vocal critics, provided brief moments of dizzy triumph and years of black despair. Before its funding was cut off in 2005 it attracted millions of pounds worth of research support which resulted in new instruments and ambitious observing programs on all sorts of telescopes and satellites around the globe. It has been completely written off several times, then each time so far has risen from the dead. Now, as the result of a casual encounter in the Caucusus Mountains, I'm pretty damned certain [my O(H) is more than millions to one on] it is right. I am. But perhaps I'm not the most reliable judge – and even I've given up in despair at times. That's life in the research trenches. You can't beat it for sheer excitement.

The hypothesis under test is: *"Most of the galaxies, even in our neighbourhood, remain to be found because they are difficult to see through our atmosphere and against the brightness of the terrestrial night sky."* See Inference Table (8:4) below. The details need not concern the reader but it is interesting to follow the roller coaster journey we rode on the clues, Weights, and Odds as I have compiled them in the

IT Please study it for a few moments to see how it was formed. I hope then to draw some conclusions of more general significance.

First of all some brief explanatory notes to go with the labels in column 4:

(A) I started with a Null Prior of about 1 in 32 i.e. 2^{-5} so that I won't fool myself with accidentally good, or biased, early results. I'm setting the bar moderately high for myself. I'm not going to believe my own results until they've overleapt this stiffish hurdle. I am going to use powers of 2 throughout because they reflect my inability to be more precise. It is much easier than using decimal numbers in this context, and one is much less likely to make arithmetical mistakes because one has only got to add and subtract the indices (powers), not carry out the multiplications and divisions. It is a very useful trick I'm going to use frequently in future (Try it). So as 2 to the power 5 is 32, then the negative power 2^{-5} is by definition one over 2 to the power 5 and thus 1/32. That's my Prior. We'll talk more about this in the next chapter. For now just think of me trying to be an honest scientist.

(B) My Weight (16) here arises from a remarkable coincidence which I discovered. The surface-brightnesses of galaxies as observed, coincide almost exactly with the optimum value prescribed by my theory. Ockhams Razor suggests we should not regard this as a coincidence but as confirmation of the theory, or at least positive support for it.

(C) A galaxy too dim to be visible itself could yet betray its presence by etching characteristic atomic shadow-marks on the spectrum of the light from sources behind it as their light passes through. I call these 'Spectral Ghosts". These are the 'shadows' referred to. Far too many are detected to be explained by visible galaxies alone. This is a moderately strong hint of more hidden ones, perhaps many more.

(D)A 'Crouching Giant' is my name for a galaxy which appears superficially as a dwarf but, when observed carefully, turns out to be a colossal dim giant mostly hidden below the sky. A spectacular specimen found by Bothun and Co.

(E)An explanation for Spectral Ghosts or shadows alternative to mine in (C) is preferred by new Hubble Space Telescope observations.

INFERENCE TABLE (8:4) : HIDDEN GALAXIES

1	2	3	4	5	6	7	8
Clue#	Date	Clue	Note	W(E|H)	O(H|E)	Q	Status
Prior	2^{-5}		A				
1	1975	Theory I	B	2^4	2^{-1}		Doubt
2	1978	Off-beams		2^{-3}	2^{-4}		Pessim
3	1983	Theory II		2^2	2^{-2}		Doubt
4	1984	QSOALs	C	2^2	1		"
5	1985	SB/colour		2^2	2^2		"
6	1987	Fornax		2	2^3		"
7	1987	Cr. Giant	D	2^3	2^6		Belief
8	1990	(2) explnd		2^2	2^8		"
9	1993	SB/latitude		2^2	2^{10}		Convn
10	1994	ESO gals		2^3	2^{13}		"
11	1995	Impey		2	2^{14}		"
12*	1995	Gnt. Halos	E	2^{-5}	2^9		Belief

13	1997	Centaurus		2	2^{10}		Convn
14*	1997	Arecibo		2^{-5}	2^{5}		Optim
15	1998	Compacts	F	2^{2}	2^{7}		Belief
16	1999	SB flucts		2^{-1}	2^{6}		"
17	2002	Sloan DSS		2^{-1}	2^{5}		Optim
18	2002	Millenium		2^{-1}	2^{4}		"
19*	2005	HIPASS	G	2^{-6}	2^{-2}		Doubt
20	2005	Starcounts		2^{2}	1		"
21	2007	DG Virgo	H	2^{2}	2^{2}		"
22	2009	Eq. Strip		2	2^{3}		"
23	2012	HSB HDF	I	2^{4}	2^{7}		Belief
24***	2013	Clustering	J	2^{5+5+6}	2^{23}		Convn
25	2013	Red nugs	K	2^{2}	2^{25}		"

(F) For various technical reasons galaxies that are too compact will be ordinarily impossible to find according to MBT. But here they are turning up in a deliberate search using optical fibres set on what looked like stars. You see!

(G) This was a huge survey specially instigated by me to search for dim and dark galaxies with the giant Parkes radio telescope in Australia. It utterly failed. That's why I've assigned it such an unfavourable Weight of 1/64.

(H) We find an apparently massive Dark Galaxy with the giant Jodrell Bank radio dish. But our colleagues violently dispute our interpretation;

that's why I've only given it a Weight of 4 for now. But it's just what MBT predicts.

(I) I am a member of the Hubble Space Telescope camera team and I was, with quite other matters in mind, studying the new and deepest picture ever taken with it of the sky. It made no sense, no sense at all. Unless! Unless the sky is full of Hidden galaxies which only come to light at high redshifts, at the redshifts where these very distant galaxies lie. No other explanation has been offered.

(J) As a result of a hint dropped to me by a fellow astronomer while we were attending a conference up in the Caucasus Mountains, I realised that three of the strongest observational arguments against our theory [asterisked numbers (12), (14) and (19)] were seriously flawed because they'd grossly underestimated their errors. Thus my Weight of 32 times 32 times 64 here deliberately reverses their earlier unfavourable Odds of (1/32) ,(1/32) and 1/(64).

(K) A 'Red Nugget' is a type of galaxy predicted by MBT but never previously seen.

NB The words like 'belief' etc are discussed in the next chapter. There's nothing sacred about them, but they seem about right to me.

What of general interest do we learn from such an Inference Table, such a tangled web of thought?

(1) Without some kind of book-keeping of the Odds I don't see how any thinker can retain some balance in the face of so much conflicting evidence. But it's got to be done. The IT, for all its obvious flaws, enables one to do just that – without a superhuman memory.

(2) There are no less than 25 clues with their associated Weights. All of those Weights are crude, all of them. Yet in their sheer numbers lies hope. Provided I am not overtly biased, the too-highs and the too-lows should roughly cancel out and lead to wisdom in the end. That's why the Detective's Equation, with its capacity to balance evidence, any amount of evidence, is the chief secret of Common Sense. Given enough evidence, and given lack of bias, there is definite hope that it will lead one towards the light. What does it matter that my final Odds are 30 Million to 1 on? A factor of 100 or even 1000 either way wouldn't affect the outcome. I could even Square Root my final Odds and still come away with Inductive Conviction in this case.

(3) As the chief protagonist I am obviously biased. I so want MBT to be true. But at least I, and anyone else, can check back over the table to look for suspiciously 'good' Weights. I've wielded my Square Root here and there, and as you can confirm, most of my highest leverages have been conferred on unfavourable clues (numbers 12, 14 and 19) – all of which subsequently turned out to be completely wrong. And I even gave up the whole project for 4 years in 2005, discouraged by the apparently unfavourable Odds. Anyway I set that cautious Prior in the first place as a remedy for any personal bias. After all I can't afford to waste my own life on a wild goose chase, any more than anyone else can. Self delusion exacts a very high price – wasted years of one's life.

(4) Scientific research generally requires funding – and mine was cut off in 2005 when things were not looking good.(The IT shows the combined Odds were only evens. And those were *my*

Odds. It was just bad luck. Also stupidity on my part. Had I been quicker to spot the egregious faults in (12),(14) and (19) there might have been less reason to cut me off. One can see why most scientists nowadays are risk-averse and want instead to join some industrial herd supported by the Conventional Wisdom of the Dominant Group (COWDUNG). That timidity could kill off science, and perhaps it is doing so.

(5) So now I am convinced – but most of my professional colleagues are not, not yet. Why they are not, and why one needs to tackle the whole question of persuading others deserves an entire chapter in itself (Chapt 12). For now note that the vital clues (23) and (24) are too recent for their impact to ripple out. And the false clues (12), (14) and (19) are still respectable among the community of astronomers for now.

(6) Less than half of the clues were gathered by my colleagues and I. The rest were gathered by others or turned up by accident. In many cases the originators had no idea of their relevance to HGs. One therefore needs to read widely and listen in. As Louis Pasteur said "…chance favours the prepared mind". I think he was absolutely right. Omnivorous curiosity seems to be an unalloyed good, when it comes to science anyway.

(7) The search for understanding can be a steeplechase ride. Breakthroughs, breakdowns, luck, misunderstandings, personnel, funding, ambition, distraction, rivalry, imagination (and its lack), stupidity, character and history all play confounding roles. To extract a clean story at the end of it, with one bright hero, would most often be to tell a pretty tale. The greatest scientific discovery of the twentieth century in my opinion – the discovery of Continental Drift – took over fifty

years to fully emerge from its chrysalis [see Ch. 3 and in particular {*}]. Alfred Wegener first saw the light back in 1912, but died a hero's death out on the Arctic ice in 1930 without convincing his colleagues. Because he'd got a crucial part of the story wrong his contemporaries dismissed him almost entirely. Anyway he wasn't one of them – that is to say a geologist. He was a bloody meteorologist, and a German to boot. As much as anything else it was the Cold War which resurrected his hypothesis in the 1960s. The US Navy, with its nuclear confrontation with Soviet submarines, needed to know what was going on on the ocean floor – and provided the funds to find out. Instead of the ancient sediments expected the sea bed was made up of young igneous rock – on the move! Wegener's continents weren't ploughing through the sea-bed – as he had supposed – but were being carried along on top of it, as it flowed. No one generation could have got it entirely right. The instruments of Wegener's day were too feeble to see the ocean floor, too insensitive to detect drift of a mere centimetre a year. Those who did claim to measure drift measured drift of 50 metres a year – because they'd wildly underestimated their errors . Their over-optimism helped to discredit Wegener – though they were not immediate collaborators of his.

(8) Yes it's those Errors again. They can be fiendishly difficult to locate. Astronomy for instance is shot through with systematic errors – as I imagine are other fields of thought. It's the unknown unknowns which get you (Chapt 10). Take the three most unfavourable clues aforementioned, one of which (19) I 'helped' to provide myself. We all forgot that galaxies are not evenly spread throughout Space, but dramatically clustered in

clumps. Thus it was all too easy to miss hidden galaxies when they huddled amongst a cluster of far more visible neighbours. Clue (24) put this to rights. Likewise the astronomical instruments of today, miraculous though they are by comparison with yesterday's, are still too feeble to find all the galaxies we would like to see (clues 17 and 18). We are still prisoners in a brightly lighted cell, impotent to see out into the vasty dark (23). The running Odds though trace the fluctuating state of play. Crude though they are my Weights follow the turbulent state of the game through no less than 25 clues and 40 years – and it's not over yet.

From this example I hope you will agree that an Inference Table, in a complex and turbulent world, is a powerful, even necessary way to hold onto common sense. Refs {2 to 8} deal with Hidden galaxies.

(8:8) DO YOU BELIEVE IN THE BIG BANG?

You may be uninterested in my hidden galaxies but what about the Big Bang– perhaps the highest profile hypothesis in all science today? It supposes that all the cosmos with its 10^{11} visible galaxies emerged from a super-dense, super-hot fireball about 14 billion years ago. Why don't we submit that spectacular theory to the searching gaze of an Inference Table in the same way as we did my hidden galaxies? After all we encountered serious worries about it in Section (7:5) where we discovered that it has more free parameters (17) than independent pieces of testing evidence (13). In other words it looked more like a Just-so story than a proper, parsimonious scientific hypothesis. It fits – but then so it damned well should . It's flexible enough to please the Vicar of Bray[5]

The table below shows my IT for Big Bang Cosmology where the hypothesis at *issue* is "The Big bang theory of Cosmology is broadly right."

INFERENCE TABLE (8:5) FOR BIG BANG HYPOTHESIS

#	Clue	W(E|H)	O(H|E,..)	
	Prior		1	
1	Nothing older than expans. age	2^2	2^2	A
2	Earlier hot dense state	2^6	2^8	B
3	U should but does not decelerate	2^{-1}	2^7	C
4	U should be but is not anisotropic	2^{-6}	2	D
5	Gals don't dim w redshift but shd.	2^{-3}	2^{-2}	E
6	B Bang could prod. Light elements	2^4	2^2	F
7	B Bang predicts structure peaks	2^4	2^6	G
8	B Bang can't produce galaxies	2^{-6}	$2^0 = 1$	H
9	But CDM variant can	2^4	2^4	I
10	But gals don't resemble CDM ones	2^{-4}	1	J
11	'Inflation' may expln. no anisotropy	2	2	K
12	Recent accelng expansn not explnd.	2^{-4}	2^{-3}	L

---

[5] The turncoat clergyman of the witty eponymous poem who altered his faith time and again to suit the turbulent times . It concludes "For in my Faith and Loyalty, I never once will falter, But George, my lawful king shall be, Except the times shou'd alter."

Big Bang Cosmology (BBC) is such a delicate creature that I've started it off in life with a Prior of 1 (as opposed to 1/32 for my own Hidden Galaxies) i.e. with an even chance of surviving infancy. As before my Weights for its 12 extant clues are given in powers of 2, and here are the explanatory notes referred to in the final column of the IT:

(A) No component (e.g. galaxy) should be older than the cosmos itself, as judged by its size divided by its expansion rate (14 billion years). Reassuringly, nothing older has been found so far (But ages are notoriously difficult to judge).

(B) We live in a bath of Cosmic Background Radiation (CBR) for which no other explanation but a hot big bang presently exists. Cosmologists have been wrong so often in the past that I've given no Weights larger than 2^6 (64) or smaller than 2^{-6} (1/64). I'm being cautious.

(C) The self-gravity of its own parts should decelerate the expansion of the universe. Difficult to measure but not observed.

(D) Light has a finite speed, therefore as every day passes light reaches us from distant regions never seen before. There is no reason why such regions should look like each other (or like us) because they've never been in causal contact before i.e. no message could have travelled between them. And yet they do look alike to parts per million (e.g. in temperature). To explain this so called 'isotropy' is a huge intellectual challenge.

(E) Galaxies should appear dramatically dimmer with redshift (recession rate) – but they don't. This should be a huge effect – a factor of $10^{-4} or 2^{-18}$ in the latest Hubble Space Telescope pictures. It was originally proposed (1930) by Richard Tolman as a test for the expansion hypothesis. The universe has failed it dramatically. {8}

(F) Stars can't generate light elements like Helium as they can all the rest. But a hot big bang could.

(G) The very slight roughness observed in the CBR is expected under BBC.

(H) It's virtually impossible to see how galaxies could form in a hot expanding universe whose radiation would tear their young bodies apart.

(I) Cold Dark Matter (CDM), if it exists and dominates the ordinary kind, wouldn't be subject to the radiation of (H) and so could form seeds around which galaxies could quickly form later.

(J) But real galaxies don't look anything like the hypothetical CDM ones. {8}

(K) 'Inflation' is a vague and entirely ad-hoc addition to BBC coined to explain the otherwise mysterious isotropy [see (D)]. There is some slight observational evidence in its favour. [We used to call these 'miracles' when I was a boy.]

(L) In 1998 observations of exploding stars designed to measure the deceleration of the expansion due to gravity [see (C) above], revealed the shocking fact that the expansion is not only not slowing down but had actually started accelerating recently. No one has any coherent explanation for this, To save BBC a new free parameter had to be taken out of the box (number 17). It's a real weirdo labelled 'Dark Energy' [Miracle number Two].

So there you have BBC as seen through the lens of an Inference Table. Of course all the Weights are mine own – and disputable. But at least I am an experienced, and I hope honest, extra-galactic observer, familiar with and enthusiastic about the general field. Indeed I originally came into astronomy partly because of a fascination with Cosmology. But BBC simply doesn't add up – or rather multiply up. If we had started with a Null Prior of 2^{-5} as I believe we should, BBC would fall short of Inductive Conviction by a factor of no less than 2^{-8} or 4 in a thousand! In such circumstances quibbling over details seems pointless.

I leave the reader to find their own way out of this morass; I can't help because I can't find one myself. Suggestions (repeatable) which have been made include:

(i) *Some* of BBC is right – but which bits.

(ii) Cosmology should be judged by different (i.e. much weaker) standards than the rest of science.

(iii) I don't know what I'm talking about.

(iv) We've got no alternative to BBC so we need to cling to it – otherwise chaos will prevail.

(v) Cosmology is a costly and prestigious profession – like banking – which cannot afford to admit that its foundations are rotten.

(vi) BBC started off healthy but when it failed its observational tests it was gradually fitted by its nurses with a set of baroque prostheses which include Inflation, Cold Dark Matter, Hierarchical Galaxy Formation, dramatic and tuned Galaxy Evolution and most recently Dark Energy. Tended by its influential friends, who naturally don't want to see it die, it survives comatose on a life-support system composed of numerous ingenious free parameters. Is it a subject for admiration, pity, disgust or despair?

(vii) BBC and Common Sense Thinking are simply incommensurate.

(viii) I haven't taken into account the likely errors in my Inference Table. We return to this possibility in the next chapter. See {1} for a good reference to Cosmology.

SUMMARY

In any complicated situation Common Sense Thinking can be guided, probably should be guided, by the use of an Inference Table. It starts with a Prior and line by line multiplies up the Weighted clues. Eventually such a cumulative process *can* lead to Inductive Conviction, sufficient conviction on which to act. (Equally it can lead to the Inductive Dismissal of an hypothesis, the direct opposite of Inductive Conviction.)

Provided there *are* enough clues those associated Weights can be rough and ready (precise to a factor of 2 only). Bully clues, clues with overwhelming Weights either way, should be examined critically, and Square-Rooted where there is any doubt about them. The Prior needs to be set low enough so that there is little likelihood of overcoming it by chance alone (we discuss such a chance in the next chapter).

As a form of Induction, Common Sense Thinking does not and cannot have a strictly logical foundation. It must assume that the universe is not infinitely complicated and that the future will be much like the past. Then it works by using association of ideas (Bayes' Rule), the multiplication of Weights (Detective's Equation) and the irrelevance of sequential order. We probably share these notions and mental tactics with other creatures. But by writing our thinking down in a disciplined way using an Inference Table, we can increase its reach, its complexity, its transparency and its reliability. Even so its conclusions are provisional, and open to further evidence. One can regard that as a fault or a virtue. For me it's a definite virtue.

Two problems have emerged which need to be tackled. Error Analysis – which can play a crucial role in setting Weights and Priors [Chapt 9]. And Subjectivity: how can a thinker convince others if all his or her Weights are based on personal judgements [Chapt 12] ?

THINKING FOR OURSELVES (Disney)
CHAPTER 9
ERROR ANALYSIS (8.8 kw)
(draft 14/9/18)

"The job of the scientist is to find out how he is fooling himself."
Saul Perlmutter, astronomer. 2011

(9:1) WEIGHING EVIDENCE

We have decided that the best reason for believing in any hypothesis is that it fits the evidence better than it has any right to do *by chance*. That immediately raises questions about the precision and reliability of both the evidence and the predictions of the hypothesis, and about the goodness-of-fit between the two. This is what we mean by 'Error Analysis' here. In the Murder-in-the Library case the forensic scientist claimed that his measured size for the footprint in the flower-bed was so discrepant from the leading lady's that the Odds against her being the murderer were 40 to 1 — on that ground alone. That handsomely overruled the Odds from her motive (2 to 1 on) and from her dodgy alibi (4 to 1 on) and so she got off — despite being the murderess. Either the scientist was too sanguine about his measurements, or he calculated his Odds wrongly and, as a consequence, an innocent party was sent to gaol for 20 years. A sounder analysis of the errors might have avoided that tragedy.

At the heart of Common Sense Thinking (CST) lies the Weight:

$$W(E\mid H) = \frac{P(E\mid H)}{P(E\mid \bar{H})}$$

which enables one to update a Prior on the basis of new evidence. If $P(E\mid\bar{H})$, the Probability of finding that evidence is high even when H is not true (i.e. $\bar{H}$) then obviously that evidence shouldn't carry much Weight — as the equation implies. Very often the most likely alternative to H, the hypothesis one is interested in, is chance. Therefore one must be able to estimate the Probability that the evidence in question could be the consequence of pure chance. For instance if we toss a fair coin 32 times what is the Probability that we throw 18 Heads by chance? If it is high (it is) 18 Heads can hardly give much Weight to the hypothesis that the coin is unfair.

The sound thinker always looks into, and if possible discounts, the possibility that Weighty clues in particular are the product of either random chance, or some other alternative hypothesis. That is what Error Analysis is about. A measurement with a large possible error could fit a wide range of hypotheses and so should carry low Weight. But if it could be made precise it will be more discriminating and its Weight should consequently rise.

The Weight Equation above also explains the philosophy behind controlled experiments. The good experimenter tries to test a single hypothesis by isolating the experiment from all but the phenomenon (hypothesis) in question, so that $P(E\mid\bar{H})$ may be driven very low, and W(E|H), if H can explain E, consequently very high. This is sometimes difficult and expensive; for instance experiments designed to detect Dark Matter particles have to be done in chambers hundreds of metres underground to shield the detectors from uninteresting Cosmic Rays that could swamp the rare looked-for events (i.e. could increase $P(E\mid\bar{H})$. Or if you wanted to test whether Chlorine in the water supply causes depression in some individuals (I suspect it does) you might have to test tens of thousands of people in various parts of the Earth, including 'controls'. One could almost characterise a good scientist as one who is obsessively interested in the weaknesses in arguments, particularly her own.

'Let's look at the numbers' is the mantra of modern man. One can't disagree with it, and broadly speaking I do not – indeed I can't because weighing evidence requires one to estimate Weights and Odds. But gathering numbers is one thing – drawing reliable conclusions from them quite another. Numbers are just as often used by shysters to peddle lies as they are by honest people trying to get at difficult truths. There is a very important general point here: animals don't use numbers, so far as we know. Indeed we didn't invent them, at least in the written form, until about 5000 years ago in Mesopotamia, where they were compiled mainly for tax gathering and shop-keeping purposes. That being so *we have no instincts for dealing with numbers*; history has not given us the time to evolve such instincts, as we have evolved instincts for judging people and spotting lies. Thus we are highly vulnerable to numerical evidence and numerical arguments – and crooks know it. For our own defence we need to learn some basic rules about the significance of numbers, and to acquire a reasoned attitude of scepticism towards them. Much of this chapter is about winnowing good numbers out of bad. Thus we cannot avoid arithmetic here because arithmetic is the very grammar of number.

As usual we develop our ideas first within the scientific context, because they can be seen more sharply there—though they are of far wider application – and in Science 'Error Analysis' has the highly restricted meaning defined in our first paragraph above. [There are lots more mistakes in Thinking than are confined within this narrow definition of 'Error' here, and we come to the worst of them later in Chapter 13 entitled 'Poor Thinking'.] But even within the restricted subject of 'Error Analysis' there are three very different kinds of Error: 'Systematic Errors' – which have their own Chapter (10); 'Counting Errors', and finally 'Measurement Errors'. Understanding that they are entirely distinct from one another is fundamental to clear thinking. Confusing a Measurement Error for a Counting Error led our forensic scientist into the tragic mistake which fooled first himself, then the detective, and finally the court.

(9:2) COUNTING-ERRORS AND SCATTER

We begin with Counting Errors because they are the simplest to analyse. Very often objects turn up randomly: raindrops in a puddle, photons of light striking a camera, voters arriving at a booth, Heads and Tails in a sequence of coin-tosses. Although there is an 'expected' or average number of such arrivals they won't in practice arrive at precisely spaced intervals. They'll be scattered about some mean or average value. Nor will Heads and Tails follow precisely one another. You might get 4 Heads, then 3 Tails, then 1 Head, then 2 Tails and so on in random order, although you expect, after a large number of tosses, roughly equal numbers of each, but rarely exact equality. How large can such random departures from the expected number be before we suspect that something more than chance is afoot — that the shower is over or the coin unfair? Many physical observations amount to counting things: electrons, stars, cells, people......and so it is important to be alert to changes that *are* significant (unexpected) as opposed to merely random effects or scatter. Our model here will be the repeated tossing of a supposedly fair coin. We'll be interested in questions like 'How many more (or less) than the expected number of Heads must turn up before we have grounds to suspect that the coin is biased?'

For that reason we need to introduce a new Statistic for measuring scatter defined by:

Scatter

$$= \frac{(x_1 - \bar{x})^2 + (x_2 - \bar{x})^2 + \ldots + (x_N - \bar{x})^2}{N}$$

where the x_i are the N different measurements of some quantity x and $\bar{x}$ their average value. The squares are just a trick introduced to prevent the positive and negative errors $(x_i - \bar{x})$ fortuitously cancelling one another out. It works because all squares are positive. One can then take the square root of the resultant scatter to get a fair idea of the *spread* of the measurements about the average value $\bar{x}$. We call this the 'standard deviation' σ_x ('sigma-sub x') where:

$$\sigma_x \equiv \sqrt{scatter} \qquad (1)$$

Next to the average $\bar{x}$, σ_x yields the most information about a set of figures. When quoting their results scientists usually write $\bar{x} \pm \sigma_x$ where the standard deviation σ_x is just as important as the average $\bar{x}$. If you haven't encountered scatter or standard-deviation before then:
EXERCISES (9:1) SCATTER

There's a fair amount of Arithmetic in this chapter, but that is all it is. This is a good place to repeat our earlier advice on reading symbolic text. Do not expect to be able to read it through from start to finish like a newspaper article. First skip fairly rapidly through a single section getting only the general sense of it and the main conclusion. It may help at that stage to jot down a summary in a line or two. Then with the main idea already defined skip back again and follow the arguments in more detail – if you are sufficiently interested. If at any stage you cannot follow a particular argument, don't wory but *carry on !* It may well clear itself up later on. But even if it does not don't despair. We scientists learn to live with half comprehending things at the first encounter or two; we know from experience that understanding often comes only after repetition and exercise. But the point is CARRY ON. If you can drive a car without understanding exactly how the ignition works…….. {1} is an readable and excellent guide to Probability.

So let's study the coin-tossing problem, not because it is interesting in itself, but because it's a good and simple model for all kinds of counting problems, problems which constantly arise in both science and every-day life. We aim to show that after Q tosses you expect to find roughly Q/2 Heads turning up (if the coin is fair) but with standard deviation about that value of about $\pm\sqrt{Q}/2$. Thus after 100 tosses we expect to find roughly 100/2 = 50 Heads but with a standard deviation of $\pm\sqrt{100}/2 = \pm 10/2 = \pm 5$. That means to say that if we carried out a series of 100-toss trials of a coin we would expect to find between 45 and 55 Heads in about two-thirds of all trials. You are going to need a result well outside those limits to claim that the coin is in fact unfair.

We don't want to do any more maths than we have to so we've drawn the branching-diagram Fig (9:1) instead. It is worth studying for ten or more minutes because then much will become clear without algebra. Each vertex represents one possible outcome after Q tosses. Thus after Q = 4 tosses we can see, under Q=4, 5 Vertices, each labelled above it with the number of Heads corresponding to that particular outcome i.e. 4,3,2,1,0 moving downwards. Below each outcome is another, more interesting number which we call the 'vertex-number'. It is a count of the number of alternative paths by which that outcome could be reached. Thus under 3 tosses there is a vertex labelled with 2 above it, corresponding to 2 Heads. The vertex-number underneath it is 3, Why 3? Because there are 3 alternative pathways corresponding to 2 Heads: HHT, HTH, and THH. Those vertex-numbers can be found by following back the two arrows which lead into it and adding the vertex-numbers under the 2 vertices from which the arrows started, i.e. 3 here = 1+2.

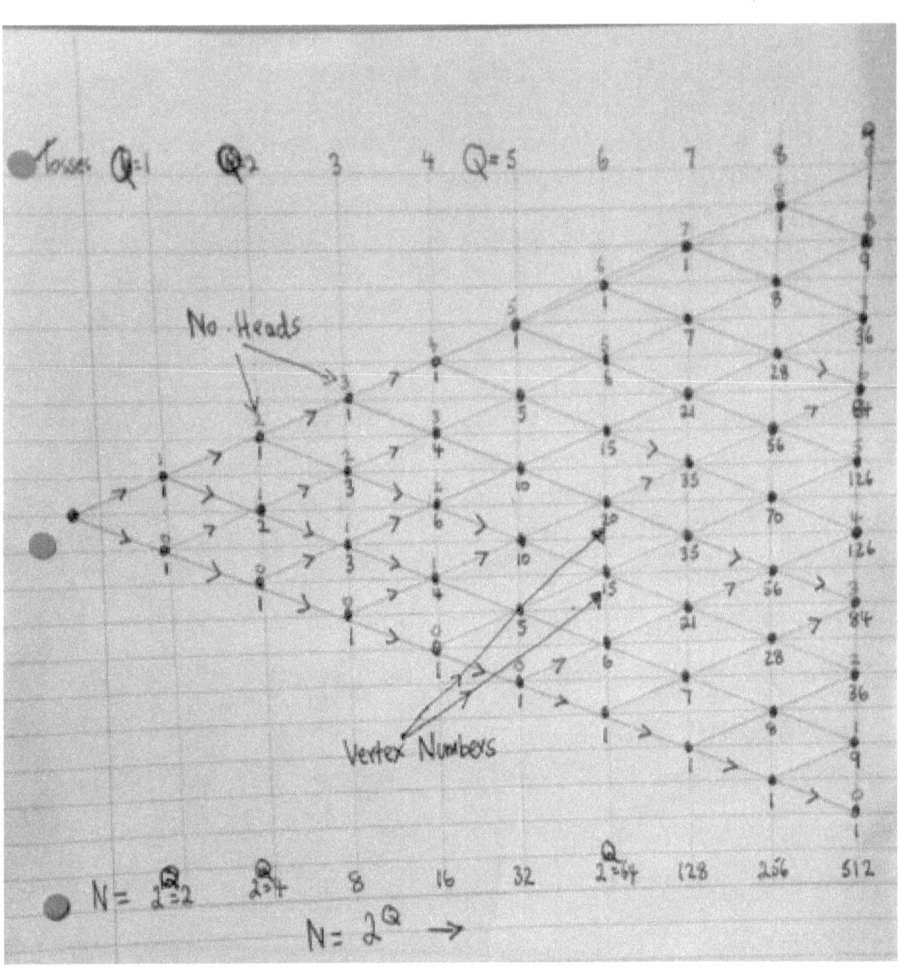

Fig (9:1) The branching diagram for tossing a coin 9 times, showing all possible outcomes, one at each vertex point. The number *above* each point shows the number of Heads at that outcome. The number *below* each point, the so called vertex-number, gives the number of alternative paths which would lead to this outcome. Thus under Q = 9 tosses the point corresponding to 5 Heads has a vertex-number of 126 indicating that there are 126 alternative sequences of Heads and Tails that would give 5 Heads. The vertex number at a location is the sum of the two vertex numbers of the arrowed paths leading into it.(Thus 126 = 56 + 70) Note how the vertex-numbers pile up in the middle of the diagram because there are more paths leading to a middling number of Heads than the extremes of all Heads, above, or all Tails, below. The number N at the bottom is the sum of all the vertex-numbers above it, the sum of all possible sequences of Heads and Tails up to that point. The important general point is that the Probability of arriving at a certain location (vertex) will be proportional to the number of alternative ways of arriving there i.e. to the local vertex-number.

Moving one's way across the branching diagram from left to right one can fill in all the vertex-numbers by such simple additions. Thus the vertex-number under 5 Heads from 7 tosses is 21, made up of 6+15. These vertex-numbers will turn out to be crucial so:

EXERCISE (9:2) VERTEX NUMBERS

Why are vertex-numbers important? Because they correspond to the Probability of each outcome. An outcome with a large number of alternate paths into it will be much more probable than an outcome with only one unique path, Thus, in the case above, 5 Heads is 21 times more probable than 7 Heads (from 7 tosses) which has only 1 path leading into it, i.e. a vertex-number of only 1.

At the bottom of each step in the branching-diagram is the sum total N of all possible pathways in the Q tosses above it. N= 2^Q because each new toss has 2 possible outcomes, H or T, thus doubling the previous number N of total pathways. Thus each N is the sum of the vertex-numbers above it; so for Q =

3, $N = 1+3+3+1 = 8 = 2^3$. This is as it should be because each vertex-number corresponds to the total number of pathways leading to that vertex.

As an exercise check that N does indeed correspond to the sum of vertex numbers above it for the case of $Q = 6$ tosses.

We can now compute the Probability of any outcome in the diagram. For instance what is the Probability of 2 Heads after 6 tosses? Go to the vertex corresponding to that outcome and read off the vertex-number — 15. Under 6 tosses the total number of pathways is $N = 2^6 = 64$. Thus the Probability of 2 heads in 6 tosses is 15 pathways out of 64, or 15/64. Easy — and you can do it for any number of Heads after any number of tosses so:

EXERCISE (9:3) CALCULATING PROBABILITIES

Once you know the Probabilities for each outcome all manner of useful quantities can be calculated: for instance '*What is the average number of Heads you expect to turn up after Q tosses?*' [Obviously Q/2]. Or '*What is the standard deviation you expect in that average number after Q tosses?*' This is not obvious but is very important and is actually

$$\sigma_x \equiv \pm\sqrt{scatter} = \pm\sqrt{Q/4} = \pm\sqrt{Q}/2 \quad (3)$$

In other words the number of Heads we expect to get, after Q tosses is

$$= Q/2 \pm \sqrt{Q}/2$$

e.g. after Q=36 tosses we expect to find $36/2 \pm 6/2 = 18 \pm 3$ Heads if the coin is fair. [We demonstrate the truth of this very important general result In Appendix A3 for those who are interested].

NB Eqn. (3) is only true for situations like Heads and Tails where the Probability of either outcome is ½ exactly. We discuss the situation where the Probability is other than ½ later in Section (11:3) on '*The*

Gambler's Secret'. A very rough and ready rule of thumb though is that if you expect *any* number x expect also a standard deviation of $\pm\sqrt{x}$ in it due to counting error alone. I call that 'The Scientist's Approximation'. It's a damned useful and easy to remember. When in doubt use it.

EXERCISES(9:4) EXPECTED SCATTER

(9:3) THE PROBABILITIES OF NUMBERS

Before they can be used in an Inductive argument Counts and Measurements must first be turned into Probabilities and later into Weights because

$$O(H \mid E) = W(E \mid H).O(H) \quad \text{(Bayes' Rule)}$$

where $\quad W(E \mid H) = \dfrac{P(E \mid H)}{P(E \mid \bar{H})}$

For the coin-tossing problem we already found the Probabilities of various numbers of Heads by using the vertex-numbers and dividing them by 2^Q (for Q tosses). Recall that the Probability of 2 Heads after 6 tosses (Q=6) was 15/64. This suggests a way to find the Probabilities, of interesting numbers, from the point of view of evidence, using the branching diagram (Fig 9:1). Let's consider the case of Q=9, at the right hand side of that diagram. Let's plot the vertex-numbers upward against the number of Heads horizontally [Fig (9:2)]. Because the vertex-numbers are proportional to the Probabilities it is a sort of Probability-diagram. The diagram humps up towards the middle because there are many more ways (paths) for getting to 4 or 5 Heads than there are for getting to 9 or 0.

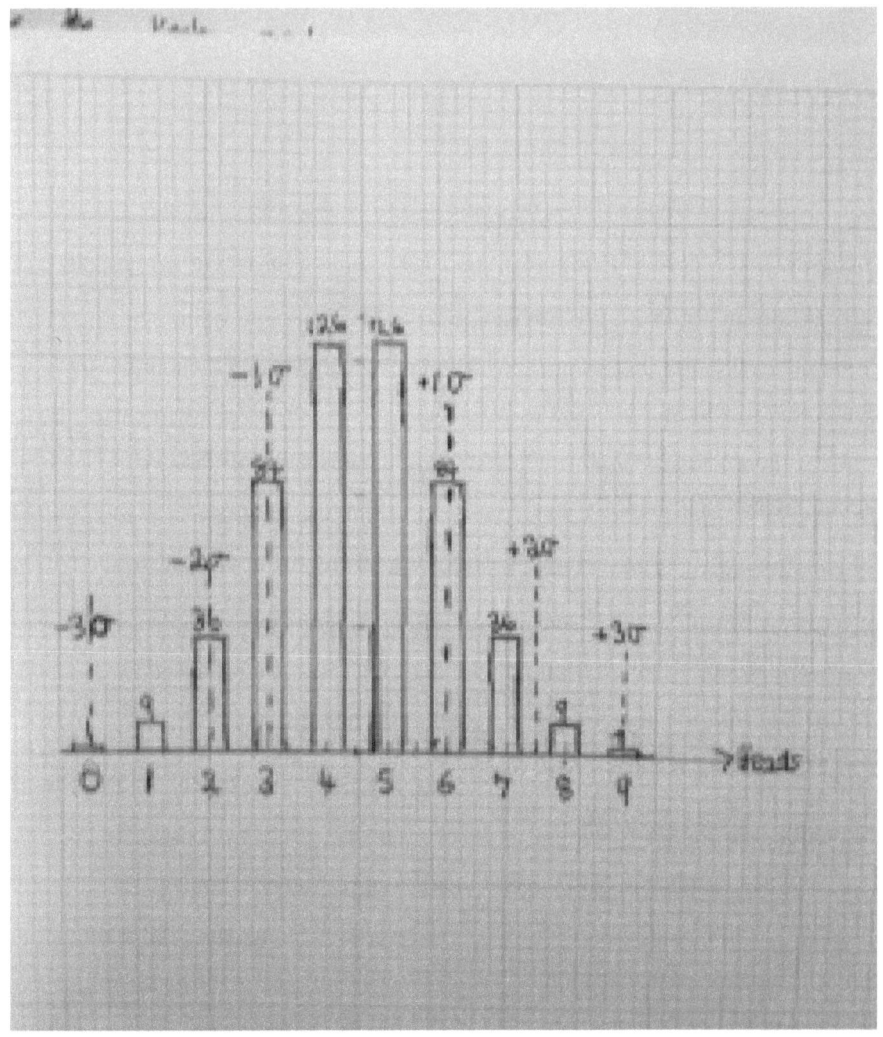

Fig (9:2) Vertex numbers plotted vertically for different numbers of Heads (plotted horizontally) for the case of 9 tosses. The vertex numbers come from the branching diagram Fig (9:1). I have marked in various σ (sigma points). Eventually for large numbers of tosses the columns all move together until they are touching to form a smooth bell-shaped diagram, and both the vertex numbers and the number of tosses can be dispensed with because all we shall be interested in are the σ values and lines and the *relative* areas to either side of them (see next figure)

We've worked out that the standard deviation σ 'Sigma' $= \sqrt{Q}/2 = 1.5$ here so I've marked in $\pm 1\sigma$, $\pm 2\sigma$, and $\pm 3\sigma$ as dashed vertical lines in Fig (9:2). We can now make some crude but useful Probability estimates. For instance what is the Probability that the number of Heads will be within $\pm 1\sigma$ of the mean value (9/2 = 4.5) ? Adding the two central columns i.e. $2 \times 126 = 252$ alternative paths, and half the two second highest (half because the 2σ line runs right through these two columns) =84 gives a total of 336 and dividing this by the total number of all such paths ($N = 2^9 = 512$ see bottom of branching diagram) yields about 66 per cent. (i.e 2/3rd of all trials lie within $\pm 1\sigma$ a result we promised to prove). Likewise the number within +/- 2σ will be [see fig (9:2)]:

$(2 \times 126 + 2 \times 84 + 2 \times [36/2])/512 = 492/512$ or about 96 per cent, whilst the number *outside* $+ 3\sigma$ will be, as we can see, (1/2)/512 or 0.1 per cent.

Such numbers would only be interesting if there was something specially fascinating about Q = 9 trials — which there is not. But it turns out that as Q increases to 20 or more trials the 'Probabilities all settle down and are *independent of Q*, depending only on σ. Indeed we can remove the actual numbers of Heads from the horizontal axis and deal only in σ s. Moreover all the vertical columns run smoothly into one another leaving the bell-shaped curve of Fig (9:3). We can even dispense with the vertically plotted vertex-numbers because in practice all we shall be interested in are the *relative* areas under the graph to either side of vertical lines corresponding to different σ values.

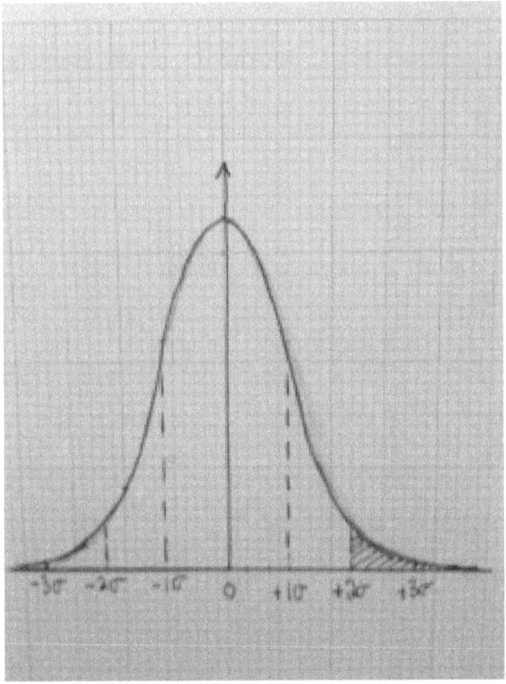

Fig (9:3) The famous bell-shaped 'Normal Curve of Errors' discovered by Abraham de Moivre in 1737 as a tool for gambling. It is very useful for estimating Counting Errors. To find Odds all you need do is estimate the relative areas underneath it to either side of your chosen σ line. Thus to find the Probability of getting a value higher than plus 2σ it is the hatched area to the right of the $+2\sigma$ line as a fraction of the whole area under the graph; its about 2 per cent. Or alternatively the Odds *against* getting such a value higher than $+2\sigma$ is the area to the left of the $+2\sigma$ line divided by the area to the right, which is about 40 to 1 against. Actually you can look these numbers up in tables because de Moivre worked them out. I have printed a short summary in Table (9:1).

Take a simple example to make things clear. What proportion of trials lead to Heads turning up 2σ or more *above* the mean (expected) level ? It's the hatched area to the right beyond $+2\sigma$, as a fraction of the whole area under the graph. We could measure that area by counting squares under a copy of the graph traced onto graph paper. But that is unnecessary because de Moivre long ago did all the tedious work for us and placed the results in easily accessible tables so that gamblers like us can use them. From such a table I find the hatched area beyond $+2\sigma$ is about 2.5 per cent of the whole, and the area beyond $+3\sigma$ is only 0.3 per cent. Likewise the fraction of area between $+1\sigma$ and -1σ is about 66 per cent, as against 34 per cent left outside. So the Odds are 66 to 34, or about 2 to 1, that a Counting Error will lie inside $\pm 1\sigma$. And as there is 2.5 per cent of area beyond $+2\sigma$, with a similar percentage to the left beyond -2σ , that leaves 95 per cent within $\pm 2\sigma$; i.e. the Odds that a Counting Error will lie within $\pm 2\sigma$ are 95/5 or about 20 to 1.

We can't avoid Counting Errors; they are, like the weather, a fact of life. We can however calculate how large they are likely to be in any given circumstance. That will enable us to avoid making claims for numerical evidence that are too strong — too Weighty in the Detective's Equation, and so running ourselves into folly. Conversely we can examine our 'opponents' claims to see if they are warranted by his numbers.

The bell-shaped curve is nowadays called, quite disastrously, 'The Normal Curve of Errors', disastrous because 'Normal' is a technical term which has nothing, *absolutely nothing*, to do with its colloquial meaning of 'usual'. 'Normal' does not mean 'normal' and strictly speaking the curve applies only to 'Counting Errors' —*and certainly not to Measurement Errors in general !* Remember how it was obtained just now. It is simply a glorified version of the counting of Heads and Tails and the number of paths leading to them, which we did in the branching diagram and later where we plotted the vertex-numbers so obtained upwards. And it is glorified only in the sense that it has been smoothed out by taking a largish number of tosses beyond 20. It was originally discovered as an aid to calculation (for gamblers) and is not, repeat not, magic. But it can be very useful — *in the appropriate circumstances* — when 'Counting Errors' predominate. The Table (9:1) below of Counting Errors (see Appendix A1 for details on how it was compiled) was inferred from de Moivre's bell-shaped curve. It gives the Probabilities of numbers N turning up discrepant from the expected value N_H Under hypothesis H) by more than 1 standard deviation σ in either direction, by more than 2σ, and so on, *as a result of Counting Errors alone*:

TABLE (9:1): NORMAL PROBABILITIES AND ODDS

	P(> ±1σ)	P(> ±2σ)	P(> ±3σ)	P(> ±4σ)
Probability	0.3	.05	.003	.0004
Odds on	1:2	1:20	1:400	1:30,000
Odds against	2:1	20:1	400:1	30,000:1

Suppose a number N has turned up in circumstances where you were expecting [on the basis of your hypothesis H] a number close to N_H. The standard deviation of N_H is known to be σ. What can you read into the discrepancy between the real number N and the expected number N_H? Does N support H, rule it out, or is it inconclusive? Table (9:1) gives the Probability of discrepancies in various ranges of σ, occurring *by chance*. Where those Probabilities are low then the discrepancy is more likely be the result of something other than chance, something interesting or 'significant' in the jargon used by scientists.

How did we arrive at the Odds in Table (9:1)? Two specific cases will illustrate the principle:

(a) $> \pm 1\sigma$: I go to the de Moivre curve and can see that the summed area to the right of $+1\sigma$ (or to the left of -1σ) is about $1/6^{th}$ of the total area (which is by definition =1). Thus:

Probability P($>\pm 1\sigma$) by chance) =1/6 +1/6= 1/3 (roughly 0.3).

So Probability P($>\pm 1\sigma$) *not* by chance) ≈ 2/3

So Odds *on* O($>\pm 1\sigma$) by chance = $\dfrac{1/3}{2/3} = 1/2$

So Odds *against* O($>\pm 1\sigma$) by chance= 2/1

(b) Outside $\pm 2\sigma$. I go to the de Moivre curve and by counting squares or whatever, can see that the combined areas lying to the right of $+2\sigma$ and to the left of -2σ come to about 5 per cent of the whole. Therefore:

P(outside $\pm 2\sigma$ by chance) =.05

So P(outside $\pm 2\sigma$ *not* by chance) = 1-.05= 0.95

Thus Odds(outside $\pm 2\sigma$ *not* by chance) 95 / 5 = 19 to 1 or roughly 20 to 1 (Great precision isn't called for here and 20 is more memorable).

(9:4) NUMBERS SIGNIFICANT AND INSIGNIFICANT

Nowadays we are besieged by numbers and numerical arguments, without having the instincts to make much sense of them, Animals don't appear to use numbers, so we haven't inherited the necessary survival skills to decide wisely when numbers are brought into an argument. Are the numbers persuasive or not? Does the person using the numbers really know what they are talking about – all too frequently not the case. Or is an attempt being made to deliberately bamboozle us? One needs to be very cautious about numbers – they are just numbers, not arguments in their own right. Ever since their invention we humans have been struggling with numbers, caught half way between perplexity and resentful reverence for that numerate priesthood who claim to understand their magic. In my experience they often don't. It is worth remembering that CST works with Weights. Evidence, and that includes numerical evidence, can not be used in the DE until it has first been *weighted*. And there lies the rub. It's far easier to collect numerical evidence, and quote it to any degree of exactitude you like, than to give it a reliable Weight. Bestowing reliable Weights invariably requires experience, judgement, wisdom and imagination – skills not guaranteed alas by the possession of a certificate in Accounting, Economics, or Statistics, especially Statistics. It's easy to babble about numbers, and many fools do, far harder to decide whether they are 'significant' i.e. Weighty. In this section we go in search of some clues as to whether numbers are significant or not.

Granted some hypothesis H, how can we use some numerical evidence to test it out? If the hypothesis predicts the number N_H, but N instead turns up, how significant is the discrepancy (N_H - N)? Is it small enough to suggest that H is likely to be true, large enough to convincingly rule it out, or somewhere in between? This is the question which turns up all the time when numbers enter a discussion. What with errors of one kind and another – Counting Errors, Measurement Errors, Sampling Errors…… – exact agreement between N_H and N is not to be expected – but what can we reliably infer from the size of the discrepancy?

An important start on this question was made by Abraham de Moivre (1733) in connection with Counting Errors with his Table of so called Normal Errors given above [Table(9:1)]. One can see that the figures in that table make some sense at least. If the Discrepancy t between an observation and the hypothesis in question is high [say $t > 3\sigma$, see column (3)] the Odds *on* it occurring by chance are very low i.e. 1/400; hence the Odds *against* it occurring by chance are very high (400 to 1). That being so the discrepancy is 'significant', i.e. it is telling evidence against H being true. On the other hand if the Discrepancy t is low [say no more than 1σ, see column (1)] then the Odds against it being mere chance are only 2 to 1 and it can hardly be regarded as significant evidence against H. The 2σ case [col.(2)] lies in between with Odds of 20 to 1 against chance being the cause of it; it looks very suggestive but not utterly compelling. Thus there has grown up a widespread convention among numerate types as follows:

TABLE (9:2) THE SIGNIFICANCE OF NUMBERS

A 1 sigma result ; "Insignificant; ignore"

A 2-sigma result : Strongly suggestive; investigate further"

A 3-sigma result: 'To be accepted unless there are special reasons to be

cautious".

EXAMPLES:

(a) (i) Of 20 Tiptuppititis sufferers (Sect 4:8) treated with Organocalm 14 get better more quickly than average. Should that evidence persuade you to part with £300 for a course of treatment? My answer: had the drug been ineffective we would expect that 10 of the 20 patients would have taken less than the average time to recover. In fact 14 Organocalm recipients seem to have benefitted. So is 14 significantly bigger than 10? Well if we use Eqn.(3) to calculate the expected standard deviation it is $\sigma = \pm\sqrt{20}/2 = 2.2$ and 14 is only 4/2.2 or less than 2σ bigger. So, according to Table (9:2), 14 is too small an improvement to persuade a rational person to part with their money

(b) In a poll of 1000 voters 481 claim they will vote Blue, and 425 claim they will vote Red. Does that really mean that Blue will win the election? My answer; the number of interest here is the difference between the 481 polled Blues and the 425 polled Reds. This difference is 56. According to Eqn (3) again we expect a standard deviation of about $\pm\sqrt{1000}/2$ or about 16 in the two polling numbers. 56/16 is more than 3σ. Thus according to Table (9:2) this looks like a highly significant difference, making it very likely that the Blues will win. On the face of it . But wait.

CAVEATS

Now for the important caveats, and there are several:

(a) The Odds in Tables (9:1) and table (9:2) refer only to Counting Errors. *They do not apply to errors of any other kind*, Measurement Errors for instance (next section), or Systematic Errors (next chapter). It is worth working out the significance category of numbers if only to see whether they are worth considering any further. If they are less than 1 sigma (or 2 Sigma depending on taste) you can dismiss them entirely as insignificant, whatever their origin. (Learn how to do that in your head, using The Scientists' Approximation – see later.)

(b) Even if they are truly Counting Errors would a small change in the numbers, such as could easily occur by chance, make a difference? Take the Tiptuppititis case. A mere change of 1 patient from 14 up to 15 would move the evidence from the 1σ into the 2σ category where they will be significant. Surely that's a warning that the numbers are too small to be reliable. It's a case of *"Small Number Statistics"* against which we all need to be constantly on guard. Had there been 40 patients in the study instead of 20, 1σ would have been $\sqrt{40}/2 = 3.2$. Now if the same *proportion* had recovered quickly (i.e. 28/40) that would be an 8/3.2 or a clear 2σ result –suggestively promising according to Table (9:2). So dismiss small-number statistics! And if you hear somebody using them then you can safely disregard everything numerical they say [Applies especially to Economics pundits who wax profound over insignificant differences in numbers, because that is what they're paid to do]. Because Counting Errors rise only with the square-root of the number Q of things counted, then as a proportion of the full number Q they rise as $\sqrt{Q}/Q$ i.e. as $1/\sqrt{Q}$ and become *relatively* negligible with rising Q. On the other hand if Q is small $1/\sqrt{Q}$ can be quite large and you will have results that may *look* interesting, but won't be '*Significant*' i.e. worth taking any notice of.

(c) You should even beware of 2-sigma results – depending on their pedigree. Supposing a drug company tests many drugs against a certain disease. In such a trawling exercise it's quite likely that *one* of the many drugs will yield a positive 2σ result – simply due to chance. So you need to know about all the negative results as well as the positive ones before deciding on the significance of one particular trial. It's obvious really but frequently neglected – or covered up.

(d) Now consider the opinion poll, which was, from a purely arithmetical point-of view pretty suggestive (3 Sigma). It could still be completely unreliable because those polled may not be representative of those who will vote in the election. That's a Systematic Error which no numbers will cure (Next chapter). Poll 10,000 bus drivers and you will get one result, 10,000 lawyers and you will get another. That's mainly why polling can be so unreliable. It's not just the numbers!

PSS Is there anything special about one-sigma or two-sigma or three-sigma results? No there isn't. It's purely a convention that has grown up – particularly among less numerate analysts who think of it as holy gospel. What only matters is the Odds-on some clue or hypothesis, but we sometimes need to be familiar with conventions that are widespread, even if they are a bit daft. Personally I would prefer to see numbers categorized not by their Sigma-class but by their Odds-on, i.e. by the Odds that they are significant e.g.:

Category Zero: Odds on of less than 10 to 1

Category Ten: Odds on of more than 10 to 1

Category Forty: Odds on of more than 40 to 1

Category Hundred: Odds on of 100 to 1 or more

I would ignore a Zero, follow up a Ten, believe a Forty and publish a Hundred.

[The problem with Significance Classes are that they are context dependent. For instance in Astronomy a 3-Sigma result is highly significant in Spectroscopy, but utterly insignificant in Imaging. It's all about the number of false-positives, compared to real sources.]

EXERCISES (9:5): NUMBERS SIGNIFICANT AND INSIGNIFICANT; USING THE TABLE OF NORMAL PROBABILITIES

(9:5) CORRECTING INFERENCE TABLES

An Inference Table may contain many clues and even more 2's and ½'s multiplied together. Thus if you look back to our Hidden Galaxy Inference Table (8:4). there were 25 clues and no less than 58 powers of 2 (positive or negative) multiplied together. So even if the final outcome *ought* to be neutral (i.e. = 1) it might instead have an artificially high (or low) value purely due to the effects of scatter which you expect in counting any number. We need to find a way of discounting this possibility – which is what this section is all about.

In that table we combined, by the end, no less than 58 2's and ½'s together. If there was no real evidence either way we'd expect exactly 29 2's and 29 ½'s — which would cancel out to give a combined Weight of 1 (i.e. neutral) for all the evidence combined in the Inference Table. Even so we could expect a scatter of 58/4 or a standard deviation σ of $\pm\sqrt{58}/2$ = +/- 8 / 2 = +/- 4 in the actual number of 2's about the expected number of 29. Thus by the end, the Inference Table could lead to Odds of up to $2^{\pm 4}$, Odds which are purely fortuitous, i.e. due to counting errors; we shall call them 'Fortuitous Odds'. Thus, by the end of the Table Fortuitous Odds could easily be as high as $2^4 = 16$ or as low as $2^{-4} = 1/16$. Earlier on, further up the table, the Fortuitous Odds would be lower. Can we do something to get rid of this unfortunate consequence of Counting Error? Indeed we can.

The obvious tactic is to go down the table line by line, count the total Q of 2's and ½'s used *up to that point*, compute $\sigma = \pm\sqrt{Q}/2$ and divide the real Odds O(H|E) already calculated there (col.4) by the Fortuitous Odds $2^{\pm\sigma}$, whichever is appropriate [i.e. if O(H|E) is 2 to some positive power divide by 2^{σ} but if O(H|E) is 2 to some negative power divide by $2^{-\sigma}$.] The adjusted Odds, I call them the 'Significant Odds', will now be largely, if not wholly, free from Counting Error.

Table (9:3) shows the old Table (8:4) thus adjusted for Counting Errors. Study Table (9:3) carefully because I'm going to suggest you adjust all your own ITs in the same way. It doesn't take much effort and it should lead to a far more convincing result, especially if you have only a small number of clues.

In Column (8) of the new Table (9:3) I've translated my 'Significant Odds' into English as follows (it's a matter of choice; if you don't like my scheme use your own) where the key is:

Significant Odds-on: 2^{10} or more :   'CONVICTION'
"   " : 2^6 to 2^9 : 'BELIEF'
"   " : 2^4 to 2^5 : 'OPTIMISM'
"   " : 2^3 to 2^{-3} : 'DOUBT'
"   " : 2^{-4} to 2^{-5} : 'PESSIMISM'
"   " : 2^{-6} to 2^{-9} : 'DISBELIEF'
"   " : 2^{-10} or less : 'DISMISSAL'

CORRECTED INFERENCE TABLE (9:3): HIDDEN GALAXIES

1	2	3	(4)	5	6	(7)	8
Clue#	Date	Clue	O(H|E)	Q	$2^{\pm\sqrt{Q}/2}$	(4)/(6)	Status
Prior			2^{-5}				
1	1975	Theory I	2^{-1}	4	$2^{\pm 1}$	2^0	Doubt
2	1978	Off-beams	2^{-4}	7		2^3	"
3	1983	Theory II	2^{-2}	9		2^{-1}	"
4	1984	QSOALs	1	11		2^0	"
5	1985	SB/colour	2^2	13		2^2	"
6	1987	Fornax	2^3	14		2^2	"
7	1987	Cr. Giant	2^6	17	$2^{\pm 2}$	2^4	Optim
n8	1990	(2) explnd	2^8	19		2^6	Belief
9	1993	SB/latitude	2^{10}	21		2^8	"
10	1994	ESO gals	2^{13}	24		2^{11}	Convn
11	1995	Impey	2^{14}	25		2^{12}	"
12*	1995	Gnt. Halos	2^9	30		2^7	Belief
13	1997	Centaurus	2^{10}	31		2^8	"
14*	1997	Arecibo	2^5	36	$2^{\pm 3}$	2^2	Doubt
15	1998	Compacts	2^7	38		2^4	Opt
16	1999	SB flucts	2^6	39		2^3	Doubt
17	2002	Sloan DSS	2^5	40		2^2	"
18	2002	Millenium	2^4	41		2^1	"
19*	2005	HIPASS	2^{-2}	47		2^0	"
20	2005	Starcounts	1	49		2^0	"
21	2007	DG Virgo	2^2	51	$2^{\pm 4}$	2^0	"
22	2009	Eq. Strip	2^3	52		2^0	"
23	2012	HSB HDF	2^7	56		2^3	Opt
24**	2013	Clustering	2^{23}	56*		2^{19}	Convn
25	2013	Red nugs	2^{25}	58		2^{23}	"

Table (9:3) shows an Inference Table, the old Table (8:4) for Hidden Galaxies, being adjusted for Counting Errors as suggested in the text. The addition of 5 new columns to the right may look complicated but it is just counting up the 2's and ½'s in Col. (5)

i.e. Q and in Col.(6) setting up the divisor needed for Col.(7) which yields the Significant Odds by dividing the old raw Odds in Col.(4) by the estimated Counting Error in Col.(6). I'm using the nearest integer for $\sqrt{Q}$ as being sufficiently accurate. They are 'Significant Odds' because the likelihood of chance has been discounted, if not removed completely (which would be impossible). Blank spaces in a column signify the value there is identical to the value above. Column 8 is explained in the text. Study for 10 minutes.

Would it have helped my research on Hidden Galaxies if I'd known about Inference Tables at the time (which I didn't)? I can see now that it definitely would — *indeed it might have saved a dozen precious and painful years of my research life*. As you can see in Table (9:3) three powerful clues, asterisked, stick out against the grain. And they all eventually turned out to be wildly wrong! Had I been running an IT *at the time* I'd have been far more suspicious and would probably have fingered the fallacies in each one — instead of much later, mostly by chance. Thus we might have got to 2012 in 2000. What a bloody waste! If you haven't got the odd decade to fritter away you might think of using an IT, and of adjusting it for counting errors.

So that's it; the (almost) complete machinery of an Inference Table with all the bells and whistles included. It is potentially a very powerful tool for hard thinking because it incorporates Bayes' Rule, The Detective's Equation and now Counting-Error Analysis as well. Of course you have still got to get your Weights roughly right, but only you can do that. Now do a couple of Exercises to make sure you've grasped it:

EXERCISES (9:6) CORRECTING INFERENCE TABLES

(9:6) MEASUREMENT ERRORS

Most measurements amount to far more than mere counting. One measures voltages, wavelengths, temperatures, distances, velocities, angles — using more or less sophisticated instruments, anything from a wooden ruler to a Space Telescope. Indeed science has largely progressed through the construction of ever more sophisticated instruments, magic carpets which can transport our minds into regions never explored before: the bacterial world, outer Space, the living brain, the atomic nucleus, DNA………. (Believe it or not Joseph Black's improvement of the simple Weighing Scale in 1750 when he was an undergraduate at Glasgow University was more important to history than Columbus' voyage. Black invented Chemistry, and discovered Carbon Dioxide! He also discovered Latent Heat, which makes him the founder of Thermodynamics, the key to improving the steam engine. He was James Watt's boss!. Poor Columbus never realized he'd discovered a new continent, while his most valuable return was rhubarb.)

How can we translate such measurements into the Weights when we need to make a reliable Inductive argument? It's not obvious — it's not obvious at all. And yet that is what we first have to do if we are going to employ our Common Sense to decide on the truth or otherwise of various hypotheses. Like all his predecessors Johannes Kepler clung to the idea that planets move in circular orbits — after all circles were simple, symmetric, aesthetic, harmonious, and therefore more pleasing to the mind of a Perfect God. But Tycho's accurate measurementsof the orbit of Mars forced Kepler, after no less than 20 years of puzzlement, to recognize that those measurements could not be fitted on a circle. The discrepancy (8 arc minutes) between the predictions of the Circular–orbit hypothesis and observations with a low margin of error (1 arc minute or less) became such that the Odds on reconciling the two became too high for an honest mind to persist. So Kepler eventually tried out the next simplest closed curve — the Ellipse, which has 2 free parameters — and everything clicked smoothly into place, not only the observations of Mars but the observations of other planets as well. Kepler was understandably overjoyed and reported: "I have attested it as true in my deepest soul and I contemplate its beauty with incredible and ravishing delight."

Tycho Brahe wasn't counting things he was measuring angles, the angles which Mars made with the 'fixed stars' as it moved through the constellations of the Zodiac. Such measurements had a limited accuracy, limited by the precision of the human eye-ball to about a minute of arc (a sixtieth of a degree). Only when Mars departed from a circular course by significantly more than such a minute could Kepler be 'reasonably certain' that something was seriously awry. Although at the time (~ 1609) he couldn't know about σ 's, or 'The Table of Normal Errors' (1733), his Common Sense, his 'instinct' if you like, would tell him to abandon an ingrained habit of thought and try an alternative.

Looking back on this great sea-change we can re-tell the story in modern words, using the vocabulary for thinking developed in the last section. Kepler would know, because his boss Tycho had told him, that there was an error σ on planetary measurements of about +/- 1 arc-minute. Tycho would have ascertained that vital number by repeating the same measurement over and over again and noting the scatter. He would further report to Kepler that discrepancies from the average value of more than 2 arc minutes (2σ) were rare (about 1 in ten say) and of more than 3 arc minutes (3σ) virtually unheard of. Thus when Mars deviated from its predicted circular orbit by more than 3 arc minutes (8 actually) Kepler could be pretty damn sure [Odds of hundreds to one] that Mars wasn't on a circular course. He'd got, as we would say in modern parlance, 'a three sigma result' [If you look in the Normal Odds Table you will see '3σ' amounts to Odds of 400 to 1 against, in this case circular orbits]. But that was for Counting Errors only! We have no warrant to extend those figures to Measurement

Errors. In modern times CERN was only willing to claim its discovery of The Higgs Boson when it obtained a '5-sigma' result (Odds of 3 million to one against pure chance). Gradually, between the time of Tycho and the time of CERN, a working philosophy has developed among scientists, for turning measurements — and importantly their errors — into decisions about the natural world. It's a philosophy partly based on gambling theory, but mostly on experience. It seems as if 3σ results, and better, *usually* stand the test of time, whereas 2σ results (Odds of 20 to 1 on) are interesting, but by no means decisive. It's become a kind of folk wisdom throughout science which might have useful applications in other fields. Whether it is any more than folk-wisdom, and has a truly rational basis, is a fascinating and momentous question to which we return in Chapter 10.

Incidentally such numbers underline the crucial importance of precise measurement to human thought, and of improving instruments to achieve it. Reducing σ by a factor of only one third could change a measured discrepancy (its 't-value') in such a way as to change the Odds for or against some crucial hypothesis from 20 to 1 (i.e. 2σ) to 400 to 1 (3σ) — see Normal Odds Table again. That is a not-obvious insight, but History reflects the truth of it. Once Joseph Black in Glasgow had devised the simple weighing balance he discovered Carbon Dioxide, Aristotle's 2,000 year old 'Chemistry' was doomed, and Atomic Theory would emerge. The Conservation of Energy, so vital to all of modern Science and Technology, awaited James Joule's development of the accurate thermometer. Galileo's spy-glass was an instrument for improving the eye's acuity by a factor of ten — but it utterly transformed the landscape of the human mind. The Hubble Telescope, raised into Space, can spy the cosmos with a precision ten times better than any telescope on Earth. The accelerating expansion of the Universe is only one of its shocking discoveries.

A scientific training rightly emphasises the importance of precise measurement and careful Error Analysis because quantitative data can occasionally deliver reliable evidence of massive Weight. Such bully clues can be decisive, as they were in the case of Planetary orbits —with Tycho's precise measurements of Mars' orbit. But we mustn't get carried away. Most of us aren't astronomers, and even if we are, most of the clues we use are not of high precision [See Tables (8:4) and (9:3)] but are not valueless on that account. In combination weak clues can be utterly convincing — if they nearly all point in the same direction. Indeed one could argue that the main purpose of Error Analysis is to subject potential bully-clues to rigorous examination. In the Murder-in-the Library a poor innocent spent 20 year in prison on account of an illegitimate bully-clue. In the Hidden Galaxy Saga 3 of the 5 heavy-weight clues (Weights either more than or equal to 16, or less than or equal to 1/16), i.e. clues asterisked in Table (9:3), turned out to be plain wrong. Getting errors wrong can prove disastrous.

We've asked the vital question; "How do we turn *measurements*, not just number counts, into Weights?" 'With considerable difficulty' is the short answer . We'll try to do better in Chapters 10 and 11. It won't be easy but we'll discover a simple and wise way out.

SUMMARY

1) Wise thinkers, aware of history and disasters past, make strenuous efforts to look for and size the likely errors in any evidence they use.

2) Counting and Measurement are such recent human accomplishments that we have no natural instincts for dealing with them. We can easily be fooled, or fool ourselves, unless we acquire such 'instincts' through practice.

3) Most care and suspicion should be directed at 'Bully Clues' — ones that could dominate the rest.

4) Broadly speaking there are three kinds of specifically Scientific Errors: Counting Errors, Measurement Errors and Systematic Errors[In Chapter 13 'Poor Thinking' we come to a much broader range of mistakes that thinkers often make]. The first two are 'random', tending to scatter and so weaken the evidence, without necessarily biasing it.

5) Counting Errors are the easiest to understand and discount. Much can be learned about them from studying the tosses of a coin. The standard deviation of number Q = $\pm\sqrt{Q}/2$. That is to say when you are counting Q things don't expect to get exactly Q every time but to get some number in the range $Q \pm \sqrt{Q}/2$ two thirds of the time and some number in the range $Q \pm 2\sqrt{Q}/2$ ninety-five per cent of the time. Only if you get some number outside this last range can you be reasonably sure (Odds of 20 to 1 on) that your expectations were wrong.

6) The Probabilities of various Counting Errors led De Moivre (1733) to his famous bell-curve (fig 9:3) and his allied Table of so called Normal Errors (Table 9:1) – which *can* be a powerful aid to reaching decisions.(Look at Appendix A1 to see how it all works). But beware of misapplying it to situations where the Errors are *not* Counting Errors (Measurement Errors for instance) when it frequently leads to horrible blunders. Charlatans love Normal Errors, which are anything but 'normal' in the colloquial sense, because they make them look decisive and 'knowing'. Just watch out!

7) When you are faced with potential evidence in the form of a number first calculate whether it is 'significant' i.e. not likely to be the consequence of mere chance. You can use the 'One-sigma', 'Two-sigma' etc convention if you like (Table (9:2).

8) Since Counting Errors increase only with the square root of the numbers involved [see 5 above], then as the numbers rise, the *relative* influence of counting errors declines. Conversely small numbers will lead to low Weights and hence the perils of 'Small-Number Statistics': numbers that look promising but are not significant. Beware. Learn to estimate the rough Significance of a number in your head using the Scientist's Approximation i.e. $\bar{x} \pm \sqrt{\bar{x}}$.

9) Inference Tables can and should be adjusted for the Counting Errors that must arise in the accumulation of Two's and Halves which make up the Weights. Such adjustment will lead to more reliable 'Significant Odds' [see Table (9:3) to see how it is done].

9) *Some* forms of Measurement Error amount to no more than counting identical particles, or photons or cells. In such *special* cases insights from Counting Theory — square-roots for instance, and the 'De Moivre's Weights Table (9:1)', *can* be reliable guides to Inference.

10) *However* most forms of Measurement are NOT mere counting. In such cases there is no obvious warrant for using 'Normal Errors' to draw inferences. There is no alternative to measuring the errors experimentally and establishing their actual distribution. This may be very hard to do out in the wings of the error distribution because such errors are rare. Thus such measurements can in practice rarely carry high Weights into any argument. As we shall argue in Chapt. 11 much of Statistics is plain wrong because it tries to apply the theory of Counting Errors to Measurement Errors.

(11) Modest increases in measurement precision can sometimes lead to spectacular increases in hypothesis discrimination. Thus the vital importance of sophisticated instrumentation in science. [Black, Joule, Frauenhoffer in Chapt.3].

(12) No number can be admitted into Evidence until it has first been given a Weight. But recall that any Weight $W(E|H) = P(E|H)/P(E|\bar{H})$,

which requires a knowledge of $\bar{H}$, the alternative or alternatives to H, as well as of H itself. If you don't know all the alternatives to H, and you probably won't, then you'd better be bloody cautious. It turns out that one can rarely afford to use Weights higher than 4, or less than ¼ (Chapter 10).

(13)   Finding the errors in one's own thinking is more than a duty, it's often the way to spectacular progress. Don't miss the opportunity! Don't let someone else have the joy of it.

> NB: Animals don't seem to use numbers more than two or three. Thus using numbers is a new and specifically human skill and so we cannot expect it to have the same kind of reliability, brought about by long usage and the cull of Evolution, that other aspects of CST can expect to enjoy. And we humans seem to advance mostly by trial and error (think of aircraft engineering or sailing-ship design or maritime navigation). So it shouldn't be surprising to find that our early essays into using numbers will turn out to be full of mistakes. Indeed over the next two chapters we will argue that most of conventional Statistics is either misapplied or wrong. On the whole this is good news not bad [though Statisticians will hate it].

Uncertainty or not, we still have to make decisions, usually by gambling on the Odds. To do nothing is a decision in itself, often not the best.

THINKING FOR OURSELVES
CHAPTER 10 (5.9 kw; 14/9/18)
SYSTEMATIC ERROR : THE ELEPHANT IN THE ROOM

"To see what is in front of one's nose needs a constant struggle."   George Orwell

(10:1) INTRODUCTION

The third major category of error cannot be discounted by taking square roots or making careful measurements − it is the 'Elephant in the Room', the 'Unknown unknown' or, as we scientists call it, "A Systematic Error". Scientific examples that have troubled history include: the world is flat; diseases are caused by witchcraft; the Earth is the centre of the Universe; continents don't move; atoms are indestructible......We may not even be aware of such Systematic Errors, let alone know of a tactic to discount them.

Such weighty elephants could have a dramatic, even fatal effect on our decision making. Is there anything in our process of Common Sense that could be adapted to help? As it happens there is. I call it PAW, standing for the 'Principle of Animal Wisdom', because animals have probably been using it for aeons to survive. Without it most or all would have become extinct. As we shall see it is very effective even in areas of cutting-edge scientific research − for which it was never evolved. But it does have some rather shocking implications. For instance it implies that sophisticated statistics is not only a complete waste of time but may be a dangerous delusion. That should come as a welcome discovery to those not conversant with higher mathematics. In essence PAW tells us that it is the diversity, reliability and concordance of evidence which should matter more than the precision of its individual clues. A combination of weak but diverse clues is always to be preferred to one strong one because that 'bully clue' may be the consequence of a Systematic Error. We all, scientists and laymen alike, desperately need to understand that.

(10:2) A GALLERY OF ROGUE ELEPHANTS IN SCIENCE.

Systematic Errors, and their effects are best illustrated for now by looking at some infamous examples that have arisen in the history of Science. Later we will look at non-scientific Elephants which are, if anything, much worse. The advantage of starting from Science is that it generally progresses so that its follies eventually come to light. That may not be so in Economics, Philosophy, History, Politics, Religion...

(A) The ancient Greeks were aware, as early as 500 BC, that the Earth is a globe. Thus Aristotle (350 BC) wrote: "Furthermore the sphericity of the Earth is proved by the evidence of our senses, for otherwise lunar eclipses (when the Earth casts a shadow on the Moon) would not take such form, for whereas in the monthly phases of the Moon the segments are all sorts — straight, gibbous and crescent — in eclipses the dividing line between light and shadow is always rounded. Consequently, if the eclipse is due to the interposition of the Earth, the rounded line results from its spherical shape." I think that was a brilliant inference.

But 600 years later the respected scholar Lactantis (tutor to the Emperor Constantine's son) disagreed: "Can any one be so foolish as to believe that there are men whose feet are higher than their heads ('Antipodeans'), in places where things be hanging downwards, trees growing backwards, or rain falling upwards?"

The Elephant in that room was Universal Gravitation — an idea that was to emerge only a thousand years later still. The force of terrestrial gravity is always towards the Earth's centre so that Australian rain falls locally downwards too.

(B) The size of the Earth was first estimated by Eratosthenes (300 BC) a Greek colonist in Egypt. From travellers he'd heard that at noon on Midsummer Day every year the Sun cast no shadows in the wells at Aswan, a city on the Nile far to the South of his own home in Alexandria. At the same date and time, as he determined by careful measurements of the shadow cast by a nearby obelisk, the Sun lay 7 and a quarter degrees South of the local zenith. Attributing this difference to the curvature of the Earth Eratosthenes could argue that 7 and a quarter degrees to 360 degrees should lie in the same proportion as the distance to Aswan to the circumference of the whole globe. But how far South was Aswan? He couldn't simply measure the enormous distance. Travellers told him that camel trains took 50 days to accomplish the journey. It followed that the whole earth was $(360 / 7.25) \times 50$ or 2,480 camel-days in circumference.

To proceed further he needed to assume something about the distance, in a straight line due South, covered by the average camel train in a day. His best estimate was 100 stadia, about eleven and a half miles. Thus his final circumference for the earth was about 28,500 miles, about 15 per cent too high, as we know today. All the same it was a transformative piece of thinking. The entire world known to the Greeks became a tiny patch [about a 1 per cent area] on a huge globe that had still to be explored.

No matter how accurately he surveyed his obelisk Eratosthenes' final estimate would contain a 'systematic error', his uncertainty as to the southward distance of Aswan. It would be very difficult if not impossible to judge the proportionate size of that error but it was almost certainly much greater than the angular precision with which he could measure the shadow angle cast by the obelisk. The naked human eye should be capable of a quarter of a per cent precision, something he could have estimated by measuring the angle several times and noting the random scatter of his values.

Here, in a nutshell, we can see the vital difference between Random and Systematic errors. Random errors are often easy to calculate or estimate whereas one might not even be aware of Systematic errors, let alone be capable of estimating their sizes. Scientists often happily quote their random errors while ignoring the Systematic variety — about which they may not even be aware. For instance the sand-conditions, or the topography on the road to Aswan, might well affect the camels' performance. So the elephant in the room was, in this instance, the 'true' distance to Aswan.

One would have imagined that Eratosthenes' brilliant thinking would have burned itself indelibly on the human mind. On the contrary. His successor the much more famous Ptolemy (150 AD), whose book '*Geographica*' summarised the discoveries of the ancient world, got the circumference of the world far too small (18,000 miles). Then came 'The Great Interruption' when men, at the behest of religious dogma, retreated to a flat world with Jerusalem at its epi-centre.

In one of the most exciting but neglected moments in history, in 1400 AD a copy of '*Geographica*' was found in the library of Santa Sofia in Constantinople by a young Florentine merchant called Palla Strozzi. What precisely happened next to 'The Great Secret' is not clear, but one way or another the knowledge came, in part or in whole, to the West. I suspect that Strozzi initially sold it secretly to the Portuguese crown, for why else would that tiny nation stubbornly spend a third of its GDP over the next century trying to find the way by sea to the fabled East?

(C) Man's persistence in believing in an Earth-centred (Geocentric) universe we have spoken of . There were at least two elephants in that room. An unwillingness to believe in our own insignificance and the existence of a learned profession (us astronomers) which made its living reconciling the observations with the Geocentric Theory. Daniel Boorstin the historian put it thus {3/1}: "Astronomers were adept at explaining away what seemed only minor problems by a variety of complicated epicycles, deferants, equants and eccentrics, which gave them a heavy vested interest in the whole scheme. The more copious this peripheral literature became, the more difficult it became to retreat to fundamentals. *If the central scheme was not correct, surely so many learned men would not have bothered to offer their many subtle corrections.*" (My italics)

In other words Systematic Errors can be self-sustaining.

(D) Circular planetary orbits were another beguiling theoretical myth almost impossible to overturn. Even possessed of Tycho Brahe's super-accurate observations of Mars it took Johannes Kepler twenty years to remove the scales from his own eyes. As Boorstin again reports {3/1}: "For most of history the human mind has abhorred a vacuum and preferred myths and factitious facts to the label 'Terra Incognita'. *How could the 'learned' be awakened to a willing confession of ignorance?*" (My italics)

The Elephant here, as it so often is, was 'Abhorrence of Doubt'. The truth here, elliptical orbits, was finally recognised through the emergence of astronomical observations of unprecedented accuracy: "Tycho made his observations with scrupulous regularity, repeating them, combining them, and trying always to allow for the imperfections of his instruments. As a result he reduced his margin of error to a fraction of a minute of arc, and provided the sharpest precision achieved by anyone before the telescope."

(E) Long before Darwin and Wallace (1858) naturalists considered the possibility of Evolution but generally dismissed it because the World was far too young. Even if they didn't believe Archbishop Ussher (1658) who, on the basis of biblical scholarship, claimed that The Creation had occurred on October 26th 4004 BC, they would have suspected that the Earth, after supposedly being wrenched from the surface of the Sun (6,000 degrees Celsius), would have taken but tens of thousands of years to cool to its present temperature.

The elephant in this room was radio-activity — discovered more than a century later (1899) (3:7), radioactivity which actually heats the Earth to keep it warm for billions rather than thousands of years.

(F) Even with the discovery of Deep Time Evolution couldn't account for the existence of evidently related large animals on different continents [see Darwin (7:3)]. The elephant here was ignorance of Continental Drift, which moves at a kilometre every hundred thousand years or so.

(G) Drift was dismissed by cognoscenti who refused to countenance the idea that rocks could flow. But if you take a walk in the Alps, or on almost any beach in my native Pembrokeshire, you will see cliffs warped like plasticene into the most spectacular folds. The elephant there was the notion of rigidity which is, in many cases, not absolute. Many materials have a 'time-constant' which separates their behaviour on different time-scales. A piece of Silli-putty when hit with a hammer shatters like glass. But left on the table it will, in a few minutes, flow onto the floor like treacle. Its time constant is a few seconds whereas mantle-rocks have a time-constant of a hundred million years.

(H) Before 1900 giving birth in European conurbia was extremely dangerous; in some lying-in hospitals as many as 30 per cent of mothers died of child-bed fever. Then in 1847 Ignaz Semmelweiss a young Hungarian obstetrician noted a startling difference between the mortality rates in the doctors' wards and the midwives' wards in Vienna General Hospital. Women begged not be admitted to the doctors' wards where they were three times more likely to die. Observations led Semmelweiss to conclude that doctors were bringing into their wards fatal poisons on their hands, poisons probably picked up in the course of autopsies on dead patients. He found that hand-washing with bleach led to a startling decline in the dread disease and recommended the practice to his colleagues. He was vilified and replaced;

the idea that 'gentlemen' could have filthy hands was out of the question. Moreover Semmelweiss had no very clear explanation for why his prescription worked — though work it undoubtedly did, with death rates dropping to less than 2 per cent.

Desperate to stop the ongoing and unnecessary slaughter Semmelweiss wrote angry letters to all and sundry. Eventually he was lured into a lunatic asylum where he was beaten to death by the guards.

The existence of bacteria had been known as early as 1674 when a Dutch draper Antoon van Leeuwenhoek devised single-lens microscopes capable of magnifying 250 times. Examining some greenish cloudy water from a lake near his home in Delft he reported : " I now saw very plainly that there were little eels, or worms, lying all huddled up together and wriggling; just as if you saw, with the naked eye, a whole tubful of little eels and water, with the eels a-squirming among one another: and the whole water seemed to be alive with these multifarious animalcules. This was for me, among all the marvels I have discovered in nature, the most marvellous of all…"

Unfortunately it would take 200 years and hundreds of millions of unnecessary deaths before Louis Pasteur established the connection between Leeuwenhoek's animalcules and infectious diseases. Childbed Fever was caused by a kind of streptococcus.

Two lessons can be drawn from this hideous story. Firstly people are reluctant to believe evidence, even overwhelming evidence, without a *backing explanatory story*. And second, *professions* (in this case the medical profession) *can set themselves up as experts, and practice profitably, on the basis of very little evidence for their claims*. If you consulted a physician before 1945 he was more likely to do you harm than good — the tide only turning with the introduction of antibiotics. Alas that same medical profession has so oversubscribed antibiotics that that they have lost most of their potency today as bacteria have evolved resistance. {3/ 3&7}

So there were two elephants in that room: lack of an explanation, and vested interest. These themes recur again and again in the history of Science.

(J) The early days of radio were hampered by the quite reasonable apprehension that electromagnetic waves couldn't travel round the curve of the Earth, couldn't compete with submarine telegraph cables, although these latter were extremely slow (a few words a minute) and hideously expensive. Fortunately Gugliemo Marconi was too ignorant to take this apprehension seriously.

The true history of radio is both fascinating and instructive. James Clerk Maxwell, using a highly dubious theory, got the right equations and predicted the existence of radio waves back in 1864. Alas his mathematics, based on quarternions, was incomprehensible even to experts. Then in 1871 David Hughes, a successful Welsh-American inventor (e.g. the microphone and telex-machine) accidently discovered that feeble electric sparks could send signals 500 yards down Great Portland Street from his laboratory in central London. The Royal Society, called in to witness this miracle, disastrously poo-poohed the whole idea, crushing Hughes. Heinrich Hertz in a far more deliberate search 16 years later in Karlsruhe Germany, established the existence of such waves beyond question— but died too young to exploit them. A self-taught telegraph operator called Oliver Heaviside then appeared on the London scene. He first invented a vital new branch of mathematics (Vector Calculus) which rendered Maxwell's Equations comprehensible for the first time, and then used them to speed up telegraphy thousands of times over. He also inferred the presence of an ionosphere high above us which could reflect radio waves around the Earth {10/1}.

One can draw several inferences from this particular piece of history:

Firstly Systematic errors (no propagation of waves beyond the horizon for instance) are often very plausible — indeed it's their very plausibility which makes them so pernicious.

Secondly Eminence, particularly eminence in *theory*, should be treated warily. Sir Gabriel Stokes from the Royal Society who turned down David Hughes' amazing discovery as 'Just Induction' was a famous applied mathematician from Cambridge. Why he ignored Hughes' conclusive evidence against 'Just Induction' isn't clear. He may have looked down on Hughes as a mere artisan whilst Hughes wrongly looked up at the 'Great Professor'. In a similar vein the abject performance of the early submarine telegraphs was owed to a wrong calculation by another panjandrum of Victorian Theoretical Physics, Lord Kelvin Ironically it transpired that he'd forgotten to include Inductance in his calculations, an omission which another 'artisan' put right — Oliver Heaviside. (Kelvin is buried in Westminster Abbey; Heaviside, who probably died of malnutrition, in Teignmouth.)

But the final inference is perhaps the most interesting. Maxwell's odd theory convinced almost nobody at the time, and would today be regarded as wrong (it was based on 'vortices' and 'idler-wheels') but it led to the *right equations*. Put in reverse the right equations, as tested in the laboratory, don't necessarily endorse the theory behind them. We discuss this surprising conclusion later (Chapter 14).

(K) Theoretical astrophysicists thought there would not be many X-rays to be seen in the universe because they couldn't think of any mechanisms to generate them. How wrong Riccardo Giacconi and his experimental space-scientists proved them to be (Sect 3:10). The systematic error here was, as it so often is, sheer lack of imagination. The Universe is always far more ingenious than the human mind can comprehend. *That's why Theory must always be subsidiary to Observation in science* — which renders it so completely different from Mathematics. Binary stars can rip the skins off one another tidally, Black Holes can swallow material, and galaxies can somehow re-ionize the gas in Intergalactic Space, all processes leading to copious X-rays. If we scientists can get a theory wrong we generally do! {3/2}.

It would appear that without constant jogging from observations and experiments, the human mind is not imaginative enough to grasp the ingenuity of Nature. J.J. Thompson, discoverer of the electron, had this to say on the matter: "In the interplay between mind and matter in scientific discovery the parts played by the two are, I think, widely different from those usually assigned to them in popular estimation. There is a widespread belief that the mind itself is desperately speculative, that it is only kept from wild imagining by the control of the stolid and prosaic partner, the physical facts. The true state of affairs is, I think, that it is the mind which acts as the brake in the combination, that the impulsive partner is the facts, and that these spur on the mind to take steps which it would shudder at when not under the influence of its stimulus. Nature is far more wonderful and unconventional than anything we can evolve from our inner consciousness." — a sentiment echoed by that witty physiologist J.B.S. Haldane who said " Now my suspicion is that the universe is not only queerer than we suppose, but queerer than we can suppose."

From these examples it must be clear that Systematic Errors can, on occasion, swamp every other kind. They are difficult or impossible to spot — they are the 'unknown unknowns' in the discussion. They can't be tamed but they shouldn't be forgotten. And there is a temptation to forget them because their possible existence will ever preclude us from reaching that iron-hard certainty that some souls seem to crave. It is always tempting to believe 'The Big Game is over; all that remains is to clear up the details.' Thus we can walk off into the sunset with anthems peeling. Such triumphalist episodes have been all too common in scientific history. For instance Lord Kelvin, President of the Royal Society no less, claimed in 1897: "There is nothing new to be discovered in Physics. All that remains is more and more precise measurement." Alas for his eminent wisdom; Radioactivity (1899), Relativity (1904), Quantum Theory (1905) and Deep Time (1907) all lay just around the corner.

We have in this section spoken only of scientific Elephants. What about the non-scientific variety: systematic errors than can arise from cultural, historic, religious and political considerations? We discuss them in a later section.

(10:3) ELEPHANT LESSONS

It is all very well listing infamous Systematic errors; what can we learn from them? If I categorise our Elephants under different headings there are three we can do little about: 'Major unknown Phenomena', 'Lack of Imagination' and 'Inadequate Evidence'. Three more need to be constantly guarded against: 'Religion' (in its broadest sense); 'Vested Interest' and 'Arrogance'. But there is one clear lesson we *must* heed; 'Don't disbelieve a phenomenon just because you can't explain it (e.g. Continental Drift, a Spherical Earth and Childbed Fever). Putting 'Explanation' above 'Observation', in other words preferring Theory to Data, was the great folly which led us into the Dark Age. For reasons inexplicable to me our human minds favour story tellers and theory makers, especially ones equipped with sorcerers' tricks, above plain-speaking messengers of the truth. It is a childish fancy we must put away if we are to grow up. More about this later.

The remaining category of common Elephant is 'Narrowness of Focus" or Tunnel Vision(Continental Drift, Child-bed fever, Quasars). If you can find evidence of a totally new and independent nature you are more likely to sidestep a particular Systematic Error. It was the sheer breadth of Wallace's and Darwin's knowledge of the natural world, acquired in years of travel and exploration in far flung lands, that gave them the confidence to ignore the then serious gaps in their argument for Evolution — for instance total ignorance of Continental Drift at the time. When I study the lives of ground-breaking discoverers, scientists included, what stands out is their 'breadth' — their omnivorous curiosity. Narrow focus easily leads to foolish 'Groupthink' (Chap 13).

The Systematic Errors of our forbears can at least act as a useful reminder to keep our minds open. And it is a very necessary one. Knowledge is nearly all provisional because it rests on Inductive Thought. To be adult is therefore to be reasonably comfortable with living in Doubt. But Doubt is painful and is eschewed by many, not only religious types. There are vested interests out there who don't want you prying into their self-proclaimed expertise — academics among the very worst of them I have to confess. We need to move on from time to time — which is difficult when you are handcuffed to absolute certainty. Had I professed Astronomy a century ago most of what I taught my students then would have turned out to be either wrong or irrelevant by now. Have we good reasons to believe the situation is any better today? I doubt it. So the vital lesson any student, any honest mind has to learn, is how to decide better on the basis of the *existing* evidence. It's the process not the product.

(10:4) THE PRINCIPLE OF ANIMAL WISDOM (PAW)

Since most animals *presumably* can't count to more than 2 or 3, can't make precise measurements and do not carry out elaborate statistical analyses, you might find it as hard as I did to understand why they still arrive at Weights good enough to make the kind of decisions which enable them to survive. Yet they must do so over and over again because so many of them lead risky and precarious lives. I imagine the best they can do is weight a clue as either 'Neutral' [W(E|H) ~1,'Weak' [W(E|H) ~ 2 or 1/2] or 'Strong' [W(E|H) ~ 4 or 1/4]. That seems too clumsy to be effective. In fact the very reverse may be true; there could be more danger in using bully clues of strong Weight and sophistication. Because of unsuspected Systematic Errors and other fundamental difficulties it is probably safer to be more vague and allow a combination of weaker and more diverse evidence to decide on the verdict. This has important lessons for us.

Assigning Weights is likely to be a trade-off between precision and reliability. More precision generally requires more ancillary assumptions and arguments each of which can unconsciously smuggle in new errors. Remember the transformation from an observation to a Weight requires at least two extra factors: an error analysis of the observation itself *and* an assessment of P(E|$\bar{H}$) – the likelihood that the observation would have turned up even if our chosen hypothesis H is wrong – by chance say. Both factors are problematical. On the other hand simple evidence may be crude but it probably will be more transparent and, as John Maynard Keynes put it 'Better to be roughly right than precisely wrong' – especially if it's a question of life or death. Subtle arguments are not necessarily wise ones.

So no wise animal can afford to give any single clue decisive Weight. That could lead to extinction. Better to rely on several clues with lower Weights in the hope that most of them will be sound.

Suppose ones Decisive Odds are about 64 (i.e. 2^6) to 1 and you want at least 3 clues to contribute towards your decision. The cube root of 64 is:

$\sqrt[3]{64} = (2^6)^{1/3} = 2^2 = 4$ (i.e. $64 = 4 \times 4 \times 4$). If, on grounds of reliability, you don't want a single rotten clue to decide for you, you cannot have any Weight as high as 8, because then any two clues of Weight 8, one rotten, could reach decisive Odds of 64 [the numbers are not much changed if you pick instead decisive Odds of either 128 (2^7) or 32 (2^5) to 1 on]. So animals don't need Weights of much more than 4, and if they don't need them why would we? Even if we could calculate them it would be unwise to use them unless we were absolutely sure they were right. Weights of 1, 2 and 4 (or conversely 1, ½ and ¼) may be all we usually need to make wise decisions. Anything 'better' might be a trap.

This exceedingly important insight comes as a great surprise to most people, as it did to me at first, but it seems to work out in practice. You will recall that in the Murder-in-the-Library [Sect (5:4)] the case against the Leading Lady was dominated by the foot-print evidence which was given a strong Weight of 1/40 by the forensic scientist. This comfortably over-ruled the other two clues Motive (W=2) and Opportunity (W = 4) so she was judged innocent and one of the servants was convicted instead. As we shall find out in the next chapter [Sect 11:6] that was a very sad mistake which led to a tragedy. The forensic scientist had arrived at his decisive Weight of 1/40 through a subtle mathematical argument that was actually wrong but which is so common in Statistics as to render much of that subject suspect. Had the 1/40 been replaced by ¼, as animal wisdom would suggest, the tragedy might have been avoided.

An even better illustration of animal wisdom, for lack of which I lost a decade or more of my research life, occurs in Table (9:3), the Inference Table for Hidden Galaxies. If I invoke the PAW to modify its Weights of more than 4 to 4, and all Weights of less than ¼ to ¼ I get Table (10:1) below. Spend a few minutes comparing the two tables because here is a true story, and not some didactic fabrication.

TABLE (10:1): ANIMAL WISDOM AND HIDDEN GALAXIES

1	2	3	4	5	6	7	8
Clue#	Date	W(E|H)	O(H|E)	Q	$2^{\pm\sqrt{Q}}/2$	O(animal)	Status
Prior			2^{-5}				
1	1975	2^2	2^{-3}	2		2^{-3}	Doubt
2	1978	2^{-2}	2^{-5}	4	$2^{\pm 1}$	2^{-4}	Pessim
3	1983	2^2	2^{-3}	6	"	2^{-2}	Doubt
4	1984	2^2	2^{-1}	8	"	$2^0 = 1$	
5	1985	2^2	2	10	"	2^0	
6	1987	2	2^2	11	"	2^2	
7	1987	2^2	2^4	13	"	2^3	
8	1990	2^2	2^6	15	"	2^5	Optim
9	1993	2^2	2^8	17	$2^{\pm 2}$	2^6	Belief
10	1994	2^2	2^{10}	19	"	2^8	
11	1995	2	2^{11}	20	"	2^9	
12*	1995	2^{-2}	2^9	22	"	2^7	
13	1997	2	2^{10}	23	"	2^8	
14*	1997	2^{-2}	2^8	25	"	2^6	
15	1998	2^2	2^{10}	27	"	2^8	
16	1999	2^{-1}	2^9	28	"	2^7	
17	2002	2^{-1}	2^8	29	"	2^6	
18	2002	2^{-1}	2^7	30	"	2^5	Optim
19*	2005	2^{-2}	2^5	32	"	2^3	Doubt
20	2005	2^2	2^7	34	$2^{\pm 3}$	2^5	Optim
21	2007	2^2	2^9	36	"	2^6	Belief
22	2009	2	2^{10}	37	"	2^7	
23	2012	2^2	2^{12}	39	"	2^9	
24***	2013	2^{2+2+2}	2^{18}	39	"	2^{15}	Convtn
25	2013	2^2	2^{20}	41	"	2^{17}	Convtn

Particularly interesting is a comparison of the last columns in each case. In practice [Tab.(9:3)] my beliefs wobbled all over the place between 1997 and 2010 and indeed I lost heart and gave up in 2005 when my Odds on H fell to evens. Had I known about animal wisdom then [Tab.(10:1)] my Odds-on would never have fallen below 2^3 (8 to 1 on). I would have stuck to my guns and probably, by now, have succeeded in finding Hidden Galaxies – if they exist. [See our exciting Appendix 10]

Why did animal wisdom 'succeed' there? It down-played the three asterisked bully-clues all of which were highly unfavourable – and as it turned out horribly wrong (They all depended on complex arguments relying on plausible but unsound auxiliary assumptions). Animal Wisdom is sceptical of individual bully clues, preferring to look for confirmation among a wider diversity of evidence. Presumably this caution could only be the outcome of Evolution – survival of the fittest – and should be respected as such. Those who were less cautious have gone extinct.

Even if you are not interested in Hidden Galaxies you should be very interested in Table (10:1) because it illustrates just how powerful CST, properly used, can be. Using very crude Weights (no more than 4, no less than ¼) it has overcome three major Systematic Errors and arrived at Odds of no less than 2^{17} (better than 100,000 to 1 on). Not bad for a superficially crude bit of gambling!

(10:5) NON-SCIENTIFIC ELEPHANTS

Humans need to be *far more wary of Systematic Errors than other animals* because our heads are full of myths, fictions implanted there by other men; it's called 'Culture'. The myths may be good for the men who planted them but whether they are good for us as individuals is another matter altogether. According to the historian Juval Harrari {*} this new human aptitude developed 30 to 70 thousand years ago during the 'Cognitive Revolution' and was crucial to get large numbers of people to cooperate in raising pyramids, taming rivers, building temples, making wars and forging empires. Here is a short list of such myths:

(a) You shouldn't mind sacrificing your life because you will be rewarded in paradise.
(b) God is naturally on our side.
(c) There is a Hell awaiting, but if you pay us we will see you are spared the tortures of…..
(d) We ………s have a divine right to rule.
(e) The (other side) are evil and must be crushed.
(f) They're savages; they don't feel pain like we do.
(g) Education is good for you and for everyone.
(h) Our little father the Czar/ Stalin/ Mao….. will look after us.
(i) Newspaper proprietors have the best interests of their readers at heart.
(j) If you work hard you'll get more stuff and that will make you happier.
(k) Our religion is the right one. Those others are blasphemers and heretics.
(l) All things bright and beautiful,
All creatures great and small
All things wise and wonderful
The good Lord made them all.

(m) They're only aborigenes/gypsies ….If you gave them land they wouldn't know what to do with it.
(n) Doctors/Lawyers/Professors/…. can't do their jobs properly unless they have much bigger houses and cars than you and I.
(o) Women are too emotional to drive vehicles.
(p) You can rely on the government News channels.
(q) Anyone who criticises our great leader is a traitor.
(r) It's our land; God gave it to us.
(s) Education is good for you.
(t) Education is good for you.

When people who we barely know give us presents we have to ask why. When people feed us information we have to ask the same question. Is it for our benefit – or theirs. If they have no good reasons to love us it's almost certainly for theirs. It's a gift-horse we should very definitely look in the mouth. Ask where the information originally came from, and who is going to benefit from it. One of the best reasons for reading history is to see how difficult it was for ordinary citizens to come by the truth. For example a major reason for the Second World War was Ludendorff's Lie; that his German army hadn't been comprehensively defeated at Amiens in August 1918 but had been 'stabbed in the back' by Germany's civilian government who had signed the armistice in November 1918. People like Hitler believed him and ….

This is most decidedly not a book about *what* you should think, but about how you might be able to think better. Having been thoroughly brain-washed three times already – at preparatory boarding school, by the family newspaper and in the army, in no case in my own interest, I'm an extremely sceptical individual now. I try to look at my own cultural prejudices and underweight them using PAW. My suggestion is that, if you want to think for yourself, you should do the same. Be warned though that it won't make you particularly popular, though it might save the lives of you and your family.

(10:6) WISDOM AND WEIGHTS

We have put Weights at the very heart of Inference, but we now have to confess that setting them with any certainty is honestly impossible. Recall that a Weight $W(E|H) = P(E|H)/P(E|\bar{H})$ where E is the evidence, or clue, bearing on one's hypothesis H. But what exactly is $\bar{H}$? It is the *collection of all other hypotheses that could account for E, apart from H itself.* And that makes sense because each such alternative ought to further weaken the case for H. But how could one know, even in principle, what all the conceivable alternatives to H might be? There was no way mediaevals could know that germs might be the cause of an outbreak of Diphtheria – and so they blamed witchcraft instead, and burned blameless old ladies at the stake. In our terms their Weight for 'H = witchcraft' was higher than it should have been because their $\bar{H}$ failed to include the sub-possibility of $\bar{H}_x$ for germs. But in the real world won't there always be such possible $\bar{H}_x$'s which we cannot imagine at the time? And if we make a very modest allowance for the entirety of such unknown $\bar{H}_x$'s by putting $P(E|\bar{H}_x) = 0.25$ we are left with a maximum value for the Weight of our chosen H of only $W(E|H) = 1/0.25$ or 4, i.e. its PAW value. So the PAW is nothing less and nothing more than the recognition *that we are not omniscient,* that there may lie causative agencies $(\bar{H}_x)$ out there beyond our present ken.[You might like to identify the equivalent argument that says that W(E|H) should never be less than ¼ either.]

So it seems we cannot avoid PAW – which is worrying. Won't it down-grade perfectly good evidence which might otherwise be decisive? For instance in Muller's first statistical investigation of smoking as a cause for lung cancer (LC) [see Sect (11:2) for the fuller story] he found that among 86 cases only 3 were non-smokers whereas among 86 healthy controls 14 didn't smoke. The difference between 14 and 3 [about 3 sigma] is formally sufficient [Appendix A1] to increase the likelihood that smoking is dangerous by a factor of no less than 100. If one started from a neutral point-of-view about smoking that evidence would surely be decisive (Odds of more than 100 to 1 on danger). But PAW would downgrade the Weight of that same evidence to only 4 – resulting in Odds too weak to even suggest a follow-up investigation. Doesn't that discredit PAW?

No, in my opinion, it does not. Formally speaking a Weight $W(E|H) = P(E|H)/ P(E|\bar{H})$ explicitly recognizes that the evidence E could arise from causes other than H (i.e. smoking), namely $\bar{H}$, which stands for "any other cause bar H". Suppose, for example, that poor people are more likely to smoke than the better off. Now there may be many other factors in poorer peoples' lives, apart from smoking, that pre-disposes them to LC. For instance they may be forced to take dirtier jobs – like working down mines where there is a risk of radon gas, a known carcinogen. Then again there is the possibility of biased mis-diagnosis: if a physician already suspects a link between the disease and smoking he may more readily diagnose it in the case of smokers than non-smokers. PAW is reminding us, as it should, about possible $\bar{H}_x$'s, and not to take any number at face value.

My second worry about the PAW is that I cannot find it, or anything resembling it, in all my most authoritative tomes on Inference. Why not? Surely the problem of $\bar{H}_x$, the possibility of unknown causes, a form of Systematic Error, is as old as the hills? But in only one of them do I find Systematic Errors even indexed – and then in connection with something irrelevant. It took me years to realize what is going on. It is not foolish, it is not sinister, but it is worryingly disingenuous. *The authors are pretending that they are dealing with CLOSED worlds where the possibility of unknown causes is supposed not to exist.* Such worlds are like card games where every possible hand can be imagined and probabalised, every possibly hypothesis enumerated: if Jones hasn't got the ace then Smith has. The possibility that someone is cheating is entirely discounted. The real world of course is OPEN, subject occasionally to entirely unexpected causes – or $\bar{H}_x$'s. But if a world is too unpredictable there would be no point in constructing hypotheses about its behaviour anyway. So what has happened historically is that scientists have concentrated their attention on worlds which appear *approximately*, if not entirely closed. Take Newton's model of the Solar System (which is light hours across) in his *Principia*. In principle it could be disturbed by passing stars (and might be one day) but even the nearest stars are so far away (light-years) that they can effectively be ignored. Thus the pretence, the approximation if you like, that the Solar system is CLOSED has proved extremely productive. More generally, science has progressed by preferentially studying systems that are practically, if not logically closed. Thus Atomic Physics could be divorced from the Nuclear and Molecular varieties, and so studied successfully almost in complete isolation from them (because each has a widely different energy). One of the chiefest skills of a good scientist is spotting phenomena which do appear sufficiently isolated to become amenable to analysis. Thus Darwin could concentrate on the micro-problem of finches' beaks among the Galapagos Islands – and so was led to vastly wider truths. Thus James Hutton (1788) could study a single Unconformity in the rock strata of the Berwickshire coast and realize that "the present is the key to the past" – making him "the father of modern `geology". And what are most experiments, if not attempts to isolate some phenomenon sufficiently from 'extraneous' causes i.e . $\bar{H}$ s ?

We thus come to understand why, from the outset, theorists have concentrated on the solution of CLOSED PROBLEMS. They may be fictional accounts of the real world – but they *are* potentially soluble. Jacob Bernoulli in the first published book on Inference ["The Art of Conjecture", 1714, Groningen] concentrated on the artificial closed problem of drawing a sample of coloured balls out of an urn, and then trying to infer something about its whole contents. That led to some very powerful and useful ideas, including "the Gamblers Secret" [see (11:3)]. Mathematical types love CLOSED problems because then they can become Certain, Decisive and Authoritative. Very likely though they are deluding themselves because in the real world very few situations *are* CLOSED. If you put your hand into a real urn, as opposed to a mathematical one, you could be bitten by a tarantula.

So where does all this leave us when we come to consider Weights? At one extreme you can adopt the Fiction that the system you are interested in is CLOSED; that there is no possibility of being surprised by an unanticipated cause. The advantage is that you might then, with all the apparatus of Probability Theory, reach a decisive conclusion from a modest amount of evidence. And with luck that conclusion might be true of the real world (i.e. no tarantula). Such fictions have sometimes worked spectacularly well in the past because some real-world systems are *approximately* closed. On the other hand many such pretenders have received venomous bites. Malthus' infamous "Essay on Population" was one such misfortune but, as usual, it was the Poor what got the bite: some 30 per cent of the British population were afterwards forced to emigrate to escape it's dolorous predictions.

The alternative is to accept that the system you are interested in is to some extent OPEN and adopt the remedial PAW. The price you will then have to pay is the search you will have to make for more *diverse* evidence (accumulating more of the same E won't hack it, because no matter how much there is it cannot increase the Weight for that particular E beyond 4).

For my money, and I can't be more explicit than that, PAW will win in the real world most of the time. See how it dealt with no less than 3 Systematic Errors, i.e. $\bar{H}_x$'s, in the case of Hidden Galaxies [Table (10;1)].

Being a sailor I would like to share with you an analogy which I find helpful here. How is one to navigate in shallow waters where there may be treacherous shoals at every hand? My belief (and experience) is that we can use the PAW to prevent ourselves from running so hard aground that we cannot float off on the next tide. Seen thus the PAW is not merely a precaution against Systematic Errors, but an enabling mechanism to penetrate sounds and estuaries that would have been entirely closed off to explorers before. In that sense Table (10:1) is the log-book of a voyage into the Dark.

SUMMARY

Systematic Errors can dominate both the Measurement and Counting varieties. They are the unknown unknowns, the hard-to-spot Elephants in every thinker's room. There are no simple nostrums for spotting them though breadth of knowledge and a knowledge of scientific history can help in scientific situations.

Common varieties of *scientific* Elephant include: unknown phenomena, lack of imagination, group-think, vested interest, arrogance, narrowness of focus, a craving for explanation, and a mistaken reverence for established authority. *Man's limited imagination means that observation and experiment should nearly always take precedence over theoretical considerations!* But judging from history the most fearsome rogue elephant, the one that most often stands athwart the path ahead, is a craving for Certainty, a cowardly fear of Doubt. Bertrand Russel confessed ('Portraits from Memory' 1958); " I wanted certainty in the kind of way in which people want religious faith." He spent 20 years trying to find it in Mathematics, but failed.

Many 'experts' and text-books on Thinking either pretend or assume tat they are dealing with CLOSED systems – ones in which all the possibilities are known so that all the Probabilities can be calculated rationally – i.e. they imagine they are playing card-games. But the real world is very rarely, if ever like that. It is OPEN, with entirely unforeseen circumstances – which must be allowed for.

To allow for Systematic Errors Nature has evolved PAW 'The Principle of Animal Wisdom': **'Never rely on a single line of evidence; even two may be risky.'** *We should adopt it too by generally refusing to assign Weights to individual clues greater than 4 or less than ¼.* Nature places the diversity, reliability and concordance of evidence above sheer weight. This caution is presumably the outcome of Evolution – survival of the fittest – and should be respected as such.

If we are going to use Weights as crude as 2, and then multiply them together with other Weights, there is no point whatsoever in estimating those other Weights to a precision much better than a factor of 2 either! Combined Odds of 13.79 are then no better than the closest binary value of 16 (i.e. 2^4). But if you look at Statistical Tables they are usually quoted to 4 places of decimals! That surely ought to make us all suspicious of the Statistics enterprise. Common Sense has gone out of the window somewhere (next chapter).

Because of Systematic Errors we must be wary of single bully clues. We should be especially suspicious of elaborate mathematical analysis of a single kind of evidence. It is analogous to locating a piece into a jigsaw on the exquisite perfection of fit of one tongue into one socket, instead of looking for a more general fit into the wider puzzle. I'm afraid much of statistical analysis is of this foolish kind (next chapter).

Powerful as it is PAW cannot entirely protect us from Systematic Errors. We need to spread our net wide and look for consensus among widely disparate evidence. This is especially true of so called cultural information which has been implanted in us by other people, particularly priests, politicians, media tycoons, schoolteachers, academics, advertisers and the government.

All the same an Inference Table incorporating the Principal of Animal Wisdom (PAW) is an exceedingly powerful way to think our way through a thicket of mixed, contradictory and sometimes unreliable evidence towards the truth [e.g. Table (10:1)].

NB: You should now take a first look at the important Appendix A8 entitled *"Certainty*

, *Falsifiability, and Animal Common Sense*". There you will find a far stronger case in support of the simple Weights advocated by the PAW; no 'better' Weights are, *or ever will be*, justifiable! Yes, that's a surprise.

THINKING FOR OURSELVES (Disney)
CHAPTER 11 (Draft 14/9/18)
STATISTICS — OR TERROR ANALYSIS (14.4 kw)

"Statisticians have already overrun every branch of science with a rapidity of conquest rivalled only by Attilla, Mohammed, and the Colorado Beetle." M.G. Kendall, Statistician

(10:1) INTRODUCTION

Many people, both inside and outside Science, feel that it is the quantitative aspect of the discipline that makes it so much more reliable than softer subjects. 'Let's look at the numbers' people say, and after doing so are inclined to issue authoritative, apparently unanswerable conclusions. The folk who gather those numbers for a living and then interpret them, are called 'Statisticians'. So influential did statisticians become that in the 20th century they rose to dominate discussions of Inference and of The Scientific Method. Students of my generation (1960s) were brought up in awe – almost terror of the subject – which became highly technical. I was myself a real enthusiast and for years taught statistics to students at both undergraduate and graduate level.

The trouble is that Statistics is far more problematical than it appears, or that its most ardent advocates will allow. To the superficial eye it looks like a branch of Mathematics, that is to say it belongs to the Deductive disciplines. But we have discovered that Science is of its very nature an Inductive subject—it attempts to draw general conclusions from a limited number of observations. By so gambling it can introduce fresh information into a subject and thus reach conclusions entirely unforeseeable without. Deduction on the other hand can only reveal conclusions that are already implicit in its agreed-upon pre-suppositions. It can only re-arrange the existing chairs – not bring in entirely new furniture. So, philosophically speaking, Science and Mathematics are a world apart from one another. Mathematicians are rigorously trained not to do the very thing that scientists try to do all the time – make Inductions.

Outsiders will be amazed that two such fundamentally opposed modes of thinking could ever have become entwined with one another – to the regular confusion of both. But so it has been. Thus the famous French philosopher Rene Descartes (1595 to 1650) imagined that the whole of truth could be spun – by deduction—from a single incontestable truth, which he chose to be *'Cogito ergo Sum'* ('I think therefore I am'). But what nonsense! How on earth could the existence of Killer Whales be deduced from nothing but Descartes' unquestionable proclivity to think? How was he allowed to get away with it? Presumably because he was, and he certainly was, a very great mathematician, but mainly because Induction, which involves gambling, was a very difficult process to justify in an era dominated by the Church. I would though put it more generally: people then demanded Certainty, as they often still do, where no Certainty can be found – only Probability. Science can deliver at best only provisional conclusions, and we have to be, indeed we should be content with that: after all better evidence may come along!

Inductive science often uses Mathematics as a tool, but that does not make it a branch of Mathematics. Alas when universities decided to set up departments of Statistics in the 20th century, they observed that both used algebra, and so embedded them in Schools of Mathematics. The mathematicians then tried to cram an Inductive subject into a Deductive drawer where it couldn't fit, and didn't belong. But the difficulties were often concealed behind a thicket of impenetrable mathematics. No wonder confusion and chaos have reigned ever since.

The course I have set myself in this chapter is a difficult and dangerous one. It is difficult because I have to argue that while the gathering and interpretation of numbers are indeed invaluable habits, many of the advanced techniques for doing so, developed by professional statisticians (mostly mathematicians), are wrong, often because they involve deduction – where it is not warranted. I aim to show, by example, that CST based on the DE, can do the job instead. And it is dangerous because I am likely to be branded as a crank, indulging myself in a private rant. So I must ask the reader's indulgence here. Let me get to the very end of the chapter before making up your mind – one way or the other. The potential reward is this: if I am right then you won't have to learn advanced statistics – when plain CST will do instead.

Statistics was conceived in a well intentioned manner. Who could argue with the gathering of numbers, and the presenting of them in the simplest, most transparent ways? The ideas of average, of scatter, of correlations between these numbers and those, became the terms of everyday thinking and debate. And such numbers were used to beneficial and telling effect, as for instance by Florence Nightingale when persuading her government that their soldiers in the Crimea were dying more from medical neglect than enemy action.

But as soon as statistics ventured beyond elementary Common Sense its troubles, serious troubles began. Even the two modern founders of academic statistics, successively professors of the subject at University College London, couldn't agree. Its first superstar, the polymath Sir Francis Galton (correlations, fingerprints, heredity, anticyclones.....), had this to say about his successor Karl Pearson : "The position now seems to be that Pearson, whom I do not trust as a clear thinker when he writes without symbols, has to be trusted implicitly when he hides in a table of gamma-functions, whatever they may be."

That early distrust was only a warning of bitter civil wars to come. Sir Ronald Fiisher, Statistics' first Pope, hated Pearson, along with almost everyone else in the subject who threatened to steal his thunder, and had this to say of a book written by his illustrious contemporary Sir Harold Jeffreys: ' …..a logical mistake on the first page….invalidates all 395 formulae in his book.'

So far as I can see the problem started when some statisticians, who mainly had a background in Mathematics, convinced themselves that they had found new techniques for getting at the truth, previously undiscovered. These new techniques were *mathematical, and therefore deductive*. But we have seen that deduction can never be the basis for Inference, a fundamental truth going back at least as far as Galileo, Huyghens and Newton. So some way or another those statisticians must have befuddled themselves before they successfully befuddled the rest of us.

Tragically, when Statistics units were set up in universities, the subsequent civil war was inevitable as academics trained in Mathematics tried to shoe-horn pragmatic subjects like science into constricted deductive channels where they wouldn't fit, and didn't belong. But the damage was done and will probably take another century to undo.

If you think I'm exaggerating let me set you a little test. Imagine that some old friends, who you haven't seen for years, are coming to stay with you. You know they have 2 children and that one is a boy. What is the Probability that the other is a girl? (It's all about bedroom arrangements).

PAUSE HERE FOR CONSIDERABLE THOUGHT

Most ordinary people quickly answer 'A Half.' Most mathematicians, statisticians and theoretical scientists, after much careful thought, conclude that the answer is 'A Third'. They often go on to say that this demonstrates the weakness of common sense, and the consequent need for ordinary mortals to consult specialists like themselves.

The problem was widely publicized and quarrelled over in the 1990s. It aroused such heated debate in our department of Physics and Astronomy (by then we'd fled from Maths) that it had to be banned from the coffee room — to prevent grievous bodily harm.

Now the fact of the matter is, despite eminent theoretical assertions to the contrary [{4:5}, {11;1}, Ian Stewart 1996] that the answer *is* actually a half — something that can readily be established by experiment [I asked my class of 99

Theoretical Physics students 'How many of you have single siblings?' 37 said 'I have'. Of these, 17 had siblings of the opposite sex, i.e. close to one half. You can work out that there is only a 2 per cent chance that this was a Counting Error away from a true value of one-third]. But try to convince a 'Mathematician' once he's made up his mind to 'One Third'. In my experience it is impossible and if you persist you will only turn a colleague into an enemy. The most they will ever concede is 'it depends how you phrase the question'. Umm... [See Appendix A4 for a discussion of the Boy/Girl problem.]

Mathematicians are not fools, and I yield to none in my love of Mathematics (see chapter 15). But Inductive thinking and Deductive thinking are very different from one another and any attempt to replace Induction with Deduction in Inference, as so many statisticians try to do, can only generate nonsense (see later).

The lesson of this chapter is that Statistics is fine — *up to a point*. And that point is precisely where it departs from common sense. Beyond that point Statistics becomes fantasy. *This is wonderful news: it means we won't have to learn, or even attend to, all that arcane mumbo-jumbo some statisticians use to terrorize peons like us.*

If you haven't been indoctrinated with the importance of Statistics then you can give this chapter a fairly light reading, skipping perhaps from here to Sects (11:4), (11:7), (11:8), (11:9) and in particular to the Summary. But if you have then you will certainly need to get some idea of how it all went so badly wrong and why there is no longer any need to be intimidated.

Certainly this chapter is more mathematical than the rest because we need to learn just enough statistical mumbo-jumbo to see through it. The most mathematical section (11:3) 'The Gambler's Secret' should at least improve your poker out of all recognition. Indeed you really shouldn't gamble without learning that secret.

(11:2) THE MAGIC OF STATISTICS

Statisticians see themselves in the vanguard of scientific progress and are not short of hubris ('pride against the gods'). I cannot do better than quote Stephen Senn, author of "Dicing with Death" a very readable modern book about medical statistics {*} by a professor in the subject: *"Statisticians are engaged in an exhausting but exhilarating struggle with the biggest challenge that philosophy makes to science: how do we translate information into knowledge? Statistics tells us how to evaluate evidence, how to design experiments, how to turn data into decisions, how much credence should be given to whom and to what and why, how to reckon chances and when to take them. Statistics deals with the very essence of the universe, chance and contingency are its discourse and statisticians know the vocabulary. If you think statisticians have nothing to say about what you do or how you could do it better then you are either wrong, or in need of a more interesting job."*

The beginning of Statistics is usually traced to a London draper John Graunt who compiled the first mortality table back in 1662 when the occupations of the dead were recorded by the 'Searchers', dames whose dangerous job it was to try and find out whether the deceased had died of Plague. He was evidently a Puritan, surprised to find:

"...how small a part of the people work upon necessary labours, and Callings, viz how many Women and Children do just nothing, only learning to spend what others get? how many are mere Voluptuaries, and as it were meer Gamesters by Trade? How many live by puzzling poor people with unintelligible Notions in Divinity and Philosophy? How many by persuading credulous, delicate and litigious persons, that their Bodies, or Estates are out of Tune and in danger? how many by fighting as soldiers? How many by trades of meer pleasure, or Ornaments? And how many in a way of lazie attendance etc. upon others? And on the other side, how few are employed in raising, and working necessary food and covering? And of the speculative men how few do truly studie Nature, and Things?" Not much change then.

Graunt was the first to discover that about 6 per cent more boys are born than girls and his percentage Survival (Mortality) table was as follows:

Age	6	16	Parent	Adult	56	66	76
100	64	40	25	16	6	3	1

Over the next two centuries Statistics grew slowly: Life Insurers were naturally interested in survival rates while governments wanted to know how many of their subjects could be taxed. Occasional attempts were made to carry out statistical experiments. Thus James Lind the Scottish naval surgeon attempted in 1747 to find a cure for scurvy, that dread disease of ocean voyagers [1600 out of 1800 aboard George Anson's 1740 -1744 circumnavigation died on passage.] Lind divided 12 sufferers aboard his ship the *Salisbury*, into an experimental group and a control group, feeding various concoctions to the former only. The results were dramatic. "The consequence was" he reported "That the most sudden and visible good effects were perceived from the use of oranges and lemons, one of the those who had taken them, being at the end of six days fit for duty." By 1800 daily tots of lemon-juice were issued throughout the Royal Navy.

Let us turn next to that iconic success story of medical statistics — discovery of the connection between lung-cancer and smoking. The first indication of such a connection came from Germany in the 1920s and 1930s, probably because lung-cancer was a rare disease in the rest of the world whereas up to 80% of miners in the Joachimstal district of Bohemia died from 'berg-krankheit'(mountain sickness) — subsequently identified as lung-cancer caused by radon-gas in the local mines. Doctor Fritz Lickint noted the close connection between smoking and his LC patients in 1929 and campaigned against the former. In 1939 Muller published the following numbers for 86 LC patients and 86 healthy controls of the same age:

	LC cases	Controls
Smokers	27	41
Non-smokers	3	14
Heavy-smokers	56	31

ANALYSIS (11:1): SMOKING STATISTICS

What does one make of these figures? Do they make a strong case for smoking as a cause of LC?

My take: (i) Clue E_1 : In the Control (healthy) group there are 14 non-smokers but only 3 among the LC cases. That is suggestive – but is it a significant difference? If smoking is not a causative agent we would have expected roughly the same number of non-smokers among the LC cases i.e. if we use the Scientist's Approximation [Eqn.(9) Chapt.9)] we would expect $14 \pm \sqrt{14} = 14 \pm 4$. But we got only 3 or 14-3 = 11 less.

Now 11/4 is a discrepancy more than 2σ but less than 3σ. If we now consult Table (9:1) of De Moivre's Normal Probabilities a 2σ in the 'right' direction (i.e. 'single-tailed'} has a Probability of .05/2 =.025 against it being chance. So if chance is our alternative hypothesis $\bar{H}$ here, our Weight for E_1 is:

$$W(E_1 \mid H) = \frac{P(E_1 \mid H)}{P(E_1 \mid \bar{H})} = \frac{1}{.025} \sim 40$$

(ii) Clue E_2 : There are 31 heavy smokers amongst the Controls, so if smoking was not causative of LC we would expect only $31 \pm \sqrt{31}$ (Scientist's Approximation) or 31 ± 5 amongst the LC sufferers. Actually there are 56 or a 56-31 = 25 or a 25/5 or 5σ larger number. Even a 4σ result in the right direction, as here, has a Probability of only (.0004)/2 = .0002 of occurring by chance according to Table (9:1) so we can compute`:

$$W(E_2 \mid H) = \frac{P(E_2 \mid H)}{P(E_2 \mid \bar{H})} = \frac{1}{\leq .0002} \geq 5000$$

Now combining both clues using the Detective's Equation:

$$O(H \mid E_1, E_2) = 40 \times (\geq 5000) \times O(H)$$

so unless you can argue that the Prior O(H) on 'Smoking causes LC' is very small (less than 1/200,000) one is almost forced to conclude that smoking is dangerous [Note how we couldn't get Weights without an $\bar{H}$].

Alas these figures never made it across the line of battle and Allied governments spent the war encouraging their troops to smoke as much cheap tobacco as possible, thereby killing far more of their own men than the enemy did! [In the FWW US General Sherman, whose wife had shares in Big Tobacco, once said :'What my men need is not more ammunition but more tobacco.' Did they kill their enemies by stabbing them in the eye with a lighted butt one wonders.]

After the 2nd World War Austin Bradford Hill and Richard Doll from the London School of Hygiene, undertook a similar but larger study using 1357 LC sufferers from 20 London hospitals and 1357 controls who were patients from the same hospitals. Their figures were;

	LC cases	Controls	Totals
Smokers	1350	1296	2646
Non Smokers	7	61	68
Totals	1357	1357	2714

ANALYSIS (11:2); MORE SMOKING STATISTICS

What should we make of these numbers?

My take: The disparity in non-smokers (61 and 7) between the two categories seems very significant to me. Were smoking not significant I would have expected the same number i.e. $61 \pm \sqrt{61} = 61 \pm 8$ in each category. Now 61-7 = 54

≈7×8 or about a 7σ discrepancy which is right off the scale of De Moivre's Table (9:1) where the highest Probability tabulated against such a pure chance was for 4σ where it was given as 60,000 (single tailed) So

$$W(E|H) = \frac{P(E|H)}{P(E|\bar{H})} = \frac{1}{1/60,000} = 60,000$$

So a 7σ discrepancy will certainly have a much higher Weight even than that. Thus according to the Detective's Equation

O(H|E) = W(E|H) . O(H)

i.e. O(H|E) > 60,000 . O(H)

So unless your Prior Odds O(H) *against* smoking being a cause of LC are very high (say 20,000 to 1 against), one is bound to concede that smoking is very dangerous. Hill and Doll promptly gave up smoking (No easy matter, as we ex-smokers all know).

Not everybody was convinced however, far from it. Here, in brief, were some of the counter arguments:

(i) "The evidence is purely coincidental."

(ii) "No physical mechanism is known for how smoking might cause cancer".

(iii) "The control sample is totally unrepresentative of the general population".(Certainly true because only 2.5% were non-smokers when 30% of the population at large were; mostly women at the time).

(iv) "There is no detectable difference between inhalers and non-inhalers", as Doll and Hill had admitted.

(v) "There may be a constitutional difference between smokers and non-smokers. Your correlation is not causation."

(vi) "There may not be a LC epidemic at all —it's simply due to improved diagnosis".

I don't know what you think of such arguments, but the editor of the famous scientific journal 'Nature' could write as late as 1984; "It will be time enough to talk about causes of LC when mechanisms have been worked out….." (I suspect the poor devil couldn't give it up. His argument is fallacious anyway: you don't have to be able to explain a phenomenon to be convinced of its existence) [6] [The weakest part of our argument is the implicit assumption that the only errors in the figures are Counting Errors so that we are justified in using de Moivre's table (9:1). It's not easy to rule out the possibility of more systematic errors – such as misdiagnosis. For instance doctors convinced of the association already might over-diagnose smokers,]

After 1950 Doll and Hill set up a 'Cohort Study' to follow the smoking habits and fates of 30,000 British males (doctors as it happened). For many reasons such a cohort study is more satisfactory: for a start it is prospective not retrospective — so the results cannot be known and therefore 'fixed' in advance. There are no problems with picking representative controls while its long term should show up effects with long 'latency' periods.

The results were conclusive and shocking. To quote from their 2004 paper; "About half of all persistent smokers are killed by their habit, a quarter while still in middle age"…… "On average cigarette-smokers die about 10 years younger than non-smokers (more by heart-disease and strokes than by cancer)" …..but encouragingly: "Stopping at age 60, 50, 40 and 30 gains, respectively about 3,6,9 and 10 years of life expectancy."

The Doll and Hill 1950 and 2004 papers can be downloaded from the web and are thoroughly worth reading, and readable. They are flagship examples of

---

[6] Much later, in 1964, probably because of the powerful Tobacco Lobby over there, the US Surgeon General released a report announcing "Cigarette smoking is causally related to lung cancer in men and points in the same direction in women." Cigarette consumption per person immediately began to fall and it is a quarter today (2014) of its value then (4000 p.a.). Hollywood had been a powerful ally of Big Tobacco.

'Good Statistics'. The right kind of data is patiently collected until the results virtually speak for themselves. No sophisticated mathematical techniques are used or are necessary. They used Common Sense working in a systematic way to speak to Common Sense. Actually Muller's 1939 data sufficed to prove the LC connection. Hill's and Dolls' revealed the even more unexpected and pernicious effects on the heart and the circulation. In retrospect one can see why no less than 97.5 % of all their 1950 hospital patients, including controls, smoked: hospitals then were mostly hospices for dying smokers.

(11:3) PROBABILITY THEORY;THE GAMBLER'S SECRET

As a student I used to play a lot of poker. Being averagely bad at it I lost more than I won. But then, as part of my course in Physics, I learned Probability Theory (PT from now on). Thereafter I won consistently, so consistently that I had to give up the game because I felt I was fleecing my less numerate friends. Anyone who gambles in ignorance of some PT is asking for trouble.

PT is a superficially simple branch of Mathematics devised in the 17th century to calculate the Odds in games of chance. What are the Odds that I'll draw more than 7 Hearts in a hand of 13 cards, and so on. It's Mathematics because it's a game; all the rules are laid down unambiguously at the start, and all the cards and dice are assumed to be fair so that the assignments of individual Probabilities (e.g. the Probability of drawing an Ace) are uncontroversial.

So powerful is the theory that there are irresistible temptations to introduce it into real life. And that's when troubles *can* arise. Real life is not a game. Nobody has laid down unambiguous rules. The Probability that it will rain tomorrow, or that your bank will go bust, *are* controversial. So to what extent is it legitimate to introduce the notion of

Probability into real life, and what meaning could Probability have in that context? In particular deductive methods, which are legitimate in a CLOSED game, have sometimes been sneaked into real-life situations where they most certainly are not. For instance some brands of statistician appear to claim deductive certainty for conclusions where scientists believe that inductive inference is the best that can be hoped for. As you can imagine this has led to colossal rows, with the field breaking up into warring sects.

As thinkers we have to be concerned with the Weights of clues where
$$W(E|H) = P(E|H) / P(E|\bar{H})$$
and $\bar{H}$ is 'not H'. The evidence could have arisen by pure chance, in which case we obviously need to calculate $P(E|\bar{H})$ — which is where PT becomes particularly important. So that's why we have this section. [It's the most mathematical in the book and can be bypassed without great loss. However Gambling Theory is potentially so useful to nearly everyone that I couldn't avoid the temptation to put it in. Ref. {9/1} is excellent on Probability.

We learned much about Counting Errors by analysing coin-tossing. We found the standard deviation $\pm\sqrt{Q}/2$ (Q tosses) and learned to compute Probabilities and Odds from De Moivre's Table (9:1) by drawing a branching diagram and adding Vertex-numbers. That worked because Heads and Tails had equal Probability — one half. Alas in most interesting problems such will not be the case. What for instance are my chances of drawing 2 Aces (Prob. = 1/13 each) in a hand of 5 cards? Or, more fatefully, what are my chances as a WWII RAF Bomber Command aircrew of surviving a Tour of 30 'Ops.' when the casualty rates are about 6% (P=1/16) per Op.? All wise gamblers must be able to answer such questions and, as we have argued that all CST involves gambling, that implies all of us. If you get nothing else out of this section it should improve your poker — it certainly improved mine. Anyone entirely ignorant of PT is going to be a 'sucker', as they say in the saloon.

At the request of gamblers PT was developed in a correspondence between two 17th century French mathematicians Blaise Pascal and Pierre de Fermat (he of 'The Last Theorem') and extended by Jacob Bernoulli (Groningen, Holland) who wrote the famous 'Art of Conjecture' (1713). We are now going to tease out Bernoulli's Theorem, the jewel and engine of gambling theory, because I believe no serious thinker can afford to go through life without it. My proof is Inductive, and would not satisfy Pure Mathematicians. So what.

The rules of Mathematical Probability can be reduced to 3, all of which seem reasonable, at least to me:

(I) P(A) the Probability of proposition A lies in the range 0 ('impossible') to 1 ('certain'). (This is just a definition).

(II) P(A) + P(not A) = 1. i.e. any proposition is either true or untrue.

(III)   P(A&B) = P(A|B) × P(B)   which means ' 'The Prob. of both A and B together is the Prob. of A *given that B is true*, times the Prob. of B' which also seems obvious on reflection.

From these 3 fundamental rules others can be deduced. For instance IF A and B are *independent* of one another P(A|B) is simply P(A) in which case (III) reduces to the very simple and useful 'Independent-Product Rule':

$$P(A \& B) = P(A) \times P(B) \text{ (A and B independant) (4)}$$

If you get stuck on abstracts (as I tend to get), turn to simple examples. Thus in throwing two dice together, one red one green, the score on one will be independent of the score on the other. Thus the Probability of throwing two 6's is, by the Independent Product Rule (4):

$$P(6\ \&\ 6) = (1/6) \times (1/6) = 1/36$$

which can be rationalised by noting that there are 36 (6 times 6) combinations of scores in all, only one of which yields two 6's.

This last suggests the usefulness of learning how to compute numbers of possible Combinations and Permutations. I like to think in terms of say 9 people hoping to get into a 4-seater car. Any one of 9 could get into the driving seat, any one of the remaining 8 into the passenger seat, any one of the remaining 7 into left rear, any one of the last 6 into right rear. In all there must therefore be 9.8.7.6 = 2924 different 'Permutations' of passengers, Permutations in the sense that they are *ordered* by seat. So we could write:

Perm.(4 from 9)= 9.8.7.6 =2924

If we don't care about the seat-ordering within the car then any of the 4 *already-chosen* passengers could sit in the drivers seat, any 3 of the remainder in the passenger seat, any 2 left rear, with the last one in right rear. Thus there are 4.3.2.1 = 24 different Permutations, corresponding to that one Combination i.e.:

$$Combs(4\ from9) = \frac{9.8.7.6}{4.3.2.1}$$

= 2924/24 = 126 combinations.

There is a convenient shorthand for strings of such descending integers multiplied together thus: 4.3.2.1 is called 'factorial 4' and written 4!(or '4 shriek')

Thus Perms (4 from9) = 9!/5!= 9!/(9-4)!

Or in general: Perms(r from N) = (N!)/[(N-r)!]   (5)

While Combs(4 from9) = 9!/(5! times 4!)

Or in general: Combs (r from N) = (N!)/[(N-r)! times r!]   (6)

These formulae can be very useful. If they are not familiar try:

EXERCISES (11:1): PERMUTATIONS AND COMBINATIONS

We can see now where the Vertex numbers came from in the coin-tossing exercise (Fig 9:1). For instance look under 5 tosses i.e. Q = 5 down to the Vertex beneath where there are 3 Heads (i.e. the number *above* the vertex = 3). The 'Vertex number' underneath = 10. But the number of Combinations of 5 successive throws that can lead to 3 Heads is, by Eqn. (6):

Combs (3 from 5)= 5!/(2! times 3!) = 5times4/2 = 10

In other words the important Vertex numbers are no other than the number of different paths by which the said score of 3 Heads can be reached. But we've argued that the Probability of getting that score must correspond to the number of possible ways of getting it *as a fraction of all possible combinations* = N =

$2^Q = 2^5$ of throws up to that point i.e.

Probability of 3 Heads in 5 Tosses =

$$\frac{\text{Combs (3 from 5)}}{\text{All possible tosses} = 2^5} = \frac{10}{32}$$

Or more generally:

Probability of a certain outcome:

= (Frequency of that Outcome) /(Number of all possible outcomes)

Or *"Probability of an outcome = Relative Frequency of that outcome"*.

So let's do a real problem. Imagine a squadron of 12 Lancaster bombers was launched against a vital target during the 2nd. World War. Supposing the probability of a single bomber hitting the target is 0.1 or 10 %, what are the chances of the whole squadron making exacly 2 hits, 3 hits and so on?

The probability of 2 *particular* aircraft hitting is obviously, by the Independent Product Rule, (0.1) times (0.1) But we don't care which *particular* 2 hit, so we should obviously multiply the above number by the total number of different combinations of 2 you can draw from 12 i.e. by Combs (2 from 12). But if we don't want any of the other 10 to hit as well we need to multiply by (0.9) ten times to ensure this, so that finally:

$$\text{Prob}(2 \text{ hits from } 12) = \text{Combs}(2 \text{ from } 12) \times (0.1)^2 \times (0.9)^{10}$$

$$= \frac{12!}{2!10!} \times (0.1)^2 \times (0.9)^{10}$$

$$= \frac{12 \times 10}{2} \times (0.1)^2 \times (0.9)^{10} = 0.21$$

while for 3 hits $P(3,12) = \frac{12!}{3! \times 9!} \times (0.1)^3 \times (0.9)^9 = 0.085$

which leads to the Bernoulli's so useful Gambler's Theorem:

Probability of (r successes in Q trials) =

$$\text{Combs (r from Q)} \times p^r . [1-p]^{Q-r} \qquad (7)$$

which we set out to find. If you are a gambler it's worth remembering.

Let's use it on another more fateful question which must have occurred to every brave airman who fought in the war on either side. "What is the chance of me surviving a Tour of 30 operations given that the casualty rate per op. = 0.06 (Typical in RAF Bomber Command)?

Well success rate p per op. = (1- 0.06) = 0.94. And using Bernoulli (7) the

$$\text{Prob}(30 \text{ returns in } 30 \text{ ops}) = \text{Combs}(30 \text{ from } 30) \times p^{30} \times (1-p)^{30-30}$$

$$= \frac{30!}{30! \times 0!} \times (0.94)^{30} \times (1-0.94)^{0}$$

$$= 1 \qquad \times 0.16 \quad \times 1 = 16 \text{ per cent or 1 in 6}$$

(Where we used the not altogether obvious fact that 0! = 1)

So bomber crews were extremely brave men who were mostly not expected to survive a tour of Ops. 55,000 aircrew in RAF Bomber Command were killed out of about 125,000, the highest casualty rate among Allied forces.

Before the days of electronic computing, calculating large factorials like 10! or even 100! was a huge practical problem. Thus the need to find approximate but practical means of computing Bernoulli Probabilities — and even today half a typical Statistics textbook is taken up with various such approximations, while once upon a time Statisticians were largely humble flesh-and-blood 'Computers' hired to do such donkey-work. Indeed that's where the word 'computer' came from. The graph or 'distribution' of Bernoulli Probabilities for different values of r [for a given N and p] is centrally peaked like the Normal Distribution in the last chapter, and for exactly the same reason. There are more pathways to outcomes with a moderate number of successes than to extreme outcomes in either 'tail'. Indeed the Normal distribution IS the Bernoulli distribution as Q rises above about 20.

Using the Expectation approach developed in Sect (9:2) we could easily show, but it's rather tedious, that the mean and standard deviation of any Bernoulli distribution are given by

$$\bar{x} = Qp \quad (8a)$$
$$\sigma_x = \sqrt{Qp(1-p)} \quad (8b)$$

The first is intuitive. For the case p = ½ the second yields

$$\sigma_x = \sqrt{Q.(1/2).(1/2)} = \sqrt{Q}/2$$

exactly the result we reached for Heads and Tails (p = 0.5 for both) when discussing Counting Errors.

Eqn.(8b) above, though accurate for all Bernoulli distributions, is a bit fiddly to remember so I will often use '*The Scientists Approximation*':

$$\text{Expected Number} = Qp \pm \sqrt{Qp} = \bar{x} \pm \sqrt{\bar{x}} \quad (9)$$

It's mostly accurate enough for practical purposes and one can do it in one's head. [For instance in 600 throws of a fair dice (8b) would predict 100 ±9 sixes whereas (9) would give 100 ±10. Practical scientists are fond of rough and ready numbers because there are so many imponderables in the real world. Mathematicians and Statisticians often don't get this.]

313

Now that we all have calculators, using Bernoulli's formula (7) and its accompanying (8) is an accomplishment we can all practice with profit so:

EXERCISES (11:4) THE GAMBLER'S SECRET

(11: 4) PROBABILITY WARS

In one sense I cheated you in section (11:3) on Probability Theory — and deliberately so. Recall that our original definition of 'Probability' as it was used in all of Chapters 1 to 9 was "..*our degree of belief in*....". But in section (11:3) Probability subtly morphed into "*relative frequency in a large number of trials.*" This second definition is not in itself wrong and in some very limited circumstances it will agree with the first. But it is far narrower than the definition we need to carry out most Inference. What possible meaning can "relative frequency in a large number of trials have to "the Mass of the Earth probably lies between 10 to the 20 and 10 to the 21 tons"?

We'd naturally like to avail ourselves of probability theory and sometimes, in limited contexts we can. But in most cases we cannot, not unless we can fool ourselves into believing that real life is a game with unambiguous rules (like Mathematics) and with values for the various Probabilities that are 'logical' and beyond contention.

The ugly wars that disfigure so much of Statistics largely stem from disagreements, openly stated or privately understood, about the particular definition of Probability in this or that context. Mathematicians will claim, and they are historically correct, that PT is a branch of Mathematics developed by them to address well defined (i.e. mathematical) problems in gambling. That is why Statisticians with a mathematical background throw fits over 'Priors' — private intuitions about this or that hypothesis which have no business in Mathematics.

Scientists however, and other practical folk, can only advance by Inference, a mode of thinking which has to grapple with a world defined by Nature and not by Man. They are trying to *find* the rules, not to stick to them.

There are huge temptations on both sides of this divide. Mathematicians feel themselves to be authoritative on Inference, because it looks like Mathematics to them, but it's not, it very definitely is not! On the other hand scientists and other practical folk would like to cherry-pick bits of mathematics, PT in particular, to do Inference. Both are blundering into a minefield. The truth is you cannot do Inference without Priors. If you try to do so you will only be deluding yourself by unwittingly adopting some sort of "Objective" Prior such as 'All hypotheses are equally likely' which is less than optimal and probably foolish. On the other hand if, as a scientist, you want to adopt some mathematical theorem to do Inference you had better be damned sure that it really applies to the case in which you are interested. If not you'll be led into folly.

The easiest way to step on a mine, and it is both easy and tempting, is to mistake a Measurement Error for a Counting Error. That's what our forensic scientist did when he condemned an innocent to 20 years in gaol.

Try not to step on mines, it's not good for you. This chapter is all about tip toeing round them, with cautionary tales of those who did not.

(11:5) THE MAGIC FADES AWAY

Numbers *should* mean little to thinkers until they can be turned into Weights; in chapter 9 we described the bell-shaped curve and in Appendix A1 tried to convert it into a table of Normal Weights which can be very useful for reaching decisions. *Unfortunately it is only valid for data with errors like Counting Errors!* What about Measurement Errors?

It should be obvious that no amount of mathematics could ever prove that the errors involved in measuring something physical, like footprints in a flower-bed, must follow a particular mathematical form. Sometimes a jumper will skid, defying all the best laid theorems of mice and men. And yet many thinkers, Statisticians in particular, came to accept that Measurement Errors should follow wonderfully symmetric mathematical forms. Here for instance is Sir Francis Galton, the first academic statistician, waxing lyrical about The Normal Law of Errors in the late nineteenth century, though it wasn't called that then;

" I know of scarcely anything so apt to impress the imagination as the wonderful form of cosmic order expressed by the 'Law of the Frequency of Error'. The Law would have been personified by the Greeks and deified, if they had known of it. It reigns with serenity and complete self-effacement amidst the wildest confusion. The huger the mob, and the greater the apparent anarchy, the more perfect is its sway. It is the supreme law of Unreason."

A century before Galton Pierre Simone de Laplace, 'The French Newton', had proved 'The Central Limit Theorem' (CLT) which even today Statisticians will quote at you as proof that errors must follow the 'Normal Form'. The text-book I learned Statistics from, written by two Statistics professors from London University, dismissed the whole matter, as if it was obvious, in a couple of lines, whereas my lecturer — a mathematician, never mentioned it at all. On the other hand two professors from Edinburgh University wrote (1929): "Everybody believes in the exponential (Normal) law of errors; the experimenters because they think it can be proved by mathematics; and the mathematicians because they think it has been established by observation."

So what is this 'Central Limit Theorem' that the great Laplace proved? He proved that *in very limited circumstances* (how could it be otherwise?) that: 'Measurement Errors follow the same rules as Counting Errors' — in which case it would be legitimate to use Normal Weights. And *sometimes* that is true in practice. For instance if I use the Hubble Space Telescope to measure the brightness of a faint galaxy I am in effect counting photons — particles of light. In that case my Measurement Errors *are* Counting Errors — up to a point. There are, unfortunately, other sources of observational error involved in that measurement that do *not* follow the 'Normal' form: for instance faint galaxies in the background, galaxies too faint to see on their own, which can screw up my measurement. I would only be justified in using Weights drawn from the Normal Law *as long as they were not too high*. Higher Weights would imply that I have a more precise knowledge of my real errors, particularly out in the tails of my error distribution, than is in fact the case. If I did a lot of hard work investigating my observational errors, making observations on the telescope itself — which costs $100,000 an hour to use, I might get believable Weights as high as 16 (2^4). I might. Beyond that I wouldn't honestly dare.

Likewise our forensic scientist; if he'd made hundreds of measurements of dozens of different people jumping out of different windows onto different soils in different meteorological conditions, he might have gained a fair idea of his true errors, which I suspect might have been quite high. But he didn't do that. He'd been told, during his Statistics Course, that using Normal Probabilities was justified by 'The Central Limit Theorem' which he'd never studied, and probably wouldn't have understood if he had. And no wonder; it has had to be 'proved' over and over again since Laplace's day, a circumstance suspicious in itself, by all sorts of characters ranging from Panufty Chebyshev to Alan Turing in the 1930s. The truth is that each proof can only have a limited range of validity — *which probably doesn't apply to the specific problem in which you are interested*. As Thomas Huxley 'Darwin's Bulldog' put it: "Mathematics may be compared with a mill of exquisite workmanship, which grinds you stuff to any degree of fineness; but, nevertheless, what you get out depends on what you put in; and as the grandest mill in the world will not extract wheat flour from peascods, so pages of formulae will not get a result out of loose data."

How could so many intelligent people have been fooled, and why does so much disparate data seem to fit the 'Normal Curve of Errors' rather well? My suspicion is that it's all got to do with curve-fitting, an important topic on which we briefly digress.

Fitting the best curves we can to the data we've got is something scientists do all the time — partly to guess at the main phenomena involved (which have characteristic shaped curves), partly to find a compact formula to describe the situation, and occasionally to extrapolate beyond the data into regions presently unexplored.

What do we mean by 'best curve' in this context? This is something of a delicate matter. We don't want the curve to follow every jiggle in the data because we suspect that most of those jiggles are due to uninteresting Noise (Fig 7:5). At the same time we don't want to miss any significant trends by fitting a curve that is too crude to express those trends. Finally we don't want a curve that pretends to more knowledge than the data seem to justify; that can only lead us down blind alleyways. Yes curve-fitting is a black art — and we shouldn't pretend otherwise. *After all it is Induction tangling with Ockham, just Plausible Inference in another guise.*

If data is really rough (noisy) the best we can hope to do is fit a straight line through it (Fig 11:1a). Why? Because a straight line pretends to the least unjustified omniscience. If we plot it in terms of its departure from the mean values $\bar{x}$ and $\bar{y}$, then only one Free-Parameter is required to describe it — its slope.[See Fig.(11:1) b]

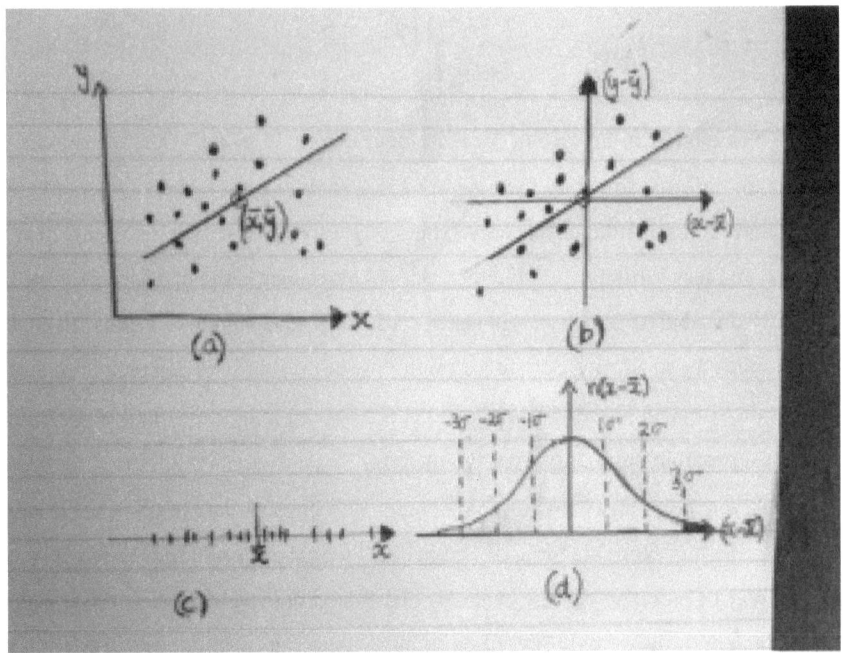

Fig(11:1) (a) Measurements of y_i against x_i for about 20 objects. The data appear to be very noisy but there is a weak trend of rising y with x. About the best 'curve' we can fit through the data is the straight line shown. (b) We can re-plot the measurements in terms of their departures from the mean values $\bar{x}$ and $\bar{y}$ as shown. Now the only Free Parameter left to describe the line is its slope. It's a 'one-parameter curve' as we say.

(c) About 20 measurements of some quantity x. They are gathered around their mean value $\bar{x}$ with a spread of about σ. (d) Now we plot how many measurements there are in each interval to either side of $\bar{x}$. They hump up towards the middle around $\bar{x}$. So we plot the simplest bell-shaped curve we can to the data. We've also marked in to either side the $+1\sigma$, $+2\sigma$, $+3\sigma$ and the equivalent negative values (both dashed). It turns out that the shape of that simplest curve depends on only the single free parameter σ. Thus the Odds on a more than $+3\sigma$ result is the area(shaded) to the right of 3σ divided by the whole area to the left of the 3σ (dashed) line. But that would only be accurate if the actual scatter of the measurements followed the precise shape of this particular, specially simple, bell-shaped curve. And why should they, anymore than the measurements in (a) should fit on a straight line? And just as it would be folly to expect the value of y in (a) out to either end beyond the line to be accurate, so it would be folly to expect the measurements in (d) to accurately follow this particular bell-shaped curve out into either wing, or 'tail' as they call it. But that is what many Statisticians presume.

Next consider the case of Measurement Errors. In so far as the measurements contain any information at all they ought to cluster round some average value $\bar{x}$, with a scatter that represents the uncertainty. There ought to be more measurements near the average, and less further away to either side. In other words the measurements ought to be 'centrally distributed' as we say, in some kind of bell-shaped curve Fig(11:1d). Obviously we shouldn't try to fit such data to a straight line but it might be reasonable (honest) to fit them to the simplest possible bell-shaped curve we can imagine — i.e. one described by the lowest possible number of Free Parameters.

At this point an awful suspicion creeps into the sceptical mind — at least into mine. Maybe the 'Normal Curve of Errors' isn't magic after all? Maybe it's just the crudest way of representing any centrally distributed but scattered data ? Maybe it's just the equivalent of the straight line fitted through noisy data that doesn't justify anything better?

Why would this be so terrible? Because just as we would be foolish to extrapolate a straight line at each end into regions unmeasured, so it would be foolish to extrapolate a crude bell-shaped curve out into either tail where little or no real information (data) is available. But those tails are the very regions capable of supplying the highest Odds to clues because:

Odds = [Large area under whole graph to left of measured point]
divided by [Small area in tail to right out beyond measured point].

And if we can't specify the *exact* shape of that error-curve, especially out in the tails, we are not justified in claiming high Odds and high Weights!

So far this is mere suspicion. To believe it we have to actually demonstrate that the "Normal Curve of Errors" is the crudest representation of any centrally distributed data that one could imagine — the equivalent of fitting a straight line to noisy non-centrally-distributed data. Such a demonstration turns out to be easy — at least if you have got a year or two of school Calculus. Since I cannot presume that you do, I have had to confine that very important argument to Appendix A5. Try to decipher as much of it as you can.

If you can't follow the appendix argument then there is an alternative. One of the most important thinkers of the 20th century was Claude Shannon who invented Information Theory. Working for Bell Telephone he became interested in the problem of sending messages down noisy telephone lines. That led him to question how you might quantify the Information content of any kind of message, be it a sound, a number, a letter, a word or even a picture. He was led to the idea of binary digits or 'bits', the least number of zeros and ones ('Ons' and 'Offs', dots or dashes) needed to transmit some Information. It's such a neat idea, and so universal, that it has come to dominate the 21st century. Texts, symphonies, phone-calls, broadcasts, DVDs, images, films.......they all store, convey or process bits of digital information. And if you know how many bits of real Information there are in a message you can compress it, and throw most of the redundant stuff away — much of the noise for instance. But what has all this got to do with Statistics, and thinking? As it happens a very great deal. We can ask Information Theory to tell us what kind of bell-shaped curve presumes the least possible Information, i.e. can be specified with the least number of bits. Yes it's de Moivre's Curve, which is also called 'The Normal Curve of Errors.'!

So our suspicion is justified. The Normal Curve is not a measure of our knowledge, it is a measure of our ignorance. It is crude; it has only one Free Parameter i.e. σ, it is the nearest thing to a straight line that will fit centrally concentrated data. And just as a straight line will fit the central regions of data rather well, so the Normal will fit the middle regions of centrally distributed data (Fig 11:1d). But it would be foolish to presume a good fit out in the tails where the real data are minimal, but where the Weights are potentially high (more than 4 say). And yet students are told that the Normal Distribution should be assumed as applying to virtually all measurements. Here is an example of that fallacy from as late as 2016 :

Taken from 'Maths for Science' the Open University, 2016, Kindle version.
"2.12 How likely are particular results?

In real experiments, as opposed to hypothetical ones, it is very rare that scientists make a sufficiently large number of measurements to obtain a smooth continuous distribution like that shown in Fig 7d. However it is often convenient to assume a particular mathematical form for typically distributed measurements, and the form that is usually assumed is the normal distribution, so called because it is very common in nature. The normal distribution corresponds to a bell-shaped curve which is symmetrical about its speak as illustrated in Fig 7d. Repeated independent measurements of the same quantity (such as the breadth of an object, or its mass) approximate to a normal distribution. The more data are collected, the closer they will come to describing a normal distribution curve."

What are we left with at the end of this tangled tale? With a scepticism for all Weights heavier than about 4 (2^2) based on measurements which have not been subjected to the most rigorous and transparent (yes transparent) Error Analysis — probably involving a deal of real measurements, not mathematical proofs based on Theorems of any kind.[7] And a healthy suspicion of Statistics in general.

(11:6) GOOD-SIMPLE VERSUS BAD-SIMPLE

There is a powerful distinction that has to be drawn when we speak of simplicity in connection with thinking: good simplicity as opposed to bad simplicity. If a situation is inherently simple that is good because we can learn much from little: a couple of measurements may suffice to predict many others, both inside and outside the present range of observation. Fig (11.2 a) is an example of good-simplicity. By some rational argument it is known that y and x are in direct proportion to one another, are 'linearly related' as we say, and so only two pairs of measurements (x_1, y_1) and (x_2, y_2) suffice in

---

[7] In many practical situations the typical measurement error δX in X is proportional to the measurement value X itself i.e. it is some fraction of it. When such errors are combined t is then easy to show that that the combined errors do *not* follow the Normal distribution but the 'Log-Normal' distribution which has a peak but is not even symmetrical and is more likely to be much larger than the Normal error, i.e. it has 'a much fatter tail' on one side. There's nothing 'normal' (i.e. commonplace) about the Normal [which originally meant 'orthogonal'].

principle to define that straight line completely and so to predict reliable values of y both within the measured range x_1 to x_2 (called 'interpolation') and out in the wings, i.e. where $x < x_1$ or $x > x_2$ (called 'extrapolation') where we have drawn a cross. On the other hand Fig 11:2 (b) shows an example of bad-simplicity. Many measurements of (x_i, y_i) show there is some kind of relationship between y and x, but with a lot of scatter [sometimes called 'noise']. The best we can hope to do in the way of prediction is draw a best-fitting straight line through the data-points as shown. We've chosen a simple straight line because nothing more elaborate is justified by the noisy data. It's simple all right but it's *'bad-simple'* because it is forced upon us and predicts virtually nothing reliable we don't know already. For instance one would be foolhardy to use the fitted line to predict y out in the wings beyond the measured range of x where for instance we have drawn a cross. Whereas good-simplicity stems from some real knowledge of the world, bad-simplicity is forced upon us by our ignorance. This may seem obvious but let's extend the argument just a little.

In Fig 11:2 (c) the vertical scale y shows the number of measurements of some quantity x falling into bins of the same size but in different ranges of x. Clearly they are centrally concentrated around some preferred average value of x = a say and with some standard deviation σ. Now IF we knew FOR SURE that the measurements had to follow a Normal distribution then we could find the best fitting such a Normal curve to the data, and draw it as shown.

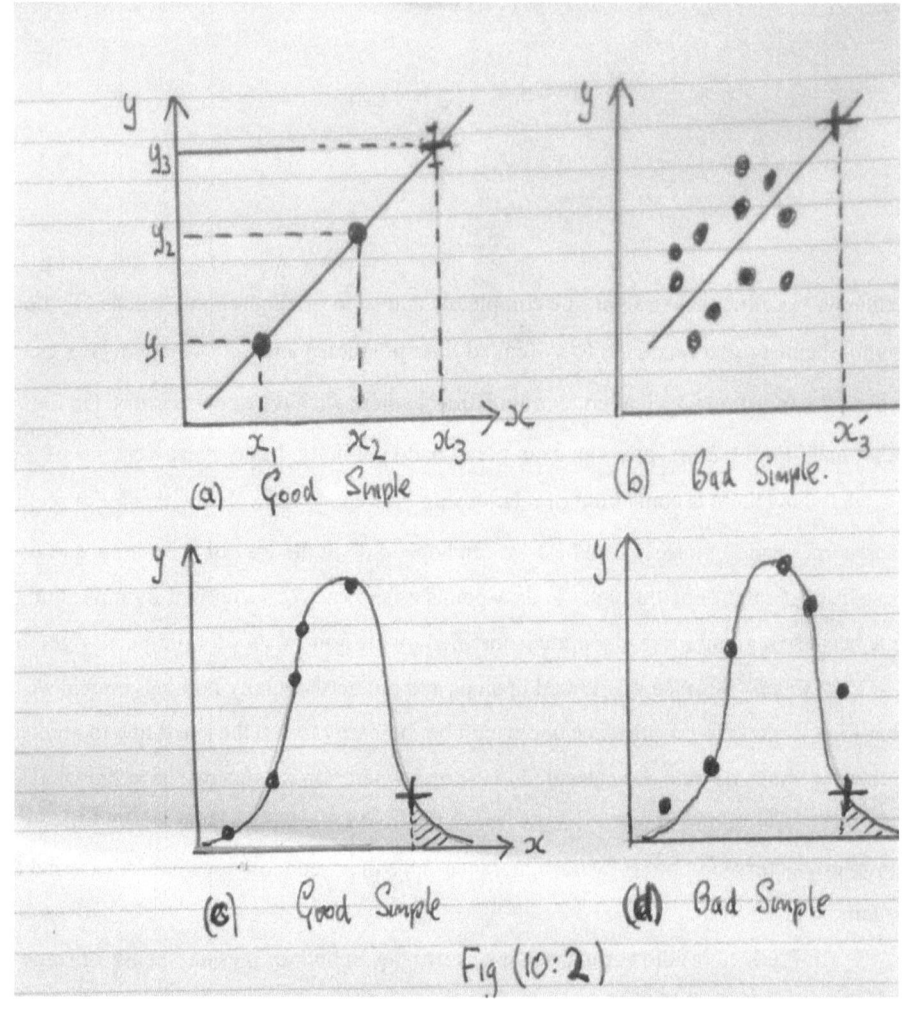

FIG (11:2) GOOD SIMPLE VERSUS BAD-SIMPLE

In (a) we have good reasons to believe that y and x are linearly related and so we are justified in fitting a straight line through some existing data and extrapolating it out to the cross and estimating the value of y at that x. This s an example of good-simple. Fig.(b) We have no such reason to believe that y and x are linearly related but the existing data suggests there is some rough correspondence between the two. We draw a simple straight line through the data because nothing more elaborate is justified. It's 'bad-simple' because it results from ignorance not knowledge. We would be foolish to use it to guess the value of y near the cross where we have no real information. Now Figs (c) and (d) show exactly analogous cases for centrally distributed data. In (c) we have independent evidence to believe the data ought to fit on a Normal curve and so we are justified in fitting one through that existing data and then extrapolating out to the cross say and using the areas to either side to estimate the Odds . In (d) we have no sound reason to expect a Normal distribution and we must not be tempted to use one. If we did we might be led to Weights which would be wholly unjustified. Thus the poor servant who spent 20 years in prison because our forensic scientist misunderstood the Central Limit Theorem.

For instance IF we knew the measurements represent counts of some sort then we know that de Moivre has proved that counting errors must follow a Normal distribution, and then we would be thoroughly justified in fitting such a Normal curve through our existing data points. Further we would be justified in extrapolating out beyond those points to say the cross and using the ratio of the areas under the curve to either side of it to calculate a reliable Odds on some quantity, even quite high Odds if the existing data is good enough to define an accurate Normal curve. This is another example of 'good-simplicity' which follows from real knowledge of the kind of data we are dealing with (in this case pure counts).

Now look at Fig 11:2 (d). It shows the number of measurements of foot-print length made in an earthen flower-bed by a single person jumping on to it many times from a window 127 centimetres above. It looks very like the data in Fig 11:2 (c) and so one might be tempted to fit a best-fit Normal curve through it which can then be used to predict the Probability of measurements way out in the wings, near the cross say – which might be of considerable forensic value. Does this jog the memory? This is a common-place example of *bad-simple*: it looks plausible but it's very very very wrong. There is no reason to believe that foot-print measurements should follow a Normal curve – what about the occasional skid? But what of the Central Limit Theorem which our forensic scientist used as his excuse to fit a Normal Curve? All it really tells one is that as you throw information away, or degrade it by blurring, what is left naturally *looks* simpler and simpler – *but is in fact further and further from reality*. No one of course sets out to deliberately throw Information away or degrade it, but they do try to cut corners – which is often the same thing. For instance they measure only a sample of the data (instead of everything), bin it into crude bins, 'round it off', throw away 'outliers', and so on. The

CLT simply tells you there is a price to pay. In other words there is no such thing as a free lunch. The CLT cannot be used to justify fitting a Normal curve to real measured data. To do so would be no better than fitting a straight line through noisy data as in Fig 11:2(b), and then extrapolating out into the 'wings'. Indeed it might be much worse because it is precisely the values out in the wings which deliver dramatic Odds and thus dramatic Weights. Thus confusing *bad-simple* for *good-simple,* which is all too easy to do, can lead to tragedy. Before he advised the detective the forensic scientist should have admitted to himself that his data in this case wouldn't justify issuing Weights of more than say 4, or less than say ¼, but unfortunately he'd been mesmerised by his misunderstanding of the CLT.

My strong impression is that much of Statistics is based on delusions of a like kind: confusing bad-simple for good. Why else is the subject riddled with Normal distributions (curves) – which are no more normal than straight lines or perfect circles in the physical world. They do occur – but rarely. The human mind appears to look for simple patterns both in Space and Time, and all too often finds them (Mathematics has sometimes been defined as 'the search for pattern'), be they in star-fields or tea-leaves. We have to be continually on our guard against seeing them where they do not exist.

A Normal distribution has tails which rapidly fall to zero as you go out a few standard deviations. Most central distributions in the real world do not have that character; they have 'fatter tails' as we say. Measurement errors for instance must allow for occasional rare events which occur way out in the wings at say 10 standard deviations. The flower-bed jumper may skid; the astronomer may bang his head against the telescope eye-piece, the sample of soldiers being measured may contain a hormonally disturbed giant, all the stocks in a sample of shares may lose value together because of a panic, a massive earthquake can make nonsense of an insurance company's actuarial calculations and so on and so on. Such is real life. Unfortunately the temptation for users of mathematics and statistics to assume that the world is simpler than it really is is often overwhelming; after all their very livelyhoods depend upon it! Caveat Emptor, or 'Buyer Beware' as they say.

Figure (11:2), simple as it is, should never be out of the mind of a thinker who wants to use numbers. *Black Swan* {11} is a wonderful polemic against the misuse of 'Normal Distributions', in the real world.

(11:7) DEATH IN THE SKY

At the height of the Second World War Britain and Germany tried to smash each other from the air. The casualties on both sides were horrendous. For instance of its 125,000 aircrew RAF Bomber Command had 75,000 casualties including 55,000 killed. Losses over Germany were typically 6% per sortie so the Odds of a crew surviving a Tour of 30 ops. were only 1 in 6, and of surviving two only 1 in 40. One Luftwaffe night-fighter crew alone claimed to have shot down 500 Lancasters (7 crew each), which were blind and helpless to any attack from behind and below, and which had so called 'escape-hatches' which were in fact criminal death traps.

To try and counter this threat one squadron of Lancasters were fitted with an electronic device, code-named Monica, designed to detect a night-fighter before it could close for the kill. It worked; several jubilant crews picked their enemies up in time to take evasive action.

After 80 operational sorties the effectiveness of Monica was assessed statistically with a view to fitting it to all bombers. The results proved disturbing. When equivalent squadrons were suffering from losses of 6 per cent the Monica squadron lost 8 crews in only 80 sorties. Their CO, Wing Commander Fox, a veteran of 25, wondered whether Monica was actually giving their positions away. He went to Group for help, and they referred him to Dr. Seeds a statistician working in Operational Research. Seeds wrote back:

"My first reaction was that these casualty data are too fragmentary. However I do realize that for your aircrew this is a matter of life and death, and so I have done my best.

Taking the Null Hypothesis H_0 to be 'Monica does not emit betraying emissions', and assuming that losses are random, I have used Bernoulli's

Theorem to calculate the probability of 0,1,2,3,up to 7 losses in 80 sorties, under the proposition that the losses are 6% under H_0, which is the case for your Group aircraft not fitted with Monica. Then

Prob (0 losses in 80) = Combs (0 from 80)$\times p^0.(1-p)^{80}$

$\quad$ = Combs (0 from 80)$\times (0.6)^0.(0.94)^{80}$
$\quad$ = 0.0071

Also Prob (1 loss in 80) = Combs (1 from 80) $\times(.06)^1.(0.94)^{79}$
$\quad$ = 0.0362

and if I add these last two Probabilities I get the accumulated Probability of one or less losses in 80 Ops of .0433. Carrying on in this vein for up to 7 losses I get the results in my table. You will see that the accumulated Probability for 7 or less losses is 0.8932 or about 89 per cent. Thus there is (100 minus 89) or 11 per cent Probability of getting 8 or more, as you did. Hence my conclusion that 'The Null Hypothesis cannot be rejected at the 10 % level'. In other words there is more than a 10% chance that your rather high losses could be the result of pure chance — with no blame attached to Monica.

Probability of bomber losses according to Seeds:

Probability	Probability	Accumulated Prob.
P(0,80)	.0071	.0071
P(1,80)	.0362	.0433
P(2,80)	.0912	0.1345
P(3,80)	0.1513	0.2858
P(4,80)	.1860	0.4718
P(5,80)	0.1804	0.6522
P(6,80)	0.1444	0.7961
P(7,80)	0.0971	0.8932

You will be disturbed to note how rapidly the accumulated Probabilities (last col.) change when a single casualty is added or subtracted; e.g. in going from 6 losses up to 7 the accumulated Probability of H_0 rises from 79.61 to 89.3, a rise of almost 10 per cent. This is sure evidence that the sample is too small — as we suspected — to be 'significant'.

How much more data would be necessary to arrive at a convincing outcome? Unfortunately these things go as the square root of the sample-size (80 in this case), so you would need to quadruple your sample to 320 sorties to double your precision. And these of course would have to be *new* sorties because the existing data cannot be used a second time after we've already used it to specify the required size of the sample. One needs to specify the stopping point of any test *in advance* otherwise there is a danger of stopping at a particularly favourable/ unfavourable point and so deriving a biased conclusion. All these precautions are as laid down in Professor Ronald Fisher's classic text "Statistical Methods for Research Workers".

I do wish I could be more helpful Wing Commander, but to say more would be to go beyond the reach of legitimate deduction.

Yours sincerely Keith Seeds PhD.

P.S. To get around the very real possibility that the losses are not random, and depend for instance on the nature of the target, then on every raid half the Monica sets should be left on, and half turned off. The off-sorties would act as 'controls'. That would randomise the test — as needed to justify a Bernoulli calculation. But to allow for this you would then need to again double the needful number of sorties to 640.

The Wing Commander was totally bewildered by this letter, and not a little angry. What the hell did a 'Null Hypothesis' have to do with matters and what import did " cannot be rejected at the 10% level" have for his poor crews, all under sentence of imminent death. If things went on as they were not one of them would be left alive after the squadron flew another 640 sorties. Not one.

"Why don't you have a word with our resident boffin?" one of his aircrew suggested "At least he should understand all that mumbo-jumbo."

A navigation expert, Dr Hanbury, who was respected by the airmen because he occasionally flew with them on ops, said:

"Look I'm no mathematician; I'm a glorified plumber. But somewhere I've got my notes on Stats. from my student days. I'll see what I can make of the figures."

Hanbury repeated Seed's Bernoulli computation for Prob(0 losses, 80 sorties), no mean feat with only a slide-rule when it meant computing big factorials like 80.79.78.77.76.75.74.73, not to say $(0.94)^{73}$, but eventually he reproduced Seed's number. But it was totally beyond him to calculate all the rest of Seeds' table.

However, in his undergraduate notes it said that for Qs above 20 (here Q= 80) the Bernoulli distribution closely approximated the Normal Curve of Errors — for which he did have a graph (fig 9:3). The standard deviation σ in the Bernoulli case [Eqn. (b) above]was given as

$\sqrt{Qp(1-p)} = \sqrt{80 \times .06 \times .94} = \sqrt{4.5}$ = 2.12 while the expected mean [Eqn:8a] was Qp = 80 times .06 = 4.8 losses. Thus 8 losses amounted to 8 – 4.8 = 3.2 or a t-value of 3.2 / 2.12 = 1.51 σ s above the mean. He then went to the Normal graph and found there was about 7% of the area under the graph to the right of 1.51 σ leaving 93% to the left. 7 /93 was satisfactorily close to Seed's 10% and 90%, given that it was only an approximation ("Anyway one might expect any approximation to be rather crude down in the tail." he muttered to himself).

"I can't disagree with Seeds," he reported "To within the uncertainties."

"So" the Wingco said "I take it to mean that there's only a 11 per cent probability that our losses are due to chance. Set that against the 100 per cent Probability of getting higher casualties if Monica is a bloody traitor and the Odds on Monica being guilty are 100/11 or about 9 to 1. We ought to switch the bloody thing off!"

"Sounds plausible," Hanbury replied "But I vaguely remember something about 'Priors'. Let me consult my notes again before you do anything."

When he returned he said:

"We've missed something. You want O(H|E) where H is 'Monica is a traitor', and you said in effect:

$$O(H \mid E) = P(E \mid H) / P(E \mid \bar{H}) \approx 1 / 0.11 = 9$$

but that's not correct because that's only the 'Weight' W(E|H) when, according to Bayes' Rule — which is what we have to use when we're doing 'Inference', we should write:

$$O(H \mid E) = W(E \mid H) \times O(H)$$

where O(H) is our Prior, our Odds on H *before* we saw the data."

"I don't follow."

"You wouldn't bet on a horse would you on the basis of a single race, not if it had some previous form?"

"Of course not."

"Well I just rang the people at Malvern who developed Monica. They say she's supposed to be fairly quiet. Their boss reckons the Odds on her giving off easily detectable radiations are 20 to 1 against. That's our Prior O(H) here. So Bayes' gives us

$O(H|E) = 9$ times $(1/20)$ or roughly a half."

"So we're stymied."

"Not completely. Bayes is telling us the Prior is as important as the Weight of evidence here. We can't do much about the Weight — unless you want to wait for another 640 ops."

"We'd all have gone West by then."

"But we can do something about the Prior. We should fly one of your machines over Malvern, switch Monica on, and they'll get one of their sniffer aircraft, tuned to the German night-fighter frequency, to see if they can pick up anything."

The experiment was tried, Monica turned out to be all too guilty, and was never employed again.

"What I don't understand" the Wingco said "Is why that statistician johnny didn't tell us all this himself."

"Good point" Banbury agreed. "I'll be seeing him at a Sunday Soviet next week. Get me half a bottle of Scotch to loosen his tongue and we might find out."

"How did it go Boff?" Fox wanted to know on Hanbury's return.

"Very revealing. Seeds can't take his drink. Was babbling by the end. Seems there are two kinds of statisticians. There's fishery ones like him who don't believe in 'guesswork' — which is what they call Priors. They don't even believe you can use evidence to strengthen belief in *any* hypothesis — because that would be 'Induction' — which Seeds considers a beastly sin. Puritans like him believe that 'hypotheses can only be falsified' — which was what he was trying to do with his so called 'Null Hypothesis.

"And he couldn't even do that could he? What about the other sort?"

"Seeds could barely bring himself to speak of them — said they were 'Bayesians' in a tone of voice implying 'rabid vermin'. They've let down the whole side apparently. They use guesswork. They happily talk about Priors. Worst of all they do 'Induction', arguing from the particular to the general — something no honest mathematician like Seeds would *ever* do. He'd rather be caught with his trousers down in the French mistress's study."

"What about you?" the Wingco laughed.

"Seems I must be one of the vermin — without meaning to be. I spend most of my time having ideas, and then trying to prove they're right — not wrong — although they often do turn out to be wrong."

"Sounds OK to me."

"But that's Induction apparently — something evil I'm supposed to give up."

"Like well ….. you know what."

"Exactly. If I go on doing it, according to Seeds, my teeth will drop out."

"You go on doing it old boy. If I'd listened to that blighter we'd never have tested Monica properly — and we'd all have bought the farm. Let me buy you a round."

"Not tonight Fox. Tomorrow perhaps."

It wasn't to be. A Junkers 88G fitted with Lichtenstein radar and upward firing cannon, sneaked up under the Wingco's Lanc over Kassel and blew him and his crew to smithereens.

The above story is apocryphal only in so far as I had to make up the actual names and numbers — which I couldn't find. Monica existed, it did betray bombers, it was switched off following a critical review. I've included it to illustrate four key points:

(i) The bizarre nature of Hypothesis Testing as preached and practiced by professional statisticians like Seeds — that is to say the great majority. They refuse to use Priors so they can never reach $O(H|E)$ — the very thing we all want to know. All they are offering instead is $P(E|H_0)$ where H_0 is their so called 'Null Hypothesis' ('Monica is safe' in this example). Since I cannot understand their logic — after 30 years of trying — I cannot explain what is in their minds. I suspect they are trying to avoid committing Inference — the very practice, the only practice that is needful in such a case. But if that is true — and I strongly suspect it is — then they have resigned from the world of Common Sense, and can be of no practical help to the rest of us. Indeed they seem to be going further and denying that evidence can ever speak in favour of any hypothesis — *only against.* In other words they adhere to the thesis of Karl Popper that "Hypotheses can only be falsified." I don't know about you but I think it is nonsense.(See Appx. 8)

(ii) The casualty figures could only be assigned Weights once the nature of their errors, their scatter was known. Sure they are numbers but they are probably not *random* numbers, as you would get from tossing a coin. Thus Bernoulli's formula would not apply to them. They were almost certainly determined by the nature of the target, some German cities being more heavily defended than others. Seeds at least acknowledged this and suggested a clever way of randomising those numbers by switching half of the Monicas off. But he still went ahead and arrived at his results by assuming that the existing numbers were random — a necessary pre-requisite to using Bernoulli's Theorem at all.

(iii) And what about the cases where Monica actually worked and crews escaped? How would you fit those into the picture? The truth seems to be that the casualty figures, though suspicious, couldn't carry much Weight – in which Seeds was right. The key insight was Hanbury recognizing that the Prior mattered, and could be tested by experiment.

(iv) The last, and not unimportant point, is that bad Statistics kills people. Statisticians who insist that data cannot be analysed until it reaches some pre-assigned stopping-point (Which rules out earlier data) kill patients, possibly in large numbers.

(11:8) AN ASIDE TO THE ANGRY READER

I expect some readers will find the criticism of Statistics here expressed incredible. I couldn't blame them; I've had otherwise courteous colleagues storm out of the room when I've dared to express them. How could anyone sane question principles adhered to by so many famous minds over the course of the last century? To them it is spitting in the face of Logic, befouling the very alter of the Scientific Revolution. Worse still, so they imagine, it is calling into question their own principles of thought on which their ideas and their reputations are founded. They no more want to hear doubts about those fundamental principles than ayatollahs want to hear criticisms of their God.

We'll return to the Wars of Thought in Chapter 14 "The Extraordinary History of Thinking" but I wanted to say just enough here to try and stay the reader's inclination to throw this book at the wall.

First of all *it's not me*, it's not some private rant[See for instance{14}] Anyone who digs deeply enough into the specialised literature will find far harsher criticisms expressed by this expert statistician or that. For instance here is Professor Edwin Jaynes, doyenne of the Objective Bayesian School (with an agenda of his own) recently criticizing, by turn, the previous Popes of American (Neyman) and British (Fisher) Statistics:

"Neyman, not a scientist but a mathematician, tried to claim that his methods were entirely deductive. For example in 1952 he states '....in the ordinary procedure of statistical estimation there is no phase corresponding to the description of "inductive reasoning"......all the reasoning is deductive and leads to certain formulae and their properties......'...".

Then: "Fisher, possessed of a colossal, overbearing ego, thrashed about in the field, attacking the work of everyone else with equal ferocity. Somehow, early in life, Fisher's mind became captured by the dogma that by 'probability' one is allowed to mean only limiting frequency in a random experiment......his later dominance of the field derives less from technical work than from his flamboyant personal style and the worldly power that went with his official position, in charge of many students and subordinates." ["Probability Theory. The Logic of Science" 2003, Jaynes E.T., CUP, pp 494 to 499].

But the paper that really made me wonder about the whole Statistical mess was written in 1991 by Colin Howson and Peter Urbach, two philosophers at The London School of Economics {6}. They demonstrated by means of simple examples that conventional statisticians could arrive at contrary conclusions simply by making apparently harmless assumptions about how they binned their data or stopped their analyses. They said: "This dependence of significance tests and confidence interval estimates on the subjective, possibly unconscious intentions of the experimenter is an astounding thing to discover in the heart of supposedly objective methodologies." And they were right; absolutely right. It was time to question the whole enterprise. As outsiders Howson and Urbach didn't have an axe to grind. And that's very important. Insiders are seldom able to see the blemishes in their own systems of thought. Such systems have to be seen and attacked from outside. The geologists did everything they could to ridicule Wegener (a meteorologist) and his mad idea of Continental Drift while it took physicists to show up Psycho-analysis for the nonsense it is.

The latest scuffle in the Statistics hen-coop has been caused by a paper written by J.P.A. Ioannidis{ with the dramatic title 'WHY MOST PUBLISHED RESEARCH FINDINGS ARE WRONG'. It argues that where scientists use a scatter-gun approach to looking for positive results [e.g. looking for new drugs] most of their claimed 'discoveries' will prove to be wrong – false-positives twisted by bias into spectacular successes. That would be worrying – if true.

However, if one reads the paper carefully – an un-refereed opinion-piece—you will find the argument, based on simple algebra, is entirely circular i.e. illegitimate. It claims that if there is such and such an amount of bias present then such and such another amount of false claims will be made. True – so far as I can see when I work through the algebra. But the amount of bias assumed is nowhere justified, and it is difficult to see how it could be. All the author can legitimately conclude is that 'foolish or dishonest research workers will make false claims'. But of course. Had the paper been refereed, and passed for publication, its title would have had to be modified to 'WHY SOME RESEARCH FINDINGS IN SOME FIELDS OF BIOMEDICAL SCIENCE COULD SOMETIMES BE FALSE' – because that's all it can legitimately claim. Then of course no one would have read it.

Given that biomedical research workers have been persuaded into using unsound Hypothesis Testing methods proposed by Fisher and Neyman I'm quite prepared to believe that many of their claims are indeed wrong. And of course some scientists *may* be dishonest. But Ioannidis's paper adds nothing solid to the discussion. And it neglects the fact that most scientists, for obvious reasons, are anxious to avoid making public fools of themselves, and will look for independent means to test a promising result before publishing. It illustrates yet again an unfortunate lesson taught us by history; "Statisticians shouldn't pronounce on the Scientific Method", for when they do they mostly make fools of themselves. Unfortunately, behind a cloak of mathematics, that is obvious neither to themselves nor to most of their un-mathematical onlookers. Ioannidis's paper does enshrine a cardinal principle though: Science relies utterly on honesty; even a modest amount of dishonesty can sink it entirely. In his wonderful scientific novel 'Martin Arrowsmith' Sinclair Lewis puts the following 'Scientists's Prayer' into his protagonist's mouth:

"God give me unclouded eyes and freedom from haste. God give me a quiet and restless anger against all pretence and pretentious work and all work left slack and unfinished. God give me a restlessness whereby I may neither sleep nor accept praise until my observed results equal my calculated results, or in pious glee I discover and assault my error. God give me the strength not to trust in God."

Your author, I'm bound to admit, has been by turn a Mathematics groupie, a naive Frequentist, a Popperite, a recidivist Baconian, a Crypto-Frequentist, an undercover Bayesian, a proud Objective Bayesian, a chastened apostate from that same currently fashionable school, a Subjective Bayesian — but all along an ardent but confused pilgrim after truth. At the end of his long pilgrimage all that remains is Common Sense. Yes, it's hard, it's very hard indeed because we're all trying to peer into minds, our own minds, forged by Evolution tinkering with our genes, crisis by crisis, over uncountable generations. We needn't feel ashamed of being confused; on the contrary: *"Men are born ignorant, not stupid. They are made stupid by Education"* so Bertrand Russell maintained, very applicable in this topic. The next section of this chapter will, I hope, convince the reader of just how hard thinking can sometimes be.

Oh yes — "Does it really matter?" You bet it does. Many of the people who've lost their jobs or their savings in the recent financial crash will find it was due to some high powered 'Quant' (They call them 'Astrophysicists' on Wall Street) inappropriately using a 'Normal Curve of Error' in some computer model run by the bank which employed him (or her). Implausible? Ask yourself what would happen to the world if a lot of experts started using 2+2=5. See {11}, {12} and [13].

(11:9) DEATH IN THE COT

In 1999 a mother called Sally Clark was gaoled for life for killing her two successive baby boys a few weeks after each was born. She was alone with each child when it died and what medical evidence turned up in court was confused. Although the judge denied it she was condemned largely on the basis of statistical evidence (there was little other), delivered by a distinguished professor of child-health Sir Roy Meadow who testified that the chance of an educated, non-smoking earner like SC having two successive innocent SIDs ('Sudden Infant Deaths') were 73 million to 1 against.

As is well known she was released following a *second* Appeal in 2003, but only on the legal point, unearthed by a volunteer solicitor, that the Home Office pathologist who had carried out both autopsies had failed to release important medical details to the defence team. The stigma remained that she was probably guilty. She had been vilified by the Press, suffered persecution from fellow prisoners, and was a broken woman. In 2007 she was found dead from alcohol poisoning.

My challenge for you is to reconsider the statistical evidence and come to a verdict of your own — *ignoring Meadow's testimony* — after all he was only a doctor not a statistician. All you need to know are:

There are 650,000 live births a year in the UK.

Approximately 1 in a thousand infants die of SID.

The chance of a second SID in the same family is 5 to 10 times higher than the first.

Thirty women a year are *known* for sure to kill their own children in the UK.

I urge you to spend at least an hour, preferably many more, trying the challenge before turning to the discussion in Appendix A6 (If you do you will learn just how hard thinking can sometimes be — even when the solution looks obvious. All I will say for now is that many august individuals and bodies have pronounced upon the matter, most of them, or so it seems to me, at cross purposes to one another.

SUMMARY

1) Drawing sound conclusions from quantitative data is sometimes extremely hard — we scientists certainly find it so, and often make mistakes trying.

2) Mathematics rarely applies to the real world; it applies to idealized *models* of it. No matter how impressive the mathematics is, any insight if offers must pass through the prism of analogy. Is the model a reliable guide to reality? That may be a question extremely hard to answer. All the more reason for raising it. In particular is an OPEN world being represented by a CLOSED model?

3) Science and Common Sense can usually only deliver provisional, that is to say *probable* answers, because better information can always turn up and it should be able to change our minds. Both are Inductive processes, and not Deductive. Beware of introducing Deduction into areas where it is not warranted. Much of modern Statistics is based on this misunderstanding.

4) Whenever one encounters or uses the word 'Probability' one must be absolutely certain of the meaning implied. Why? Because Inferences that might be sound under one particular meaning may not be valid under another.

5) Collecting and summarizing quantitative data is a valuable craft. But drawing conclusions from it that go beyond Common Sense is a black art. Beware especially of confusing Bad Simple with Good Simple [See Figure (11:2)] Statistics has not made the case that it is privy to insights beyond the reach of ordinary Common Sense. Much of it is based on confusing bad-simplicity for good simplicity. Think babies and bath-water.

6) If advanced Statistics has any accomplishments to speak of they are modest — at best. It usually relies on assumptions and approximations not valid in the very application in which you are interested; for instance The Normal Curve is not normal and strictly applies only to *random* Counting

Errors. If it is fitted to anything else, Measurement Errors for instance — *then beware*.

7) Mathematics on its own can never be used to prove the form of an Error distribution. Other presumptions about the Natural world will be required. This is absolutely fundamental.

8) The Probabilities associated with Measurement Errors can be found in general only by EXPERIMENT OR OBSERVATION. High Probabilities require precise knowledge of errors out in the tails of that particular error distribution, that particular measurement. Since those out in the tails will be rare it may be extremely difficult (expensive) to obtain such Probabilities.

9) A number cannot be used by a serious thinker until it has been turned into a Weight. And that can be very tricky. We have learned that Weights require a knowledge of $\bar{H}$ ("all hypotheses, other than H, that could account for E") as well as H, and so they are doubly fraught. Unless we have a very exact knowledge of *both* we should use the simple PAW Weights: 4, 2, 1, 1/2, 1/4.

10) Common Sense is a process of adjusting ones earlier (prior)ideas so it must start from a first Prior. Beware of any process which claims to do without Priors.

11) If a School of Statistics cannot accommodate Clues with rough and ready Weights, then it is useless to us.

12) Probability Theory is a valuable branch of *Mathematics;* invaluable in games; of limited application to Nature. If you use it, or others use it against you, be *very sure* that its use is legitimate. [Recall the bombing casualties where the data were not random.]

13) We can't assume that the bizarre recipe for Hypothesis Testing proposed by Statisticians is sound. On the contrary. It seems a confused attempt to pretend that Inference can be practiced without Induction. It can't.

14) The Statistics profession exhibits most of the characteristics of a nascent and insecure priesthood: sectarianism, intolerance, unconvincing claims to authority, gurus, obscurantism, self-reference ....

15) In my experience most people who use any more than rudimentary Statistics, including scientists, don't appear to understand it. (If the subject is incoherent why should they?). Most poor devils who practice Statistics have been terrorised into doing so by, in my opinion, illegitimate claims for its power and objectivity. Others apply it as a form of woad — for impressing their colleagues and frightening their opponents. Don't be fooled; it's probably a sign of insecurity.

16) We shouldn't use, or accede to, any statistical argument unless we thoroughly understand it *ourselves*. It's quite likely to be wrong. So we must do our own Statistics! *Statistics is too important to be left to Statisticians.* (In my experience the majority can't even get the Boy/Girl problem right).

17) This is very good news for the rest of us. We can carry on doing our serious thinking using Common Sense alone — without listening for the tread of an approaching Statistician. Statisticians are NOT scientists and one should be very wary of attending to their opinions on the Scientific Method, or Common Sense – which they have not been shy in expressing. Those with a Mathematics background find it hard or impossible to concede that science must rely on Induction and Induction must involve gambling. The habits of thought of a mathematician involve Deduction acting on axioms taken to be self-evident. The gulf between the two modes of thought is enormous to begin with and is emphasised by training.

These last three chapters illustrate, if nothing else, the fallibility of individual human minds, even when they are honest, reasonable, educated and conscientious. In the next chapter we examine how science, as a collective, tries to filter out most such failings.

THINKING FOR OURSELVES (Disney)
CHAPTER 12 (14/9/18)
PERSUASION (4.7 kw)

" .... *Science, so often seen as a way of arriving at clear-cut answers – is itself a process of muddling through to the truth.*" Economist newspaper, 2008, July 5th., p 101

(12:1) INTRODUCTION

We are social animals. In our context that means that reaching a private decision is all very well – but if you cannot convince others then that decision may not do you – or anyone else, much good. Thus thinking and persuading are inextricably linked in human life. Even in science, that supposedly objective activity, the honour often goes not to the one who first had an idea, but to the one who convinced the community that the idea was right. Thus Einstein eclipsed Poincaré, and Darwin eclipsed all his predecessors and contemporaries, including Wallace.

Precisely how and when a scientific discipline changes its collective mind is a big but unanswered question. Science has no formal political or judicial structure, so change is a messy business. As it is individual scientists change their minds, at least in principle, one by one. There is no overt coercion.

Persuading other humans to believe in a controversial hypothesis with possibly fateful consequences is, *and ought to be*, a challenging process. CST has so often been dismissed – by Philosophers and Statisticians in particular – because of its inherently personal (subjective) nature, making it, so they argue, unsuitable for convincing others. One chooses a Prior (which can seem arbitrary), selects ones evidence, assigns each clue a personal Weight, and arrives at one's decision by compounding one's Weights and one's Prior together using the DE. Could anything be more personal and private than that? How on earth can we hope to transform convictions reached in such a private way into public concordance? That's a tough question which this chapter attempts to answer.

(12:2) BEYOND REASONABLE DOUBT

Does the undoubtedly personal nature of CST really disqualify it as a tool for persuasion? I think not. I will argue that there is a model here – Trial by Jury. Each juror makes up his or her own mind on the likelihood of guilt. X thinks the Odds are 50 to 1 on, Y thinks a dozen to 1, Z 1000 to 1. They differ wildly as to the actual Odds on guilt but that doesn't prevent them reaching a verdict. The judge has explained to them the notion of 'Blackstone's Ratio' [8:6] – the conventional definition of 'Beyond Reasonable Doubt' in the legal business. She has told them that if their Odds are higher than 10 to 1 on they should bring in a verdict of 'Guilty'. As it happens all 12 jurors have Odds higher than that so they can bring in a unanimous guilty verdict. It is not of course fool-proof but absolute certainty is not to be expected in a human court of law. If such was demanded then any legal system would be unworkable. {4/6}

Formally speaking there is no such thing as a 'Blackstone's Ratio' in Science, but there could be, and informally speaking there probably is such a 'Tipping Point'. Let's call it the 'Minimum Consensual Odds' or MCO for now. As the evidence accumulates, and as individual scientists assign their own Weights to each clue, then because their Odds combine multiplicatively (using the DE) one can imagine an abrupt transition as a rather convincing new clue tips the compounded Odds of many individual scientists over the MCO at much the same time. Around coffee tables and over the phone the 'beyond reasonable doubt' signal reverberates outward and the community, without saying as much in a formal way, actually changes its collective mind. It would be like tripping the switch which empties a dam.

Note that if the compounding of Odds was additive then that tripping would probably be a hesitant and rather messy business but, thanks to the DE, it is not additive, but *multiplicative*. To see how it might work in practice consider a numerical example. Imagine the MCO was 128 (i.e. 2^7) to 1 on, and suppose a pretty convincing new clue came in which was generally conceded to have a Weight of about 8, then all scientists with previous Odds on the hypothesis of between 16 to 1 and 127 to 1 would be tipped over the MCO boundary into the 'beyond reasonable doubt' category.

It seems to me that this judicial method of collective decision making, when combined with the DE, would give a perfectly sound way for Science (and much else) to proceed. It allows for individuals to choose their own Priors, make up their own Weights and arrive at their individual Odds. And yet they are not prevented from reaching a consensual verdict just because their combined Odds do not precisely or even roughly match. And there is no pretence at certainty – because there is no certainty available. Nor should there be because there ought always to be room for new evidence – if and when it should turn up [That is why any decent judicial system has to have provisions for Appeal].

The one embarrassment is the lack of a formally agreed MCO ['Minimum Consensual Odds' remember] – Science's hypothetical equivalent to the Law's Blackstone's Ratio. Indeed even the Law is rather coy about this vital ratio, which is not widely talked about in public. The Law, when it is sentencing a defendant to life in prison, would rather insist on its absolute majesty than admit that it has reached its verdict on Odds no better than 10 to 1. Likewise scientists, or some scientists at least, would prefer to believe in the absolute nature of a scientific verdict. As we explain more fully in Chapter 14 ['The Extraordinary History of Thinking'] that preference dates from a long-lasting misunderstanding of Science. Up until the 18th Century men could believe that Science was a branch of Mathematics, Geometry to be precise, which *is* capable of delivering absolute proofs using Deduction – but only if the fundamental laws of its world [its 'axioms'] are agreed upon in advance. But in Science we are trying to *find* those laws – we are not free to insist upon them in advance.

The retreat of the 'Absolutists' in Science has been stubborn and bloody. They fought back in the 20th century under the banner of 'Statistics'. They wanted to insist that there was some 'rational' way of deciding arguments which was objective, and not the result of gambling. A sort of messy compromise has been reached where no one actually admits that there is a MCO, or if there is, what its value might be. And perhaps the actual figure doesn't matter all that much because in Science, unlike the Law, the circumstances of the case are nearly always *repeatable*. If there's doubt you can nearly always observe more atoms, more DNA, more galaxies to push the Odds in favour (or against) to more convincing levels – provided you make the effort. Thus CERN wasn't prepared to announce the existence of the Higgs Boson until the Odds on it reached " ten million to 1 ('5 sigma')" although of course that estimate didn't cover Systematic Errors [Chapt. 10)] which could be vastly higher.

Personally I think Science should be less coy about its MCO. On the one hand it would admit to all concerned, as it should, the provisional nature of Science. On the other hand it would deter those silly 'Relativists' who want to claim that all hypotheses, scientific or not, are 'social constructs' and therefore equally plausible. If the MCO were say 128 (2^7) to 1 then any hypothesis which had Posterior Odds of less than that value in the minds of the majority of experts, would not be respectable.

In my opinion MCOs (you notice I do not mention a single 'holy' value) should be the subject of constant and active debate. They might vary from field to field, from issue to issue. They need to be 'in the air' to give controversy a sense of reality. One might argue that the MCO should be higher when the issue is more momentous. But I'm not sure about that. Take Global Warming: if we insist on too high an MCO before we take action we might be condemning most of life on Earth to extinction. That's why the MCO, rather like the Burden of Proof [Chapt. 13], has to be in the forefront of any significant debate.

The MCO seems such a vital issue that surely it has some pedigree? All I can find in the scientific literature is a 'Two-sigma Result' and a 'Three Sigma Result' which are strictly relevant only in cases where the Errors have a 'Normal' (i.e. not normal) distribution [Ch 10/11] and also when the amount of data is limited. In that context a single-tailed '2 Sigma result' has Odds of 40 to 1 against it being the result of chance alone, and is said to be 'significant but not conclusive' while a '3-Sigma single-tailed result' has Odds of 800 to 1 against it being due to chance and is said 'to be believed except in unusual circumstances'. It sounds to me as if there's a shadowy MCO of between 40 to 1 and 500 to 1 operating in such fields. That reinforces my impression that MCOs of between 64 (2^6) and 512 (2^9) to 1 on trip many scientist's switches.[8] In Chapter 9 you might recall my translation of Odds into words:

Significant Odds-on: 2^{10} or more : 'CONVICTION'

" " : 2^6 to 2^9 : 'BELIEF'

" " : 2^4 to 2^5 : 'OPTIMISM'

" " : 2^3 to 2^{-3} : 'DOUBT'

" " : 2^{-4} to 2^{-5} : 'PESSIMISM'

" " : 2^{-6} to 2^{-9} : 'DISBELIEF'

" " : 2^{-10} or less : 'DISMISSAL'

which was and is nothing more than a suggestion. Amend the translation if you like, but try thereafter to remain self-consistent.

P.S. The 'n-sigma' terminology should, in my strong opinion, be dropped because it is context-dependent. For example in astronomy a '3-sigma' result is

---

[8] The International Panel on Climate Change has laid down some wise guidance on how to write about Uncertainty. For instance they translate Odds of 100 to 1 into "Virtual Certainty"

significant in spectroscopy but utterly insignificant in imaging – where you are looking for '10-sigma'. The difference is explained by the number N of independent pixels (picture-elements) in the two cases. In spectroscopy N ~ a thousand and so the number of false-positives using a 3-sigma criterion would be roughly 1, which is acceptable; but in an image, with typically ~ 10 million pixels, it could result in 10 thousand false positive images which would be utterly confusing, and probably more than the number of real sources (e.g. stars) in the image. Thus it is much less confusing to use 'Odds' not 'Sigmas' in describing the significance of any result.

(12:3) SCIENTIFIC DEBATE IN PRACTICE; PEER REVIEW

In practice scientists do not admit to reaching collective decisions by using a Judicial process as outlined in (12:2), nor to the use of MCOs (Minimal Consensual Odds). Most of our decisions are private because we are working by ourselves, or with only a handful of colleagues. And when consensual decisions have to be made they are usually made behind closed doors 'in committee'. A communiqué or paper finally emerges, and what went on behind those doors is nobody else's business. Like most senior scientists I've sat on many many committees and with a few exceptions have found them to work rather well, particularly if the chairman is not a crook and the same evidence is available to everybody round the table – which is not always the case. Some members, often including the chairman, may have a private agenda which they will not declare – then look out for trouble. But otherwise, in my experience, most scientists will make up their minds quite rationally, and not too selfishly, on the basis of the evidence. Some are even prepared to have their minds changed! As a species we're rather good at reaching consensual decisions – when we are allowed to.

Note however that such committees usually have narrow 'terms of reference'. They are formed to disburse monies between projects, design a satellite, or write the paper summarising a particular observation or experiment. They are generally not convened to decide on "Does Dark Matter Exist?" or "Is Big Bang Cosmology credible?" – the kinds of broad issue which drive the whole scientific enterprise and which interest the general public. Such decisions *are* made – for instance between 1920 and 1930 Physics switched from Classical Mechanics to Quantum Mechanics – but *how* they are made is still something of a mystery. In fact they are not made in any conscious fashion – they just 'happen'.

Formally speaking Science proceeds through the publishing of papers in learned journals where they can be read by anybody interested. The process works like this:

The author(s) write the paper and send it to the editor of a scientific journal. The editor dispatches it to one or more 'Referees', supposed experts in the field, who send back their comments. On the basis of those comments the editor will either accept the paper as is (rare), reject the paper, or (most commonly) ask the authors to make modifications (often extensive) before re-submitting. If the authors modify and re-submit there may be a second round of refereeing, even a third, and at some point the frustrated authors may demand a new referee, which the editor may or may not agree to.

This 'Peer-Review' process has many merits, the chief of which is its gate-keeping value. It warns: 'Don't submit shoddy work; it will be rejected.' This is some re-assurance to potential readers of the journal that they won't be entirely wasting their time if they read its papers. This is more true of prestigious journals where the standards of refereeing are generally higher.

The standard of refereeing is however extremely variable – even at the same journal. Some capable and conscientious referees can add immeasurably to both the science and the presentation of a paper and – in extreme cases – save the authors from making public fools of themselves. Others are lazy, incompetent, prejudiced, even vindictive – hiding behind a cloak of anonymity to stab their more productive rivals in the back. Referees aren't paid so the whole business is highly capricious. Terrible stories can be told of careers ruined and ideas stolen. Ultimately, dissatisfied authors can pull a paper from one journal and submit it to another.

Refereeing and publishing are emotive subjects in Science because the number of 'published peer-reviewed papers'(PPRPs) is the benchmark by which a scientific reputation is most often made and a career is launched. Junior scientists who omit to publish regularly will 'perish' – or at least fail to get promoted or even to stay in their chosen profession. At the other extreme are 'turbo-scientists' who pride themselves on publishing a dozen PPRPs a year – which is ridiculous considering that even great scientists have only a handful of good ideas over their entire lives. It is confusing busyness with productivity and suggests a lack of self-confidence. But I suspect that no one was failed promotion for publishing 'too much'.

Refereeing is one form of 'Peer-Review', which is generally a good idea. Even those demi-gods Newton and Einstein made serious mistakes [Newton opined that marine chronometers, achromatic refracting telescopes and man-made flight would all be impossible]. We all need some one, and preferably more than one, to sift through our thinking for errors. But the reality of peer-review is very imperfect. Too many scientists misunderstand the Scientific Method to do it well. And those that do have to squeeze it between a host of other pressing commitments – with sometimes lamentable consequences. And even competent referees are captives of their time, soaked in the prejudices of their age. For example today it would be hard to find a referee who wasn't prejudiced in favour of Big Bang Cosmology. Thus people rarely write papers against BBC because it would be too hard to get them published.

And even when they do get published it's my impression that most astronomers don't have the time to read many PPRPs, not deeply; a handful perhaps per year. So precisely how they are persuaded to change their minds is an interesting question. They talk to colleagues over tea, attend seminars and conferences, they read review articles and semi-popular magazines and are in constant touch by e-mail with collaborators. It's an imperfect, gossipy market-place with many channels of communication. There are far more people than good ideas about, with the result that some bad ideas receive far more attention and linger on longer than they should. On the other hand good ideas are sometimes picked up fast and widely disseminated. It's human and imperfect – what else could it be?

Ironically the general public imagines that Scientific Arguments are far more certain than they often are. They don't understand the limited role of Deduction in Science, the necessity for Priors, the fallacy of much statistical thinking, the subjectivity of Weights, the capricious nature of refereeing and, above all, the difficulty of discounting Errors, particularly the Systematic variety (Chapt. 10). Almost by definition a scientific argument is *Provisional*, it is hopefully the best that can be managed *at that time*. The Odds on it may be high, higher by far than on most non-scientific arguments, but they are nevertheless usually far from certain; they may need amending as new evidence comes in. That has been the fate of many famous theories in the past from the Inverse Square Law of Gravitation to the Immutability of Atoms (before Radio-activity was discovered).

Although Peer Review is sometimes held up as a particular glory of Science it is not a practice of great antiquity, nor is it above reproach, corruption or change. Max Perutz arguably abused his position as a reviewer to 'leak' the work of Wilkins and Franklin at Kings College London to his colleagues Watson and Crick at Cambridge – which may have materially helped them to snatch the structure of DNA and thus a famous Nobel prize. Darwin used his position as secretary of the Linnaen Society to ensure he wasn't pipped to publication by Alfred Russel Wallace. Luckily for Einstein his first Relativity paper was never refereed, enabling him to get away with no acknowledgement of his predecessor Henri Poincaré, as well as permitting him to beat Poincaré into print. King George the Third had to intervene to see that John Harrison, inventor of the Marine Chronometer, wasn't cheated out of the huge Longitude Prize by the Astronomer Royal Nevil Maskelyne and some of the other Prize referees. And so on. Richard Horton the editor of *The Lancet*, the prestigious journal of medical research put it thus: "We portray peer review to the public as a quasi-sacred process that helps to make science our most effective truth-teller. But we know that the system of peer review is biased, unjust, unaccountable, incomplete, easily fixed, often insulting, usually ignorant, occasionally foolish, and frequently wrong."

Today the whole process of refereeing is under challenge from the Internet. Anyone can set themselves up as the editor of an on-line journal with the aim of making money by charging would-be scientific authors. Then there are authors so anxious to 'get out there' that they publish electronically on some website even before their papers have been refereed at all. The excuse is given that they get better 'feedback' from their readers than they would get from any official referees and perhaps they do. My impression is that it has more to do with establishing priority, spiking rivals, and making dubious claims for 'glory'. I know 'hot' papers in astrophysics that have been written even before the observations have been made – to save time in some unseemly race. Recently a collaboration centred on Harvard claimed to have found 'Inflation' in the Cosmic Background Radiation. You will recall 'Inflation' as a notoriously shameless Free Parameter that had to be introduced in the 1970's to save Big Bang Cosmology. Their un-refereed paper was released, press conferences were called, champagne bottles uncorked in anticipation of Nobel Prizes. Insiders were however aware that a much weightier result was anticipated from the European Space Agency six months later. And when it came there was no sign of the 'Harvard result'(Almost certainly a Systematic Error that had been widely anticipated). You can draw your own conclusions; as far as I am concerned it only underscored the precarious state of Inflation and of BBC in general. Groupthink?[See Chapt 13].

My expectation is that eventually the Internet will transform peer-review. The author will launch his/her un-refereed paper out on the net, and every subsequent reader will have the opportunity to attach non-anonymous comments at the end, to which the author can respond. Thus a paper will become an open debate. What more could anyone want? It has taken much longer to come about than it might have done because big money is involved – funding for privately owned scientific journals and for the scientific societies which often own them and depend upon the journal subscriptions.

I don't want to denigrate conventional scientific debate. I just don't want those who rubbish Common Sense Thinking to claim that they have a much better alternative mode of persuasion and consensus.

What so often saves the day in Science is its cut-throat competitiveness. If you come up with loopy ideas or poor data you can be sure your rivals won't be shy to point them out in public. In the long run that is probably a far more effective censor than official Peer Review. None of us want to look ridiculous.

(12:4) A HESITANT GUIDE TO PERSUASION

If you are, or aim to become, a professional scientist you will have to stick to the debating style of that culture, as laid down in the last section, quirky as it may presently seem. Science is far too big a Juggernaut for either you or I to change that style, and it would be foolish to even try. So this section is not for you – it is for the rest of us who must use CST in everyday (non-scientific) life.

I suggest that the best way to persuade others is to lay out as faithfully as possible the very same arguments that persuaded you. They need to know your Prior (and why you chose it), the Evidence you have selected, the Weights you attached to each clue – and your running Posterior Odds. In other words they need to see your Inference Table (IT).

Be sure to include, and fairly explain, the strongest arguments against your hypothesis. It was Darwin's willingness to do that [Chapt 7] which disarmed so many of his potential critics. If your listener finds you've left something unfavourable out they'll only be on the alert for more such lapses.

Don't use Weights more precise than a factor of 2 because the more precise they are the harder they will be to defend. Anyway if the difference between 2 and 2.6 is going to change a decision then it's generally unwise to decide.

At the end mildly explain your own MCO – your 'Tipping Point' – but don't insist that your listener(s) adopt it too. It has to be rather vague, rather intuitive, so arguments over its precise value are seldom productive.

I have found the advantages of this style of Persuasion to be:

(a) It is non-confrontational; you are explaining your own thinking rather than overtly trying to change anyone else's mind.

(b) It invites the listener to construct his own IT and thus to defend his own POV instead of just rubbishing yours.

(c) It avoids wasting time on points where there is broad agreement.

(d) It draws immediate attention to the most contentious (weakest?) points in your own argument – a valuable thing for you to know – after all changing your own mind has to be more important than changing anybody else's. It is surely better to lose an argument than to continue on in folly.

That's all the advice on Persuasion that I can give, and I give it hesitantly because I haven't had as much experience as I would like to test it thoroughly. All one can say for sure is that it is *maximally transparent* – which cannot be bad—and it gives both 'sides' a good chance to change their minds.

(12:5) AFTERWORD

I left the writing of this chapter until last, imagining it would be straightforward. How wrong I was. Many attempts had to be thrown away before some sort of sense emerged.

The problem is the fundamentally provisional nature of Induction (CST) and therefore of Science which is based upon it and is acting as our model for Decisive Thinking in general. For Science to progress we have to presume that most of the scientific knowledge we already possess is right while recognising that our aim is to show that some of that very same knowledge is either wrong – or in serious need of revision. In other words Science has to act as both an Established Church – and a Reformation – at one and the same time. That is not an easy act to perform, or describe, because it involves a fair degree of constructive hypocrisy. If most scientists didn't act as if they believed that most science must be right they wouldn't be able to collaborate to build the new instruments and new theories that are aimed at showing that some of that very same science must be wrong! If they each pursued their own private agenda based on purely personal beliefs the whole constructive edifice would collapse. Yes the habit of 'Provisionality' – which is the central tenet of Science, as of all Inductive thinking, is hard. And for many thinkers it has proved simply too hard. Like children they search for and insist upon a certainty which is not honestly to be found.

      The result of this provisionality is the ill-understood process by which science *appears* to change its collective mind. It is informal, often hypocritical, sometimes pretentious, haunted by history, personality and place, occasionally haphazard and usually difficult to pin down. But that doesn't mean to say that it is wrong most of the time – or that a much better decision-making process could be readily devised. The conventional process is saved, above all, by the *perdurance* of Nature. Atoms, species and stars seldom change overnight. They are there in their vast numbers to be tested and re-tested again and again and again. In the long run they cannot be made to lie. Credible science must be above all *repeatable* and that is an area where Peer-Review comes in. Authors cannot just declare results without acknowledging earlier measurements, or earlier interpretations, which may differ from their own.

A third assurance is the interconnectedness of Nature, which is very hard to fool. Change this measurement or that hypothesis and it will likely have repercussions all over the place. If it does not then the change is not very convincing. It falls into the category of a 'fine-tuning parameter' [a barely justified new Free parameter], or a Just-So story, which Friar Ockham has taught us to despise [Chapt.7].

The weaknesses of the collective decision-making process in Science include its obscurity, its occasional pretence at deductive logicality, and its impotence in the face of a widespread prejudice. Thus The Argument by Design, like Aristotle's 'Four Elements', was infinitely flexible, while today's Big Bang Cosmology and the Standard Model in Particle Physics aren't much better.

I don't pretend that CST is going to get us round 'The Problem of Provisionality' which is fundamental, unavoidable, indeed it is not a problem, but the glory of progressive civilization. It can however help to make the process of collective decision-making more *transparent,* both to those inside the process, and to the general public without. Understanding the necessity for a Prior, seeing the Inference Table laid out with all its Weights (and their adjustments for chance [9:3]), and realising that a collective decision cannot be reached without some MCO (Minimum Consensual Odds) can only help all parties to act in a more civilized and more constructive way. Nothing arouses more distrust than the suspicion that one is being coerced into agreement by underhand or unsound arguments. Laying it all out is not just common sense, it is common politeness. And if you are not prepared to do so then, it seems to me, you don't deserve to win an argument, scientific or otherwise.

THINKING FOR OURSELVES (Disney)
CHAPTER 13 (17.2 kw)
POOR THINKING. (Draft 14/9/18)

"Finding the occasional straw of truth awash in a great ocean of confusion and bamboozle requires vigilance, dedication and courage." Carl Sagan, in 'The Demon Haunted World', Random House, 1991

(13:1) INTRODUCTION

In the past thinking for oneself was too often a forlorn hope. One simply couldn't acquire the information needed to make a wise decision. For instance it more often paid to go to a certificated physician, however inept he was, than diagnose one's disease for oneself. The Internet is in the process of changing all that. Yes a physician, because of his training and experience, ought to know far more than you do about *diseases in general.* But is it likely that he will know more about *your particular disease* than you do after you have put into it the amount of internet research your personal interest in it would justify but which his personal interest would not? It is debateable and we all know cases to the contrary. The general point is that thinking for oneself is becoming far more advantageous, as compared to relying on experts, than it was in the past. However, before we start to make fateful decisions for ourselves, we need to become thoroughly familiar with some of the all-too-many subtle ways there are of thinking poorly – the burden of this chapter.

There are obvious ways to think badly – being prejudiced for instance, narrow minded or illogical. However they needn't concern us so much precisely because they are obvious – in others at least, if not always in ourselves. It is the more subtle ways of thinking badly that we need to worry about the most because they can ensnare us all without us having the least idea of what is going on. If you are not thoroughly aware of the following traps then thinking for yourself could become highly dangerous:

(i) The Fallacy of Falsifiability. (13:2)

(ii)   The Principle of Limited Variety. (13:2) and (13:3)
(iii) The Demarcation Problem (13:3) and (13:4)
(iv)   The Burden of Proof issue. (13:9)
(v)   Tunnel Vision.((13:6) and (13:7).
(vi)   Confirmation Bias. (13:8)
(vii) Mistaking Cleverness for Wisdom.(13:10)
(viii) Misplaced Deference.(13:10)
(ix)   Group Think. (13:12)
(x)   Resurrective Thinking (13:13)
(xi)   Uncritical use of Analogy (13:14)
(xii) Forgetting the Tarantutula (13:15)
(xiii) The Pygmalion Complex (13:16)

These are not simply traps for fools which the intelligent will automatically avoid. On the contrary, as we shall see, these are traps which can ensnare anyone who hasn't been trained to spot them and who isn't on constant alert. Brilliance of mind is not enough to either think well or avoid thinking badly; animal instinct alone is not much help with the subtleties of modern thinking when it has to deal with cultural, i.e. unnatural concepts. As an example consider the famous philosopher Rene Descartes (1596 – 1650). unquestionably a great mathematical mind – he invented Coordinate Geometry, the vital precursor to Calculus. At the beginning of his famous philosophical treatise "A Discourse on Method" he wrote: "Good sense is the most evenly shared thing in the world, for each of us thinks he is so well endowed with it that even those who are the hardest to please in all other respects are not in the habit of wanting more than they have." At first I thought he was being ironic, but reading on one finds he was not. He was arguing that sound thinking must be a matter of learning as well as ability. The real irony is that the method he went on to teach is, judged from a modern point of view, total nonsense. He argued that everything important, including the existence of God, can be

Deduced from a single primary certainty: *'Cogito ergo sum...'* (I think therefore I am). But we know today that Deduction can do no more than re-arrange its initial pre-suppositions. All its conclusions must be implicit in those presuppositions. Thus nothing really new can arise out of it. You will never prove that buttercups are yellow from the laws of Euclidean geometry. That is why Induction is needed to introduce new information into science, as well as into everyday life. But Descartes *didn't know that.* And so, brilliant though he was, he devised a preposterous system of philosophy which is, I believe, still taught in certain establishments. Descartes makes a better warning than a guide.

If I become rather passionate in this chapter it is because I have become entangled in nearly all of these snares myself – and have in consequence lost years, amounting in the end to decades of my working life as a scientist. Nature has bequeathed us no instincts for avoiding these nightmares because they are entanglements of the silken cultural web which we have spun around ourselves, without understanding all its potentially fatal consequences.

We human beings hate to be discovered in bad thinking. Either we won't admit to it, or we blame the unfortunate who points it out. If this chapter makes you angry with the author – then good. It probably means you are learning something useful.

(13:2) UNDERESTIMATING THE SOPHISTICATION OF COMMON SENSE: POPPER AND THE FALSIFICATION FALLACY.

When you show the Detective's Equation to 'experts' on thinking their first reaction is very often "So what: everybody knows that!" But when you ask them for chapter and verse silence ensues – or at best some reference is forthcoming to an aside in an obscure paper where very little was made of it. In fact Common Sense Thinking [CST] is a subtle matter which has evaded some of the very best minds – as we shall now see.

Sir Peter Medawar Nobel laureate in Medicine (for graft rejection) and Director of the National Institution for Medical Research (1962 to 71) was one of the smartest and wittiest scientists to write about the Scientific Method – which he did very explicitly in three books, most explicitly in 'Advice to a young Scientist'. As you will gather from the number of quotations from him in this volume I admire his style and agree with many of his conclusions, for instance: "The generative act in science, I have explained, is imaginative guesswork. The day-to-day business of science involves the exercise of common sense supported by a strong understanding, though not anything more subtle and profound in the way of deduction then will be used anyway in everyday life..." Nevertheless the Scientific Method (SM) he advocated, borrowed as he was proud to admit from Sir Karl Popper, was wildly wrong. Since Popperism has cast a long shadow it is worth analyzing its fallacies in some detail {1}.

Medawar's explication of Popperism begins with the famous Syllogism:

Major premise: all men are mortal.

Minor premise: Socrates is a man.

Inference: Socrates is mortal.

If Socrates turned out to be mortal – which he was – being murdered by the first democratic government in history – that tells one nothing about either of the premises in the syllogism (he might have been a fish). But if the inference is *falsified* by experience then one at least of the two premises must be wrong. Thus there is an *asymmetry* to Inference: falsification appears to be a logically stronger process than confirmation. And it was from that point that Popper's argument sets off: he maintains: 'Hypotheses cannot be proven, but they can be falsified'. Thus his Scientific Method is one of 'Conjecture and Refutation' – generating hypotheses, then looking for evidence to refute them. All we are allowed to conclude then is that the truth must lie among all those hypothesis that have so far not been refuted.

Yes I know it sounds useless and unnatural but Popperism gained a surprisingly wide following, particularly in Britain. So what's wrong with it?

Firstly a Syllogism is a *Deductive* process and, as we have argued, Science and common sense are *Inductive*. Thus the asymmetry of Inference referred to above is *irrelevant* to science and hence to common sense in general. Secondly falsification is *not* the clean process which Popper imagined. To see why consider two examples, one ancient, one modern.

As long ago as 300 BC Aristarchus of Samos suggested that the Earth orbits the Sun – not vice versa. His hypothesis predicts that in that case stars ought to exhibit 'parallax' – that is to say the nearer ones ought to nod back and forth relative to the further ones on an annual basis (as you can see with trees when you nod your head back and forth). For 2000 years astronomers tried to detect that parallax – but even the most careful observations failed. Thus a key prediction of the Heliocentric hypothesis was falsified, and so it was concluded that the hypothesis must be wrong.

Actually (as Aristarchus himself had guessed) the stars are all so very far away from us that the parallax motions were too small to detect with the instruments of the day, and it wasn't until 1848 A.D. that Frauenhoffer was able to construct telescopes of sufficient precision to detect parallax – which was immediately found. So it wasn't the hypothesis in question *but the process of refutation* that was flawed.

In a more modern context you will recall (Chapter 4) that my pet Hidden Galaxy hypothesis asserts that many, even most galaxies, are hidden below the night sky – because it is so optically bright. But if that is true then those Hidden Galaxies should still show up in the radio where the sky is much dimmer. So we built a special receiver and surveyed the whole southern hemisphere with the great radio telescope at Parkes in Australia. Sure enough we found 4000 radio sources, but *every single one of them turned out to have a bright optical counterpart*. Thus my 'Conjecture' was dramatically refuted, my hypothesis apparently wrong. And so I conceded at the time (2005).

Then in 2009 my suspicions were aroused by something I was told at a conference on galaxies held in the high Caucasus Mountains. Valentina Karachentseva from Kiev University told me she had looked at the Parkes radio positions on deep photographic plates and in many cases had found dim (almost hidden) galaxies much closer to those positions than the bright ones selected by the Parkes team (of which I had been a member). Immediately alert I went back over the Parkes identification process and discovered a fatal flaw. We had grossly underestimated the effects of 'Galaxy-clustering' – the very strong tendency for galaxies to cluster together both in position *and* recession-velocity. If hidden galaxies were clustered with bright, as seems natural, it is very likely that close to every dim or dark one there will be a bright one with the 'right velocity' and which looks to be the plausible source of the radio signal – but is not. And my detailed calculations showed this to be the case. Thus the whole process of refutation was in this case flawed and the Hidden Galaxy hypothesis still stood {8/3}.

The general point, as W.V.O. Quine argued long ago, is that in scientific practice refutation is no cleaner than confirmation. It generally requires a *chain* of argument, any link of which might be responsible for an apparent refutation, not necessarily the hypothesis under test. That is why I think of it as *The Fallacy of Falsification*. Many other such examples can be quoted. Popperism is plain wrong.

Popper didn't believe in the Inductive process (i.e. CST) and, in an infamous paper {13:1}, claimed to prove that Probability (Odds) couldn't be used in science. Like many philosophers, mathematicians and statisticians he yearned for the simplistic Yes/No logic of Deduction and thus tried to reject the more powerful More/Less logic of Induction. As far as I can see his Conjecture/ Refutation manifesto is a resurrection of Aristotle's discredited Hypothetico / Deductive approach translated into modern phraseology. But I was persuaded, by Medawar, to think I was a Popperite for more than a decade. Thus I came to believe that a single 'crucial observation' was enough to scupper any hypothesis, whatever other evidence there might be. This entirely undermined my own, and other peoples', confidence in the Hidden Galaxy project.

Peter Medawar's mistake, and mine, was not to know *enough* about the Philosophy of Science. Had he read more widely he would surely have come across W.V.O. Quine and would presumably have rejected Popperism himself. As for me I hadn't read Quine either and was too easily seduced into abandoning my own beloved hypothesis because I didn't question the 'refutation' as thoroughly as I would today.

And had I known of the Detective's Equation at the time that refutation wouldn't have troubled me much anyway because it would have had to overbalance all the positive evidence.

If a scientist as brilliant as Peter Medawar could laud Common Sense and then so blatantly misunderstand it, there is a lesson for us all: Common Sense Thinking is not simple – it is subtle, far more subtle than logical deduction – which can be programmed into a simple computing machine.

In chapter 14 we go into the fascinating and literally incredible history of Thinking.

NB There is a much stronger case against Falsifiability in **Appendix 8** "*Certainty, Falsifiability, and Common Sense*" which you should look at now.

(13:3) THE PRINCIPLE OF LIMITED VARIETY (PLV); WHEN CST WON'T WORK.

Nearly every subject it seems wants to call itself a science nowadays. Since (some) science is manifestly successful and widely respected it is hardly surprising that it should be surrounded by such wannabes. How is the layman to judge the claims of Psychology, Sociology, Economics, Dianetics, Orgonomy, Haemeopathy, Astrology, Osteopathy, Politics, Phrenology, Telepathy, Management, Christian Science …..to be sciences? This is no small matter because illegitimate claims have led to some of the greatest tragedies in history. For instance the Communist Party claimed that The Communist Manifesto was 'scientific', therefore unarguable, giving it the right to eliminate or re-indoctrinate anyone who disagreed with it or stood in its way. Thus tens of millions of poor devils were sent to their dooms in gulags across the Soviet Union, China and Cambodia. And on the mundane level we are daily faced with prodding to purchase this service or that product because their effectiveness is 'scientifically proven'. Thus nobody can afford to ignore the 'Demarcation Problem' – when is something a science and when is it not? Think of all those spas which made a handsome living out of selling filthy-tasting water to gullible people who had no idea what the

Demarcation Problem was. No less a person than Charles Darwin, whose later life was plagued by illness, spent much time and money in such spas.

As you might expect, our approach to this problem will appeal to the Detective's Equation. A science establishes general hypotheses (laws) on the basis of limited evidence. It does so by gambling on that evidence, eventually hoping to find Odds $O(H \mid E_1, E_2 ...)$ so high in favour of (or against) H that nobody sane could disagree. The mechanism for so gambling is the DE whose secret is its *multiplicative* nature. Thus a limited number of clues with modest Weights, but mostly working in the same sense, can lead to Inductive Conviction, i.e. Odds high enough to take action. But there is an important proviso here in the form of the Prior Odds O(H). Unless these can be assigned a finite form no amount of evidence can overcome them, and there exists no possibility of reaching a decisive Posterior $O(H \mid E_1, E_2 ...)$. Take as an example the interpretation of a dream. In the dream you are playing the piano, with your fiancée turning the pages. A lion enters the room, eats your fiancée, then turns into your Aunt Mary, who you hardly know. What on earth does it mean? That you really fancy Aunty Mary? Or that your fiancée resembles her in some unpleasant way? Or that Mary is trying to warn you ? Or that she has evil designs on your marriage? Or that you had strong cheese for dinner last night? Or that you mixed your drinks? The possibilities are literally endless. You could go on writing them down almost for ever. But let us say, for the sake of argument, that there are just 1 thousand possibilities. The Prior Odds O(H) on most of them therefore cannot be more favourable than a thousand to one *against*. Now ask yourself the question; 'Is there any probability of assembling evidence on any one of these hypotheses, whose combined Weights, when multiplied together, would exceed say 10,000 to 1, for only then could a Posterior value of $O(H \mid E_1, E_2 ...)$ of 10 to 1 on or better be reached when they are used to multiply the Prior. I would submit that such a probability is vanishingly small. That being so the field of dream interpretation cannot be a science! And yet that is how Sigmund Freud started off to pioneer the field of Psycho-analysis and Psychology{4} . And Jung's ideas of the Collective Unconscious had, if anything, even less support. They both wanted to be scientists, and huge numbers of educated people regarded them as such; but they were not, not by a country mile.

The general point is that the DE won't work unless there is a roughly denumerable, finite number of hypotheses to be tested so that the Prior O(H) can be comfortably overcome by the likely evidence. John Maynard Keynes, who seems to have worried about this most in modern times, called it *'The Principle of Limited Variety'* (PLV). Induction will only work in situations where the hypothetical possibilities are limited. Those branches of science which seem to work rather well are of this kind. There are for instance only 92 kinds of atoms – or Elements – which makes Chemistry possible. Had there been 92,000, as there well could have been, then Chemistry might have defeated us. And common atoms are much less in number – which greatly restricts the number of common compounds (molecules). Proteins are made up of only 20 amino acids; DNA has only 4 bases and so on. The sciences where we have had the most success do indeed exhibit Limited Variety: Electromagnetism has only 4 governing equations, as has Fluid Mechanics. But it does not follow that all physical phenomena will be so limited in their nature that they will be amenable to Compound Induction. One thinks of the medical field with its innumerable proteins, compounds, cells, bacilli, viruses and so on. And what about brain research? The human brain contains ten billion neurons, each connected to others by about one thousand dendrites. There is no good reason to assume it will yield to science however big Neuroscience's ambitions and resources might be. Outwardly, with all its machines, it might seem very like a science – but is it really? Don't be taken in by externalities. Could the DE be made to work in that almost infinitely complex field?

(13:4) WHEN A SCIENCE IS NOT A SCIENCE: ECONOMICS.

Moving away from the Physical sciences towards the Social Sciences so called, the difficulties become greater still, perhaps insuperable. I'm now going to concentrate on one such field, Economics, to examine the challenges. I've chosen Economics because I know something about it and because it claims to be a Science – the 'Dismal Science'

{2}. Economists even win Nobel prizes *of a sort*. [They don't come out of the same stable as the science prizes, are not comparable, and are something of a dangerous sham.{11/11}]

For Economics to claim that it is, or could become a science, it must demonstrate that the Principle of Limited Variety applies to it. But how could it do that? Take the recent financial crash of 2007/8. Practically nobody foresaw it, but dozens of books and thousands of learned papers have been written about it since, pointing to different culprits which *include*: greedy bankers, toxic mortgages, opaque financial instruments, over-leveraging, vast international imbalances (China saving versus US borrowing), auditors in cahoots with the companies that paid them, Fanny Mae and Fanny Mac (you don't want to know), the scuppering of the 1944 conference on international banking at Bretton Woods, Nixon refusing to back the US dollar in the aftermath of the Vietnam War(1971), over-saving, poor wealth distribution, flash trading, inadequately financed pension funds going in search of unrealistic returns, poor or non-existent supervision of the system by financial supervisors, the Euro, hubris following the collapse of Communism, a naïve belief in 'perfect markets', the inappropriate use of 'The Normal Distribution' by financial 'Quants', insurers ignoring the possibility of correlated market movements, extremely foolish advice given by the actuarial profession, dishonesty on the part of politicians willing to buy votes by offering unaffordable utopias and raising government debts, house owners foolishly believing they were rich because house prices were rising.....and so on and so on. When I read and try to understand the various hypotheses, they all carry a degree of plausibility to me. Moreover they can interact with one another in a whole variety of plausible and dramatic ways. Thus H_1 plus H_2 plus H_3 is really a new hypothesis H_{88}, so the hypothetical possibilities are uncountable— or nearly so – abrogating the Principle of Limited Variety.

There is another way to look at the matter. Imagine that Economics *is* a science capable of generating sound principles and accurate predictions. Suppose that it predicts that farmers will make more money from selling beef than selling milk. Then smart farmers will switch from dairying to beef production. Through scarcity the price of milk will rise; through oversupply the price of beef will fall. The very prediction of the allegedly sound Economic theory has proved to be *self-defeating* ('reflexive' in the jargon). And it seems to me that any 'science of human behaviour' would be self-defeating in the same way.

Because a subject is not a science doesn't necessarily make it worthless. Take painting, one of my hobbies. There are no hard and fast rules on how to paint; nevertheless there are rules-of-thumb which it is generally unwise to ignore because when you do so you usually end up with paintings which satisfy neither you nor your 'customers'. At the very least it is better to know those rules than to be ignorant. When one of my paintings has 'gone wrong' they sometimes help me to diagnose the problem and put it right.

In my opinion Economics is not a science, and never can be, but it is probably better to know its rules of thumb than not. For instance 'Discounted Cash Flow Analysis' is a useful accounting technique we could all learn with benefit when we are at school (if we have the necessary maths). It won't make us 'winners' but it might save us from becoming unnecessary 'losers'. It's a useful tool but not a branch of science. Having it to your hand will help – but not turn you into either a master-craftsman or a millionaire.

Banking is one of the most valuable – and yet most mysterious activities devised by Man. Imagine trying to live without it. One couldn't borrow to buy a house say or save for one's old age. Life would literally be hand to mouth. At the same time banking is of its nature fundamentally flawed because a bank borrows 'short' (from its depositors) and lends 'long' – to its borrowers. If all the depositors take fright and demand their money back at once (a 'run on the bank') the bank won't be able to oblige – unless it can quickly borrow large sums of money from somewhere else – usually the government in the last resort. So banking is really a huge 'confidence trick' – which works to the benefit of nearly all of us, nearly all of the time. It's like Induction in that regard – a form of gambling we could never do without. It is valuable but risky, and we should never allow ourselves to forget that. When we do we can run into big trouble as we did in 1763,1772,1792, 1796,1819,1825, ...1884,1890,1896,1901......1929....2007, crashes which ruined lives and changed the course of history. And worse, far worse, could lie in store. The whole international banking system presently runs on a wing and a prayer. Countries can run up huge debts, and what is worse – huge surpluses. Governments can fix their currencies, and do fix them, to beggar their competitors. In other words it is a complete shambles.

Surely this is not an argument for us to give up on trying to regulate financial affairs; on the contrary! Temporary fixes can probably be bodged together from time to time, and an honest attempt to do so was made by John Maynard Keynes at the Bretton Woods conference in 1944. Unfortunately the US delegation, headed by John Dexter White (who was probably a KGB agent) – jibbed at one essential condition – that countries in too much financial surplus– as the US was then – would have to pay penalties – as would countries in too much debt. Selfishness and short-termism left us with the dangerous, and it is dangerous, shambles we somehow survive in today. The idea that there are brilliant scientists out there who can understand it all and can redesign it permanently, after all they have 'Not-the-Nobel-Prizes' in Economics, is a complete delusion. They're *not* scientists – it's *not* a science, and it never can so be. Mankind has to find the resolution, the wisdom and the enlightened self-interest to try and re-adjust international banking, from time to time, before it takes us all down into the abyss. Only the Second World War got us out of the Great Depression. If we don't recognize the danger, and it's an unnecessary danger, of trying to continue on a wing and a prayer, then we will crash and burn – it's only a question of when. In banking there is and has to be a 'lender of last resort' – usually the government. That's what keeps the magic going. In international banking there is at present no such agency with the means to save the day. International imbalances have to be regulated in some way, and governments running huge surpluses are not to be admired but punished – or at least penalised severely. International trade can create enormous wealth – but is capable of destroying it too, and has done so many times in the past.

I'm not a financial expert (I'm not convinced that anybody is – yes there are winners and losers – but that may all be chance). The point of this digression into finance is to recognize that it is no bad thing to realize that Economics is not a science. On the contrary. It is no great turbine running on diamond bearings that engineers have designed and can control with certainty; it's a dodgy business desperately dependent on wisdom, luck, unselfishness, regular supervision, and occasional re-design. Honesty is probably far more important than 'smarts' while recognition of the vulnerability of the whole rickety machine is a necessary precondition to getting round a table to try and keep it going. As it is many of us are being ripped off much of the time.

The folly of supposing Economics to be anything like a science was illustrated by the Long Term Capital Management fiasco. LTCM was a sort of US investment fund based on 'The Black-Scholes Equation' and fronted by two 'not-the-Nobel-Prize winners'. Set up in 1994 it at first earned stellar returns (40% per annum after fees). But after haemorrhaging over 4 billion dollars in 4 months in 1998 it became a danger to the entire international financial system and had to be bailed out by a consortium of big banks organised by the US government. Some estimate that it cost society a Trillion dollars, or 3,000 dollars for every man, woman and child in the United States. {*}

If you still have any illusions about Economics you should read *'Capitalism and Freedom'* by Milton Friedman "the most influential economist of the second half of the 20th century" and a ' Not-the-Nobel Prize' winner to boot.{3} To call it 'naïve' is far too polite – it is counter factual and foolish. He confuses 'free enterprise' with 'freedom' and ……well read it. Either he's not an economist or Economics isn't a science.

(13:5) PSY*******Y : WHY IT CAN'T HELP US TO THINK BETTER.

How on earth could one write a book on Decisive Thinking without incorporating extensive knowledge from the fields of Psychology and Psychiatry. Psychology is defined to be 'The study of the mind especially as it affects behavior' while Psychiatry is 'A medical (i.e. certificated) specialty devoted to the study, diagnosis, treatment and prevention of mental disorders'. My answer is: "Quite easily because we describe CST in terms of an *Algorithm* – the DE, which cannot depend on the underlying hardware (in this case the human brain) on which it runs." That is just as well because both Psychology and Psychiatry are highly controversial subjects and certainly not sciences.{4,5 &6}

An algorithm is a step-by-step procedure for carrying out a calculation – in our case for arriving at the Odds on some hypothesis. It can be described in human language because it operates at a far higher level than a computer language and can run on a variety of switching systems, provided they have sufficient memory and power. You can use an algorithm without having to know anything about electronic circuitry or physiology. An algorithm is analogous to a recipe. The recipe for Beef Stroganoff is the same in English or Russian, and the dish can be cooked using gas, electricity, wood or coal. Likewise I would argue that CST can run in a bird's brain as well as in our own even if the precise physiological mechanisms are entirely different. *Our fundamental assumption is that CST is a survival mechanism that developed under Evolution and no species can afford to run a less than optimal algorithm* – subject to its own limitations in memory and processing capacity.

The human brain certainly does have odd irrational quirks: like children we sometimes believe what we want to believe, rather than what the evidence supports; or again patients will undergo an operation with ' a 90 per cent survival rate' but refuse one with 'a 10 per cent death rate'. Even so such quirks cannot be all that important because, in the long run, they would have been eliminated by the cruel sword of Evolution. So worrying about odd psychological quirks must be of marginal interest. And that is just as well because Psy*****y has an appalling record (I use ***s to denote both fields). We need to go into that somewhat in order to defend ourselves from those who would accuse us of neglecting their subject.

We've touched upon the difficulty of Psy*****y in connection with the interpretation of dreams; Psy*****y there ran slap bang into the Principle of Limited

Variety (PLV) – there are usually too many plausible hypotheses for CST to be of much help. This is a tragedy because we really need to know how the mind works, if only to treat those of us who become mentally ill (about one third) during the course of our lives. But aspiring to understand – and succeeding in doing so – may lie continents apart. If, because of the PLV, we cannot advance using CST, then we have no means of crossing the ocean in between. No matter how sophisticated our machines, no matter how sublime our biochemistry, if we cannot eliminate uncountable hypotheses we can't even leave the beach. The brain has ten billion neurons, each with about a thousand dendrites reaching out to make connections and fire synapses all over our heads. It seems to me there is no compelling reason to believe that we will ever 'solve' the human brain. Consider an infinitely simpler electronic computer working steadily away on some physical problem for days. Suppose an army of technicians, mathematicians and scientists was funded to work out what it was 'thinking about'. After a millennium they succeed in tracing every single electric pulse in every transistor in the system from start to end of the long calculation. Then what? They set out to try and guess what that calculation was about. They could do so by trying out many different calculations to find out which, if any, reproduces the pattern of pulses mapped in the machine. And let us say that after a million trials occupying 10,000 years they finally succeed. They find that their computer has solved a certain Elliptic Partial Differential Equation called Poisson's Equation. But that is as far as they will ever get because there are a large number of physical phenomena [see (15:10)] described by that very same equation. For instance there is no way of ever finding out whether the machine was working on a problem in electrostatics or in gravity. This an example the so called 'Mind-Body Problem': deciphering the body is never to decipher the mind living in it (and vice versa). When I hear some neuro-scientists talking about the success they anticipate – if only we give them enough money – they sound exceedingly naïve to me, exceedingly. To quote the famous neuro-surgeon Henry Marsh "Despite the gargantuan scale of modern neuro-scientific research our basic understanding of how the brain works, and how the electro-chemical activity corresponds to thought and feeling, remains deeply inadequate and incomplete."

Now for a little sad history. Let's start with Sigmund Freud (1856 – 1939) "the father of psychoanalysis". As a young doctor in Vienna he diagnosed the effects of sexual abuse (often by family members) on many of his young female patients. But when he tried to publish he was met by the stern disapproval of his senior medical colleagues and so, in order 'to get on' he reversed his diagnoses and attributed their behavior to 'hysteria', 'transference' and 'wish fulfillment' {4,5}. In so doing he sentenced generations of young people to the guilt of having been abused. If this accusation is true – and it's in his private diaries apparently – see Masson {6}, this makes Freud one of the greatest monsters in medical history. In any case Freud went on to fame and fortune as his invented therapy 'Psycho-Analysis' spread around the globe like wild-fire while his phraseology ['e.g. 'inferiority complex', 'unconscious mind' etc.] became buzzwords among the fashionable intelligentsia. The problem was that Freud had no inkling of the Scientific Method and so he never tried to find out whether his methods were actually effective. Yes some of his patients recovered after many expensive years 'on the therapist's couch' but no more than would have recovered spontaneously – as tests were to prove 50 years later. It was the Snake Oil Delusion SNOD once again – operating on a colossal scale. According to a curator of the Freud museum in London, the only people who take Freud seriously nowadays are "members of the liberal arts faculties in universities".

Our next villain is Sir Cyril Burt (1883 to 1971) the first knighted psychologist and professor of that subject at University College London. Burt was entrusted to analyze the IQ tests of identical twins separated at birth by the turbulence of the Second World War. Such twins are extremely rare (only a few dozen were located across all Europe) but are potentially very valuable for distinguishing between the effects of heredity and nurture. Burt's results were unequivocal: 75 per cent of IQ – whatever that means – was inherited. On the basis of earlier research by two of Burt's assistants, who turned out to be fabrications of Burt's diseased mind, sweeping changes were made to the British educational system – the notorious 'Eleven plus exam' was introduced to select those few who were worth educating from the many more who were not.

Much later alas (1974) and after his death, Burt's figures were found by an American researcher to be 'far too good to be true'. Burt evidently didn't understand 'scatter in data' [Chap. 9]. Increasing numbers of twins ought to have led to more *absolute* scatter but less *relative* scatter – but Burt's figures did not. His ignorance convicted him as a fraudster – but only after the great harm was done. Ironically, modern figures tend to support Burt's 75 per cent figure for the inheritability of IQ. But so what? IQ is no guarantee that you will be able to think or decide in a more common-sense way – nor would you expect it to be. In a phrase 'IQ is not Common Sense'.

Our last villain Dr. William Sargant (1907 to 88) was for quarter of a century Director of Psychological Medicine at the prestigious St Thomas' Hospital in London. As a very young man I spent 2 years National Service as an infantry soldier in the British Army. During our first few months of basic training we were so effectively brain-washed that I'm quite certain that if ordered to do so most of us would have proudly leapt out of a trench into a hail of machine–gun fire as so many did in the First World War. That instilled discipline, that 'esprit-de-corps' is the key weapon in any fighting force.

It was only after I left the Army that I began to question how it had been done. How had I so completely surrendered my free will to a barking corporal on the drill square? I felt ashamed and needed to find out – if only to make sure it never happened again.

Eventually I came across a remarkable book ' Battle for the Mind' by William Sargant who claimed to be an expert on 'brain-washing' then a hot topic after so many US soldiers had succumbed to it as POWs in North Korea. Sargant ascribed our all-too-effective indoctrination to 'conditioned-reflex' a phenomenon investigated originally by Pavlov in the training of dogs and circus animals. It's all about the 'immediacy' of reward and punishment, as opposed to its quantity. I was convinced by

Sargant's book. I felt everybody ought to read it as an act of self-defense. Did you know that if you can be induced to lose your temper you become far more susceptible to persuasion?

All would have been well – but then the Internet came along. To my horror I found that Sargant had been some sort of a monster who had inflicted a great deal of harm among his many patients (see Wikipedia). Despising psychotherapy he believed in more dramatic physical interventions including psycho-surgery, electro-convulsive therapy, deep-sleep treatment and insulin-shock therapy. In doing so he relied on personal opinion and dogma rather than on clinical evidence. He was even prepared to hoodwink his patients into undergoing drastic and now discredited treatments when they were too drugged to understand the possible consequences. Many felt afterwards that he had ruined their lives.

Every profession has its rogues and the existence of three such prominent villains does not, of itself, prove that Psy*****y as a whole is discredited. Mental disease can cause so much suffering that the temptation to intervene, even when the outcome cannot be accurately foreseen, must be overwhelming at times. Certainly the 'safer' alternative of doing nothing cannot always be recommended as more humane.

What I do criticize Psy*****y for is letting 3 such prominent rogues get away with their malpractices for so long.[9] All three 'died in their beds' without being exposed. That suggests to me that the methodology underlying Psy*****y is at best very weak. It occasionally works on the 'suck it and see' principle (e.g. Prozac) but it doesn't have, and perhaps never can have, a strong theoretical basis – on account of the PLV. It's rather like banking – a tool we badly need but which we must regard and regulate with great suspicion. Banking has its spectacular rogues too – like Bernie Madoff who

---

[9] Things do not appear to be getting any better. In 2012 Diederick Stapel a Dutch professor of Social Psychology was caught out for having published at least 55 academic papers and supervised 10 PhD theses based on fraudulent data he had 'made up'. The 3 investigating committees found many bad practices in general use in the area including: omission of inconvenient data; the re-running of experiments until the 'right' result turned up; misunderstanding statistics; private(unshared) data……

stole 65 Billion (!) dollars before being apprehended.

I rest my case. Psy*****y is quite definitely not a science and I don't believe it has much to teach us about CST{5,6}. Just because Psychologists like to think of themselves as scientists, and write books with titles like "*How the Mind Works*" doesn't mean we have to believe them. In my opinion they are as naïve as the Phrenologists who, not so long ago, claimed they could tell whether a man had committed murder by feeling the shape of his head. Ugh! If you think I've been too harsh read Freud's famous book '*The Interpretation of Dreams*' {4}. I'm damned if I can find an ounce of CST or science in it.

(13:6) TUNNEL VISION I: MISSING ALTERNATIVE HYPOTHESES AND AFFIRMING THE CONSEQUENT

is one of the most frequent causes of failure in science. The astronomical 'experts' clung to their narrow expertise in epicycles and so denied the Heliocentric truth for centuries [6:3]. Evolution seemed impossible so long as geologists ignored the idea that the Earth's interior might be heated by its own radio-active minerals. Present day astrophysicists cling with increasing desperation to 'Dark Matter' and 'Dark Energy' when the observations might have quite other explanations.

And it's not just science which suffers from tunnel vision. I was recently watching a heated debate on Scottish independence. Absolutely no one mentioned the inconvenient fact that because of its immensely long coastline (6000 miles) Scotland on its own is virtually indefensible. Were Russia say to land a division on the lower Clyde who but the rest of the Brits could or would save it from enslavement? *It is far easier to spot something wrong that is present in an argument than to spot something right that is missing.*

Because

$$W(E \mid H) = P(E \mid H) / P(E \mid \bar{H})$$

there are two ways to get the Weight W(E|H) of a clue far too high: to overvalue

P(E|H) or to undervalue P(E|$\bar{H}$) the Probability of the alternative to H, and in particular the probability of chance. Since the latter is all too easy to do, and all too difficult to spot, it is the more pernicious of the two. Thus we start with that and tackle the alternative in the next section.

We have encountered several examples of undervaluing P(E|$\bar{H}$) already:

(i) The Tiptuppititis sufferer (4:8) felt it unlikely that his cure (E) could be spontaneous and not the result his Organocalm treatment (H) so he found a Weight

$$W(E|H) = \frac{P(E|H)}{P(E|\bar{H})} \approx \frac{1}{small} \approx large$$

and argued (in a circle) that therefore O(H|E) was much larger than O(H).

(ii) The prosecutor in the student murder case [Chapt 8 and Appendix 2] automatically assumed that P(E|I) – the Probability of the right blood-group (E), if the accused was innocent (I), was so low that the Odds on Guilt:

$$O(G|E) = \frac{P(E|G)}{P(E|I)} \times O(G)$$
$$= \frac{1}{very\ low} \times O(G)$$

and hence that O(G), which is actually very low(Millions to 1 against), could be ignored (the 'Prosecutors Fallacy').

(iii) In the Sally Clark case [Chapt 11 and Appendix A6] the Judge failed to point out to the jury that that they should assume that P(E| Innocence) should be 1 because she's only in the dock because she has two dead babies (E). But in that case

$$W(E|I) = \frac{P(E|I)}{P(E|G)} = \frac{1}{1} \text{ in which case O(I|E)=1} \times \text{O(I)}$$

which is very high because the Odds on a random woman being innocent are huge. She's innocent so far as the evidence is concerned.

There was a very similar miscarriage of justice in the Netherlands when a children's nurse Lucia de Berk was found guilty of multiple murders in 2004 and sentenced to life in prison because it seemed statistically unlikely that she could have been present at so many unexpected deaths. But of course improbable things happen all the time! Fortunately she was eventually set free in 2008 though again, like Sally Clark, on grounds other than poor thinking. We are all reluctant to own up to that. There is something of : "Constable Jones arrested the prisoner in the dock because of his low forehead and slitty eyes. And just look at him!"

(iv) The lady with the apparently cancerous woozle (4:6) grew unnecessarily alarmed because she assumed that P(+test| not ill) was so low that

$$W(+test \mid ill) = \frac{P(+test \mid ill)}{P(+test \mid not\ ill)} = \text{very high}$$

so high indeed that it would overcome the Prior Odds O(ill) which are actually very low. And in this she was fortunately quite wrong.

When we first encountered this widespread error in thinking [Sect.4:8] we christened it 'SNOD' standing for 'The Snake Oil Delusion'. On reflection that was unfortunate because it implied some mental slight-of-hand imposed upon us by a malicious fraudster. But in none of the cases instanced above was malice involved. This fallacy is far more pernicious than that *because we practice it so regularly upon ourselves*. Logicians call it 'The Fallacy of Affirming the Consequent'. I have been among whole lecture theatres of learned professors falling for it like young children entranced by a fairy tale. Because hypothesis H implies consequence C, and C is observed (i.e.'affirmed') does NOT imply that H must be true. Why not? Because any number of alternative hypotheses ($\bar{H}$), some not even dreamed of, could be the cause of C instead. The learned professors are not fools – or generally not more foolish than the rest of us – but still they fall. Here is a typical example. Professor X gets upon the stage and describes an elaborate computer simulation he has just made of how galaxies form in the early universe. Then he shows some dazzling coloured slides, or better still movies, of his hypothetical galaxies co-mingling out of the primeval dark. The audience is wowed, some even break into spontaneous applause. The Truth at Last – now we know! Dark Matter really does exist!

But hold on. Wait a mo'. X has had to make a dozen or two assumptions (hypotheses $H_1, H_2, H_3...$) before he wrote his colossal computer program – for instance all the laws of dynamics he has used and the properties of his 'particles'. Even supposing he's got his sums right (by no means always the case), all his simulated 'galaxies' demonstrate is that such constructions are not incompatible with his particular hypothesis set – as a whole. (He probably wouldn't have been showing his slides if they were not). More crucially they do not demonstrate that such plausible constructions would not arise from alternative and very different hypothesis sets – unless he had computed them *all*. [See '*Computer Modelling and its Pitfalls*' on my website, if you are interested].

So why do we (all) consistently fall for this fallacy – 'Affirming the Consequent'? I wish I knew. Partly wish-fulfillment I suppose – we all want to believe an attractive fairy story. Partly lack of imagination—and I don't mean that insultingly. How could the ancients have imagined bacteria, or we moderns anticipated AIDS or the Internet?

Personally I wouldn't want my children to go into the sea before they could swim – or venture into adulthood naïve enough to Affirm the Consequent.

Don't assume that flawed thinking of this kind is either rare or inconsequential. Millions of us are getting, or will get, dramatically lower pensions than we have saved for because of just such muddled thinking by pension-fund managers.

Because we are living longer than our parents a Defined Benefits Pension Scheme is likely to be in deep trouble today and, to meet its obligations, must look for its investments to return 6 to 8 % annually. This is more than twice the historical average and so can only be reached either by taking serious risks – or by believing in magic. The 'risky' tactic was investing in bank shares which were returning 30 % on equity in the days of crazy leveraging (15:4). When those shares inevitably sank governments mostly bailed the banks out, very largely using the money of those not fortunate enough to have a private pension (which strikes some as unfair). The alternative, i.e. magic, came in investing in Hedge-Funds run by Managers who claimed that their wizardry could get returns way above the average 3 %. Any basis for this had to rely on past success which – as all stock-market surveys show – is no guarantor of future success – but was almost certainly luck. But a lot of pension funds fell for it. Now here's the horrible sting. The Hedge-fund manager will typically charge '2 and 20' – a 2 per cent annual fee on your investment *and* 20 per cent of any profit. Since they can't expect more than 3 percent return on average *they're skimming off two thirds or more of your returns*. There are hedge-fund managers making Three Thousand Million dollars a year – for doing nothing which the office-boy couldn't do. Enjoy your retirement – if you can afford to.

The more I learn about 'hard-headed business executives' and 'hard-nosed money-men' the more I have to laugh. The crazy towers you find in downtown New York and London are the outward symptom of the craziness going on inside. It's largely science, technology and innovation that create new wealth (potentially up to 7 per cent annually) – only a fraction of which finishes in our pockets. My advice, when it comes to money, is 'Do your own thinking'. But it's too late for me now; some incredibly rich Hedge-fund guy has no doubt used my savings to purchase the cocktail cabinet on his super-yacht. Serves me right: I should never have relied on a 'financial expert' to do my thinking for me.

(13:7) TUNNEL VISION II : FAVOURING PET CLUES ; ANIMAL WISDOM

The easiest way to make a huge fool of yourself in science is to overweight a single clue, and we have touched upon several such instances already. Thus Sir Harold Jeffreys, who was really a mathematician, came to think of himself, and was accepted by others, as a world authority on Geophysics, bringing out numerous editions of his weighty book 'The Earth' from 1924 onwards. At an early age Jeffreys fixated on a single 'fact': rocks are solid. He was then able to 'prove' that continental drift was 'out of the question', an opinion he stuck to for over 40 years. The truth, as anyone who has walked on a Pembrokeshire beach or been to the Alps can see, is that rocks can fold dramatically – given enough time. On the long timescales relevant to drift (> 10 million years) they are obviously plastic while over the short periods relevant to earthquake-waves they are indeed 'rock solid'. Jeffreys' mistake was to overweight the earthquake evidence. That was the elephant in his room {3/11}.

Lord Kelvin, another distinguished mathematical type, held up Evolution and Geology for a generation by insisting that his cooling calculations precluded the Earth from being more than 20 million years old. As it happens, though he couldn't have known it, the Earth was actually being heated by its own radioactivity. Had Kelvin been wiser he would have granted his single cooling argument a far lower Weight (That is Animal Wisdom).

Sir Fred Hoyle, a great astrophysicist, and one of my mentors, threw much of his stellar reputation away later in life by overweighting a single argument. He calculated that the mathematical probability of creating spontaneously a single simple protein here on Earth during its entire history is infinitesimally small. Therefore, he argued, life must have been created by some super-being out in the cosmos and brought to us by a meteorite or comet. He and his collaborator Chandra Wickramasinghe even claimed that the pattern of a 'flu epidemic among school-children in Cardiff demonstrated that fresh 'flu virus must be coming in from Space.

Alas the core calculation was very dubious because it implicitly assumed that only one particular set of proteins (our own) is capable of sustaining life. There may be trillions of billions of trillions of alternative viable sets. And the 'flu argument was manifestly wrong: if it were true isolated tribes like the Inuit would have developed resistance to such cosmic epidemics – which sadly they had not. Had Hoyle given a modest weight to his calculation, and combined it with all the other clues on the Origin of Life (i.e. PAW) he might have avoided his agonizing pitfall. [10]

---

[10] Add Sir Gabriel Stokes, Sir Arthur Eddington , Sir Ronald Fisher and Sir Hermann Bondi (all Cambridge) and it seems that be-knighted theoreticians are especially prone to making disastrous arguments (Bondi got the dark-sky problem quite wrong

The easiest way to fool oneself is not to even consider the right hypothesis. It's the easiest thing in the world to do because the right hypothesis may lie entirely outside and beyond one's present ken. Thus bacteria were inconceivable to the mediaeval mind which instead settled on witchcraft as the likely explanation for an outbreak of sudden deaths in the community. CST can construct Odds on any hypothesis and we will tend to settle on the one with the best Odds – even if it is quite wrong. We will be especially vulnerable if we allow a single clue to generate such a high Weight on a particular H that we will look no further. As Boorstin said " The greatest obstacle to discovery is the illusion that we already know." That is the whole point of the PAW, the 'Principle of Animal Wisdom' (10:4). It disallows from setting such high (or low) Weights on a single clue in order to save us from narrow mindedness, from folly, from Systematic Error, even from death or extinction. Instead it urges us to look for concordance among a broader collection of evidence – because that should be more reliable. However convincing a single clue may be to you, history suggests that a combination of weaker but disparate evidence is what we should all be looking for instead. In his riposte to people like Jeffreys Alfred Wegener, a much broader and wiser man, said: "Scientists still do not understand sufficiently that all the earth sciences must contribute evidence towards unveiling the state of our planet in earlier times and that the truth of the matter can only be obtained by combining the evidence." Alas he'd been entombed in Greenland ice for 40 years before history proved him right. {3:11}

There is a general point here: no matter how conscientious the thinking, nor how rich the evidence, if the correct hypothesis isn't in the field any decisions we make will usually be wrong. For that reason alone wise people must remain generally skeptical – and tolerant. How can Sunnis and Shias, or Protestants and Catholics kill one another if, as seems likely when you consider the number of mutually exclusive religions about, none of their creeds is right? [Sect 16:6]

---

in his text-book on Cosmology). Could it be that knights forswear Animal Wisdom when they kneel before the monarch, or do they mistake mathematical precision for sound physical ssumptions?

(13:8) PREJUDICIAL THINKING

CST tries to reach sound judgments on the *balance* of the evidence. The DE is nothing more or less than a balancing mechanism to find out on which side of the question the preponderance of weighted clues should tip the scales: fight or flee, guilty or innocent, right or wrong, submit or rebel, embrace or reject, do it or don't…….. . Anything which interferes with that balancing mechanism can only lead to bad decision-making and ultimately to self-destruction.

At one end of the prejudicial spectrum lies wishful thinking: looking for and weighting evidence so as to back up a decision already made on emotional or other grounds. Children may believe in Father Christmas because if they don't he won't come down the chimney and fill their stockings. Or as Francis Bacon put it: "The human understanding when it has once adopted an opinion draws all things else to support and agree with it. And although there be a number and weight of instances to be found on the other side, yet these it either neglects and despises, or else by some distinction sets aside and rejects, in order that by this great and pernicious predetermination the authority of its former conclusion may remain inviolate."

Scientists like me are all too prone to pounce upon and delight in evidence which supports their favorite idea, ignore or underweight evidence which undermines it; that is sometimes called 'Confirmation Bias'. But one painfully learns that such childish behavior exacts a terrible price. It steals away one's life in large chunks, years of it, even decades. Growing up is all about learning this bitter lesson. Some of us never do. For instance most of us prefer to read newspapers which support our political point-of-view – a wholly self-destructive indulgence. If we gave that up, or took at least two papers with disparate points-of-view, we probably wouldn't find ourselves in so many unnecessary wars.

The reverse of wishful thinking is negativism, often to new ideas or to other peoples' points-of-view – especially if they are, in your judgment, the 'wrong' kind of people : foreigners, disbelievers, social or racial inferiors……sectarians……. you name it. This too exacts a heavy price – for both sides of the sneer. In the last chapter we talked about one of Science's supposed glories – Peer Review. The scientist submits his ideas, his plans, his request for funds to an independent panel of expert referees, hoping for a positive response. Now in my experience most referees, including myself, look for the unsound aspects of the proposal. In a competitive atmosphere one such discovered flaw will suffice to kill a proposal dead.

But that can't be right. Indeed it is fatally wrong. We have already seen that so many great scientific ideas had damning evidence against them early on, and some still do. Heliocentrism had lack of stellar parallax, Evolution had the short Age of the Earth, Newtonian Gravitation had the ridiculous notion of 'Action at a Distance", Continental Drift had the supposed rigidity of rock while Relativity still has the unexplained constancy in the Velocity of Light. It's not the flaws alone in a hypothesis which matter – but the *balance of the evidence*. Looking for flaws alone is negative thinking of the worst kind, one which holds back progress on a massive scale. We mustn't think like that.

How are we to avoid the worst effects of Prejudicial Thinking? Above all by self-awareness. If you find yourself prejudiced one way or other before you come to make up your mind (assemble your Inference Table) then lean the other way. And the easiest way to do that is to adjust one's Prior to counterbalance ones' prejudice. For instance [Table (8:4)] I introduced a prior of 1 in 32 so that I would not be carried astray by self-indulgent affection for My Beautiful Theory about Hidden Galaxies. Today I would probably be harsher still on myself.

Prejudicial Thinking is probably the commonest form of bad thinking there is: we constantly need to be on our guard against it in ourselves. If we can't change our minds then we cannot, by definition, use common sense. It is dangerous to associate only with like-minded people; get our information from only one source; appoint only biddable subordinates. So many leaders fail or fall precisely because they tumble into such self-indulgent traps. Juries, boards, committees, cabinets and parliaments are mechanisms to try and prevent personal prejudice leading towards disaster.

(13:9) THE BURDEN OF PROOF

Many hypotheses come to light, even come to rule the world, long before conclusive evidence arrives to settle their validity one way or the other. In those circumstances general opinion and decision will tend to rest with the side already 'in the saddle' – who very likely got up there through an accident of history or power. The rulers of the roost are then able to rest 'the burden of proof' on the other side and insist that they themselves are right—'until proved wrong'. Quite apart from the moral asymmetry, thinkers should be seriously worried by the opportunity this offers to delay progress. If both sides are not researched with equal vigour and equal dispassion reactionary opinions could hold sway almost indefinitely. Thinkers need to be very aware of the Burden of Proof (BOP) issue so as to discount it, or even resist it, when it arises.

Think of some purely scientific examples: the Earth is flat, the Sun orbits the Earth, Creation occurred a very short time ago; there are only four elements, organisms do not evolve, animals cannot think, rocks are too hard to flow, continents do not move, radio waves will not bend round the Earth, animals cannot live in the deep, physicians do not cause child-bed fever, the World must be cooling down,....None of these hypotheses was foolish, most were very plausible, yet all, in the fullness of time, proved to be wrong. In retrospect – it's always much easier in retrospect – we can see it would have paid to examine them all more dispassionately and researched them with more vigour—though in many cases constructive progress would have been hard without modern tools. But it would certainly have helped to know how little *positive* evidence some of them had in their favour, and how many relied on fable-telling of various kinds, including religious authority. In some cases sound negative evidence was available but the burden of proof required of it was too much to bear, the habit of the times too prevalent to overturn. The Ancient Greek astronomers (500 BC) pretty much knew the Earth was a globe but when Pytheas (325 BC) sailed out through the Pillars of Hercules, then on through the British Isles to discover castles of white crystal afloat under the Northern Lights nobody believed him when he got back to Marseilles. And the same thing happened to Admiral Cheng Ho (1430 AD) when he arrived back from East Africa with a giraffe (and many other wonders beside). The courtiers convinced the Chinese Emperor that it was actually a k'i-lin, a figment of the Imperial imagination.

This is no place for criticism. Anyone can be taken in by the BOP issue – I certainly have been myself. But those who want to think clearly must be constantly aware of the issue, and be willing to take steps against it tripping them up more often than it should

How can one avoid BOP blunders for oneself? First of all constantly bear in mind that CST is (nearly) *always provisional*[I insert 'nearly' because there is hardly any more doubt that the Earth is round is there?] New clues can (nearly) always be added to sway the scales the other way. It may not happen all that often but it can happen, it does happen, and it will happen. The history of science leaves us no room to doubt.

Secondly try to be alert to the BOP issues in one's own field. If a controversial issue comes up, briefly list the main evidence on both sides and ask if BOP is involved.

Thirdly ask yourself where the COWDUNG ('Conventional Wisdom of the Dominant Group') comes from. Is it well founded, or merely historical inertia masquerading behind an asymmetrical BOP campaign? A constructive way to proceed is to assemble the evidence both for the hypothesis and against, weight as fairly as you can with binary Weights – bearing in mind PAW— and feed it into the Detective's Equation to see which end of the scales come down. I recently tried that with Big Bang Cosmology – very much the COWDUNG in my own field – to discover some rather surprising, not to say dismaying results [see my website].

There should be no pretending that the BOP issue is simple, or the choice in many cases obvious. Before Darwin it would have been easier to insist that a gulf divided human minds from animal: afterwards to insist the very opposite. But if there is a BOP question it is always wise to be aware of it.

Let us conclude with a final example from my field of Hidden Galaxies because it illustrates how easy it is to slip into trouble. We set out to look for HGs with a giant radio telescope. But every one of the 4000 odd radio-sources we found appeared to have a bright optical galaxy right next door, one with much the same recession velocity as the source in question. What were we to conclude? 'That Hidden Galaxies therefore do not exist.'? Or 'Galaxies always huddle close to one another and so you would expect to find a bright galaxy right up close to every hidden one, and this is merely evidence of that.' Most of the team, including me, swallowed the first interpretation, unthinkingly accepting the burden of proof. But we needn't have done. We could as well have insisted that the 'Identifiers' actually 'prove' that the bright galaxies they had identified in their neighbourhoods were indeed also the sources of the radio emission. And they would have failed some of the time, proof in this instance being often hard to find. Where should the BOP lie here, and why?

I have confined attention to scientific issues, but BOP is even more insidious in political, social, religious, racial and gender areas where it is often manipulated deliberately. Many of the strongest opinions only survive because asymmetries resting on BOP hold sway.

(13:10) MISPLACED DEFERENCE; HOW TO SCREW YOURSELF

We humans are all too easily cowed – after which we can be fleeced out of most of the wealth we have created. History is very largely the record of small groups of humans cowing much larger groups into surrendering their wealth, their labour, their liberty and even their lives. Think of the Pharaohs, the Romans, the European aristocracy, the Church, the Nazis, modern African and Arabian kleptocrats, and the Communist Party. You don't have to use force, though the odd crucifixion or auto-da-fe can help. We humans are all too easily awed by outward manifestations: huge cathedrals, pyramids, tall hats, military bands, extravagant facial hair, munificent uniforms and jewelry, sonorous oratory, anthems, sky-scrapers, mansions, landed estates, exclusive accents, titles, decorations and Prizes, ludicrous vehicles by land sea and air……..and why not? They may be ridiculous in themselves but they can be used to exert power and steal even more wealth from one's fellow-beings. Churches and aristocracies are very successful extractive industries based on outward pomp. And what about all those financial types with their skyscrapers?

Even in modern times we need to acknowledge this weakness in ourselves, this proclivity to deference, which is very likely genetic because primate societies are strongly hierarchical. The silver backed gorilla is still with us, as this recent comment from 'Schumpeter' in the Economist testifies: "The typical Chief Executive is more than six feet tall, has a deep voice, a good posture, a touch of grey in his thick lustrous hair and, for his age, a fit body. Bosses spread themselves out behind their large desks. They stand tall when talking to subordinates. Their conversation is laden with prestige pauses and declarative statements."

What the heck has this rather touching, rather comical hangover got to do with thinking? A very great deal I'm sorry to say. Maybe we moderners are less taken in by archbishops' hats and Rolex watches, but we still defer to Professors and slick computer-generated slides. It's a defect in our genes – which may once have been of evolutionary benefit but is almost certainly self-destructive now. Today we defer to 'experts', very often experts accoutered in modern-day manifestations of brain-power; these include certificates, degrees, doctorates, professorships, fellowships, memberships of this or that prestigious or professional body, knighthoods and Nobel Prizes. When it comes to thinking straight about a particular problem many of these manifestations are worthless, or relevant in only strictly limited contexts. A Nobel Prize in Solid State Physics is all very well and is probably proof that you did something frightfully clever with electrons 30 years ago. It's no entitlement to an authoritative opinion on global warming (see later).

Remember:

1 Passing exams is evidence of very little but the ability to pass exams.

2 Practically anyone can get a degree from some university in some subject – 'Media Studies' for instance from The University of South Central Rutland.

3 PhD's (Doctorates in Philosophy) may be awarded for little more than doing a Supervisors dirty-work. Practically nobody fails – even in supposedly tough subjects like Physics (I could tell some stories….).

4 Almost anyone with a PhD these days can become a professor by hanging around some university for long enough. All you have to do is apply to your Appointments

Committee and demonstrate due diligence. For every professor in Britain 30 years ago there must be at least a hundred now, at least.

5 Fellowships and knighthoods are harder to acquire but not if you come from the right institutes and hang out with the right sort of people – particularly doing government-inspired business. You may have been a crack-shot research worker in your day, but you probably wouldn't have given it up to become a VIP if you were still good at hard thinking – because hard thinking carries its own incomparable rewards.

When university dons had to give up holiness as the chief justification for their handsome emoluments they settled for 'cleverness' instead. It was a cunning move because neither can be easily questioned. Hierarchies set their own standards and make their own appointments – "Jones got a double first at Saint Botolph's College, so obviously he's a clever fellow."

Now it might be true that there really is a general quality of cleverness. I certainly believed so when I was a young scientist and feared to meet somebody so much cleverer than myself that I would have to give up astronomy in despair. It never happened because, I find, the people who make discoveries in my profession are driven, work extremely hard, know a lot and have had good luck. Apparently cleverness won't do – even if it exists.

Within CST as we have described it, it is difficult to quite see where 'cleverness' would in any case be much help. Evidence will come from curiosity, knowledge, research and drive. The judgment to assign good Weights comes from – well where? I'm not sure: experience, and making previous mistakes might help. The one opportunity for imagination lies in constructing new hypotheses, though in science at least the problem more often lies in choosing among a multitude of hypotheses that are either obvious – or already exist. Even 'geniuses' like Darwin and Einstein became famous for hypotheses that were already 'out there'. Darwin's grandfather Erasmus Darwin was one of many who discussed Evolution at length while 'Einstein's Special Theory of Relativity' was really constructed by Henri Poincaré, though not precisely in its final polished form. Einstein's contribution to the Special Theory was more in exposition than creation. The later General Theory was his own {3/10}.

There are great dangers in denying the importance of 'expertise' and I certainly don't mean to do that. Many of the greatest achievements of modern man depend on blending together many very different skills. For instance constructing the Hubble Space Telescope required the combined skills of rocket scientists, opticians, astronomers like me, computer scientists, celestial mechanicians, atmospheric physicists, detector engineers, aircraft designers, cryogenic experts, atomic physicists, gyroscope engineers, materials experts, mathematicians, radio engineers, system managers, programmers, archivists and so on and so on… You need so many specialists because knowledge is specific and not generic. I might be able to give an authoritative opinion on certain aspects of galaxies but not on planets and certainly not on optics – or many of the other manifold skills required by that project. And I imagine the same is true of many other fields. Your doctor knows very little about *your* body, your bank manager even less about the trouble *your* pension pot is in.

The point I want to make is *that you should rarely try to escape the responsibility of thinking for yourself*: to do anything else is too risky, though it is tempting at times, certainly for some one as lazy as I can be. Nowadays, with adequate diets, there is no calorific excuse for avoiding hard thinking anyway. Beware! There are all too many 'experts' out there trying to tell you they know best. They may indeed know better about something but it's rather unlikely that they will know better than you do about something close to your own heart. Nor are they so much cleverer than you that it will make up for lack of specific knowledge about your particular problem. Only defer as a very last resort and when you are pretty darned certain that the expert in question knows whereof they speak. *The internet has altered the whole balance between experts and the rest of us. Before the internet it was very difficult for us laymen to find many of the facts needed to make an informed decision.* The doctor who had access to a medical library really did know far more than you or I. That's not necessarily true any longer. One of my heroes is an eminent astronomer I know whose wife Elaine was dying. When the doctors gave up on her and said she had only days to live Bob went to his university medical library with all her records, diagnosed her problem within 48 hours, and told the doctors how to save her life. She's hail and hearty 25 years later. Bob wouldn't need a medical library today. The internet has really moved the ground so that more of us could well imitate Bob.

Trading on misplaced deference is as old as witch-doctory – and one of the easiest ways to make a handsome living. No matter how often it is stamped out it will rise again in another guise. Henry the Eighth dissolved the monasteries (which had sequestered a third of England's wealth) only for his very own Church of England to later seize the riches created by scientific agriculture. Victorian parsons were some of the wealthiest, idlest fellows in the land. Now the so called "professions", particularly medicine, education, the law, management and 'finance', have discovered that they have a right to much of our wealth. We mustn't let them steal it.

If man-made Global Warming takes down the environment, and most of us poor creatures with it, Misplaced Deference will probably be responsible. In '*Merchants of Doubt*' Oreskes and Conway {8} carefully document how a tiny but fanatical group of US scientists orchestrated a sinister campaign against the notion that carbon burning industries could be destroying our atmosphere. Led by the former President of the US National Academy of Sciences, Frederick Seitz (1911 – 2008), half a dozen men, none of them atmospheric experts, have used every dirty trick in the book to prevent all effective climate-change regulation. These include:

1 Using the uncertainty in any honest science (which is provisional remember) to argue that nothing is certain enough to take action.

2 Ignored, or failed to understand, that the uncertainty in this clue or that can still be overwhelmed by the multiplicative nature of the Detective's Equation.

3 Exploited the fallacy that a scientific expert in one branch of science can be authoritative in others (i.e. Misplaced Deference).

4 Disguised their secret cabal behind a smoke-screen of 'think-tanks' funded by self-interested and ruthless industries like coal and oil.

5 Funded journals and committees to give the appearance of disinterested peer-review when the very opposite was intended. They used journalists and the right-wing media to further their ends.

6 Selected evidence that supported their case; neglected or even denigrated evidence that did not.

7 Exploited US insecurity about the Cold War (they were all ex defense-scientists) to undermine opposition spokesmen and opposition arguments.

8 Demanded and got 'equal time' from much of the media, a concept which has no place in scientific debate.

And so on and so on. Why they did it remains a mystery. Political naivety, of the kind peddled by Milton Friedman, is the probable answer because the same gang also campaigned with equal ruthlessness, against tobacco regulation, nuclear winter, CFC regulation to preserve the Ozone layer, the anti Star-wars campaign, DDT control….. Their naivety, like Friedman's, lay in confusing 'free-market' with 'freedom' so that, to them, any government regulation was a crime which justified the dishonesty of their own side. They relied especially on Misplaced Deference – and have largely got away with it because lay people, and many scientists alas, are ignorant of the Scientific Method.

There is no end to misplaced deference – some of it very comical. In Britain nearly everybody would like their children to go to Oxbridge – but why? Just go and look at the buildings. Until very recent times they were not centers of learning but seminaries for training priests – and proud of it, made extremely rich by selling 'Indulgences' to wealthy 'sinners' so that they could avoid the agonies of Hell. When they did study the natural world it was mainly to look for more proofs of the existence of and wisdom of their particular God – it was called "Natural Philosophy". Their corrupt tentacles reached everywhere, even to controlling the Medical Schools in London {16/6} as well as half the ecclesiastical 'livings' in Britain. Yes of course they had some famous alumni – but in those days there was nowhere else for them to go, nowhere because they prevented the setting up of any proper university in England for another 600 years!. They were centers not of progress but of reaction and privilege, which contributed almost nothing to the rise of science, industry, egality and commerce going on in Glasgow, Birmingham, Manchester and London. As late as 1920 Robert Clifton Bellamy, who had held the post of Professor of Physics at Oxford for 50 years, wrote only one significant paper during that time on the grounds that "research betrays a certain restlessness of mind."

Even more comical, tragi-comical I would say, is the modern tendency to set up business–schools inside universities, and in ivy-league universities at that, ivory towers which, by choice, proudly know little of the stresses and concerns of commercial life. Why? Because there's a lot of money in it — one hell of a lot (tuition fees are typically a thousand dollars a week). In return 'students' get a certificate from a prestigious school which will help them to intimidate others and to schmooze with influential people. They're clubs for would-be silverbacks. The only justification that I can see for them must be that misplaced faith in 'cleverness' adverted to above. It has little to do with learning or thinking. Had James Watt or Henry Ford gone to an academic business school then God help us….. Selling certificates to Heaven – whether it's supposed to be up there on down here – is an endlessly profitable way for parasites to gorge on human aspirations.

(13:11) CAREERISM; CONFORMING FOR ADVANCEMENT

We might as well admit it: we humans are an hierarchical species. We can't survive without very close collaboration among ourselves. Living out on the open plain among larger and fiercer animals demanded a degree of self-discipline which is not always conducive to independent thought. The truth is that most young scientists don't fight their way to the top – they creep to the top by pleasing their elders and betters. To my mind this creeping – though it does have its positive side as we shall see – is a major cause of poor thinking in astronomy, and probably in most other intellectual endeavors beside. Whereas Misplaced Deference is from outsiders toward insiders, Careerism is misplaced deference of some insiders towards others.

Hierarchicalism in academia works at all levels from undergraduate to senior professor. The undergraduate may have to sacrifice understanding for high exam marks. The Ph.D. student cannot afford to clash too often with her Supervisor. The post-doctoral fellow must research on fashionable topics if his ambition is to work in the 'right' places with the 'right' people. The junior academic must fight for tenure by pleasing her senior colleagues. The tenured researcher with a growing family needs a better-paid professorship which means buttering up the appointments committee. The professor is expected by the university to bring in handsome government funds to pay for assistants. The observer must fight for highly competitive satellite or telescope time by pleasing the Time-Allocation-Committee. The successful professor will look to get on prestigious committees. And even then there are Directorships, Fellowships and Nobel Prizes to be coveted. Science nowadays is a Big Church with a few popes, a School of Cardinals, much room for preferment, and with colossal funds flowing into its coffers. Being a curmudgeonly atheist is no way to get on.

It's all very well deploring such a scientific church – but churches are good at raising money and erecting cathedrals. What would we be without the Large Hadron Collider at CERN, or the Hubble Space Telescope? Erecting such cathedrals requires the dedication of thousands of superb craftsmen for decades of their lives. Such toilers require unambiguous goals; you cannot expect them to be awkward heretics as well. Such projects take 20 years to 'launch' and another ten to exploit. Not many cathedrals have been raised by disbelievers. All the same, scientific misbelievers are needed too. That's what makes Science such a fascinating and unpredictable mix. It's not a story of heroic individuals versus dumb crowds. For instance:

(1) Inside a cathedral with the anthem swelling it is difficult (by design) to pick out the defects in the Church's creed. Outsiders like Darwin – a gentleman's companion, and Wallace – a humble specimen-collector, could see Evolution – to which the bishops of biology were blind.

(2) The Geological establishment ridiculed Continental Drift. But 50 years after Wegener's book on the topic the US Navy, in its submarine conflict with Russia, needed to know what the ocean floor was like. Sea-floor spreading was discovered, and with it Plate Tectonics, the modern version of Continental Drift.

And so on. The unoriginal point I'm trying to make is that a Big Science requires a Big Church in order to build the Big Cathedrals it always covets and sometimes needs. But Big Churches are hierarchical and conservative by nature and so nurture a wealth of poor thinking and a congregation of humble souls[11]. And what is certainly true of Astrophysics is probably to be expected in other big fields too, from Neuro-science to Economics. After all we all need to earn a living, and careerism and independence of thought seldom mix.

The essence of hierarchical thinking is conformity, conformity to the fashionable paradigm of the day. The irony is that such conformity isn't so much enforced by autocratic grey-beards from the top as embraced by younger people trying to ascend the slippery career ladder from below. They are too focused on the next trembling step to worry about the big picture. How many have either the motive, the knowledge or the character to dissent?

This sad state of affairs leads one to question why society funds any pure science such as astronomy. I'd like to think it was natural curiosity, and there probably is something to that, but I suspect it has more to do with exercising a cadre of storm troops who can quickly be mobilized to tackle the next potential human disaster. It worked in the Second World War, it is working with AIDS and Global Warming now. But if that is the case then conformity of mind is not what society should be principally looking for amongst its scientists. Perhaps we should be erecting less cathedrals and encouraging more skeptical individualists. Working in large 'teams' [10,000 worked on

---

[11] Our witty friend Medawar wrote "Humility is not a state of mind conducive to the advancement of learning."

the search for the Higgs Boson] shouldn't be a refuge for the faint of heart or the weak of mind. Perhaps.

(13:12) GROUPTHINK

There is a phenomenon called 'Groupthink' in which the desire for harmony and conformity within a group results in poor decision-making. Members minimize conflict and reach consensus without critical evaluation of the alternatives and/or by isolating themselves from outside criticism. Its influence is not easy to 'prove' in any particular instance but it has been invoked to explain historical events from the failure to anticipate Pearl Harbor, the Bay of Pigs fiasco, and the escalation of the Vietnam war. There is persuasive evidence to argue that it played significant roles in both the Challenger and Columbia Space-Shuttle disasters and in the launching of the Hubble Space Telescope with a hopelessly flawed mirror. One can certainly sense it within committees – especially where there is a dominant chairperson, and at large scientific meetings where a majority of junior personnel are looking for the 'right' silverback to follow. Reluctance to be first in speaking out against an (often unstated) consensus view can then become stifling, especially for junior people. And of course the longer it goes on the harder it is for dissenters to get to their feet. Moral courage is a rare virtue at the best of times and, in my experience, is rarer still among young people and academics. Perhaps BBC ought to refer to 'Big Bloody Church' instead of 'Big Bang Cosmology'.

(13:13) FORGETTING FRIAR OCKHAM: RESURRECTIVE THINKING

Although it stinks, a dead hypothesis can be kept alive almost indefinitely by attaching to it more life-support systems in the form of 'Free Parameters' (FPs). The famous mathematical physicist Johnny von Neumann claimed "With 4 FPs I can construct an elephant; with 5 I can make it wiggle its trunk." Thus Greek chemistry and Greek medicine were kept alive for 2 millennia although they were mostly nonsense. Some great theoretical edifices of modern science have so many FPs that I for one feel queasy in their presence. Thus the 'Standard Model of Particle Physics' has an inelegant 18 – oddly about the same number as Big Bang Cosmology [Sect 8:8]. Particle physicists were hoping to find a better theory (Super-symmetry) but have been disappointed by the Large Hadron Collider's failure so far to find the corresponding super-symmetric particles. Big Bang Cosmologists are far more brazen. Whenever the patient looks sickly, and to my mind he's looked very poorly of late, they whip out a new FP, give it a sexy name (how about 'Inflation' or 'Dark Energy'?), and jab it into the patient's vein. You can't say they are wrong for certain, but you might prefer, as I do, to leave the room.

The hardest maneuver for any army or any general to perform, is a successful retreat. Most professional thinkers prefer to soldier on towards the grave. As the founder of Quantum Mechanics Max Planck put it "Science advances funeral by funeral." Maybe that is too cynical because most of us at least pay lip-service to Ockham's Razor with its insistence on simple hypotheses – because they should be far easier to test and reject. But as science becomes more complicated, more interconnected and more expensive, it becomes much harder for the marching army to retreat. All those useful assistants would have to be forgone, all that travel money returned, all those superb craftsmen laid off, all that self-publicity retracted, all those dreams of a Nobel Prize laid aside. How many of us have the courage and the energy to admit and retreat?

My advice is "Count the Free Parameters!" Marching armies may not retreat but they can be routed – as history documents. It is generally not wise to forget Old Father Ockham.

(13:14) UNCRITICAL USE OF ANALOGY

The human mind, so long as it is awake, tirelessly searches for analogy, where 'Analogy' is defined as "an inference that if two things agree in one respect they probably agree in others". Analogy can be a wonderful source of new hypotheses but it can be a treacherous advocate when it comes to judging whether those same hypotheses are sound. Scientists exploit analogy all the time – and we have encountered many instances:

- Darwin and Wallace spotted the analogy between the human "struggle for existence" and the wider struggles amongst plants and animals (3:11).
- Bradley on his boating picnic spotted the analogy between the flag flying at his masthead and incoming starlight, and so proved that the Earth is orbiting the Sun.
- Maxwell arrived at his Electromagnetic Equations using an analogy with fluid vortices – which was incomplete but nevertheless very provocative.(15:10)
- Naval radar scientists realized that the interference effects of the sea's surface could be exploited to identify cosmic radio sources. The analogy there was between moving aeroplanes and apparently moving 'radio-stars'.(3;11)
- Mouse models are widely used in medical science where it is hoped that the analogy between mice and men will lead to new cures (for men!).

The problem with analogies is of course that they are only partial, invariably incomplete. Thus Science has committed many blunders in their name. For instance:

- Descartes, thinking there was an analogy between Mathematics and Science, imagined he could spin off all the truths of the world by Deduction from a single incontestable truth, namely 'I think therefore I am'.
- Politicians saw a faithful analogy between the breeding of real human populations and an almost childish mathematical model proposed by the Reverend Thomas Malthus (1798) – with tragic results for the British poor.
- Marxists were deluded into an imagined analogy between the 'The Communist Manifesto' and Science – resulting in tens of millions of

deaths.
- For two millennia we saw an analogy between Mathematics and Physics and so were hogtied by the belief that both were deductive ; even Newton fell for that.

The above blunders all fall into the category of mistaking truly OPEN systems (i.e. ones that may contain unsuspected causes) for CLOSED ones where unsuspected causes are ruled out by fiat (i.e. games). An Urn Problem (a la Bernoulli) is all very well – but not if there is a tarantula lurking in the urn [See Sects (10:6) & (13:15) below)].

It seems to me that analogies are fine for suggesting new hypotheses, but not fine for judging whether they are sound. One still has to weigh the evidence for and against them using the Detective's Equation. And if we are wise we will always observe the PAW which, to employ yet another analogy, is donning a gardening glove before daring to stick one's hand in that urn.

(13:15) FORGETTING THE TARANTULA

The easiest, most treacherous and nastiest trap into which any of us thinkers can fall is imagining that there is no hypothesis, beyond the possibilities already in our mind, to explain the existing evidence. Thus otherwise perfectly kind mediaeval people blamed and burned alive harmless old ladies because they could not imagine that germs might be responsible for outbreaks of diphtheria among their children. It is such a pervasive trap that that it deserves a dramatic name, so henceforth I am going to label it " *Forgetting The Tarantula* ". Why? Because the philosophy of Inference owes so much to Jacob Bernoulli's book "*The Art of Conjecture*" (Groningen, 1714) in which the thinker was challenged to infer the whole contents of an urn full of coloured balls by withdrawing and inspecting only limited samples of them. But in real life urns can contain all manner of unsuspected and nasty surprises, including tarantulas. Ordinary people fall for it because our imaginations are naturally limited. Experts in particular fall for it because it bolsters their claims for Certainty, claims on which their influence, to say nothing of their incomes, may largely depend. Priests who claim privileged access to (some) god, financial wizards who know which way the market is going to jump, economists who know exactly what the outcome of Brexit will be, physicians who learned the Latin names of all the bones in the skeleton at medical school, academics who've been consorting with other academics for far too long, logically inclined philosophers who've been writing vast tomes about The Scientific Method without ever getting their hands dirty, Statisticians who are really mathematicians who don't believe in Induction, and dogmatic people of all kinds who find they cannot live without Certainty, are all dupes of The No Tarantula Fallacy.

"But" one might reply "Suppose there *is* a Tarantula in the urn – doesn't that make serious thinking impossible? What's the point of worrying about it if there's nothing one can do? Many times there won't be a bloody Tarantula, so one might as well go ahead and ignore its putative existence. Anyway look how Newton worked out the detailed dynamics of the Solar System, and Einstein deciphered the secrets of Space-Time – precisely because they ignored possible tarantulas, and got on with it. Your tarantula phobia is just another primitive superstition that has held up, and will ever hold up progress. We bold thinkers have to get on with the business, taking into account the evidence that is actually to hand."

We of course, who know and understand the PAW, recognize that Certainty is unattainable and we know how to avoid the worst dangers of the No Tarantula Fallacy by never using Weights in our thinking outside the range of 4 to ¼. That always leaves room for hidden Tarantulas. It cannot find them directly of course but it forces us to look for such a range of diverse and coherent evidence that their presence in a given situation can eventually be made to seem most improbable, or most probable – as the case may be. Thus in the search for Hidden Galaxies the tarantula turned out to be galaxy clustering – which was eventually forced into the open by all the other diverse but cohering evidence.

You can often spot scholars suffering from the No Tarantula Fallacy by their fussy insistence on their own precisive logic. They try to establish their credentials with some elaborate mathematical proof of some apparently obvious point and you only find out much later, to your horror, that their entire scheme relies on the unstated pretence that the world is a closed system, a mathematical urn sterilized of all tarantulas, 800-pound gorillas and six-and- a- half- ton bull elephants {11:9}. Good luck to the poor deluded creatures – but don't pay any attention to them. Nicholas Nassim Taleb has written a most readable polemic "*The Black Swan*" {11:11} against The No Tarantula Fallacy – which he calls "The Ludic Fallacy" because 'Ludos' is the Latin for 'game' and such people are playing closed games in preference to treating with real open Life.

But it's not just 'them' it's 'us' – all of us. Just in the course of writing this book I've fallen heavily into this trap at least a dozen times and no doubt I shall continue to throughout my life; after all it's so much more comfortable to forget the tarantula.

One of the great benefits of being a practical, as opposed to a theoretical scientist, is that one continually comes upon tarantulas (Theoreticians can and frequently do ignore their existence). There's always some awkward and unexpected happenstance which makes life interesting for us. In fact it's the bloody tarantulas which often lead to the greatest discoveries (Think of Oersted, Becquerel, Kepler, Faraday, Snow, Darwin, Bradley and Poincaré in Chapter 3) .

My best qualification for writing this book is that I have fallen helplessly into almost every pitfall that Thinking could set across one's path. There's a subtle version of the No Tarantula Fallacy out of which I have just crawled, severely bruised in self-esteem: I label it WIMP standing for "Weight is More than

Probability". My fictional agents – the Forensic Scientist and the Detective in the Murder in the Library case (Ch. 5) made *two* fundamental errors, not just the one I had intended to illustrate. Yes the scientist quite wrongly supposed his measurement errors (of the footprint remember) had a Normal distribution – when they should have been *measured*. But even supposing he had rectified that, *he still couldn't get reliable Weights*. It was assumed on reasonable grounds that only a woman from the household could be responsible for the murder. And suppose further that all those women, bar the actress, had feet larger than 18 cm., then not one of them, apart from her, could possibly have left a footprint so small. So if $\bar{H}$ was 'One of the other women in the household is guilty' then $P(E|\bar{H}) = 0$, in which case $W(E|H) = P(E|H)/P(E|\bar{H})$ is infinitely large, no matter how low $P(E|H)$ or $O(H|E)$ is. The vastly important general point is that *without an $\bar{H}$ you cannot estimate a Weight*! Thus there are no such things as 'Normal Weights', even if there can be Normal Odds. So, beware of WIMP. Here is yet another argument for PAW – which can prevent us from falling into many a subtle trap. PAW would here have enjoined all parties to set $W(E|H) = ¼$ here, not 1/40. Then justice, even right, might have been done. Of course the foot-print size should have ruled out the parlour-maid – whose feet were too big. But being poor she didn't have, like the other suspects, a decent lawyer, and so finished her life in place of the guilty actress – who became a Dame. ['Tis the rich what gets the pleasure, tis the poor what gets the blame'.]

(13:16) THE PYGMALION COMPLEX

We all know the story of the sculptor Pygmalion who fell in love with his own creation Galatea to the extent that he could not, or would not, see her stony nature. Scientists, particularly theory builders and computer modelers, are sometimes bewitched likewise. What they have created becomes too valuable to abandon, even when every one else can see her blemishes. Thus Arthur Eddington, who had made a brilliant start to the theory of stellar structure (1913) covered up its cracks (1917). It was the start of a disastrous downward spiral. We saw how next he twisted the 1919 eclipse observations to proclaim The General Theory of Relativity – which admittedly was Einstein's creation, not his own. It was as if he couldn't bear to be wrong. His hubris was punished in the 1930s when he developed his "Fundamental Theory" in which the entire cosmos was controlled by the mystical number 137 – revealing that brilliance had descended into priestliness, if not mania. Likewise Lord Kelvin in the late 19th century saw evidence for his young (20 M y old) cooling Earth wherever he looked and Sir R A Fisher couldn't abear the thought that anybody's ideas on Probability but his own were sound. In later life all held positions of immense influence and hordes of uncritical disciples who held progress back.

Computer simulators are particularly prone to this mental disease. For my Ph.D. I constructed a beautiful computer model (which I called 'Ethel') of a forming star – which provided the most startling insights. But gradually I realized that Ethel was a treacherous creature, prone to telling beguiling lies. Imagine my astonishment therefore when an identical twin of Ethel appeared from Caltech and was hailed as a great breakthrough. Its young sculptor, who shall remain nameless, was appointed to a professorship at a famous New England university. Rather than shoot him down in public – after all I'd fallen for Ethel myself – I approached him privately – but it was far too late. He angrily rejected any notion that Ethel might be a worthless trollop. The net result was that Star Formation Theory in the US was held up for twenty years.

A far more serious obstruction of the same kind today concerns Cold Dark Matter or CDM. Conceived by half a dozen ingenious people in the 1970s to explain some major puzzles in extra-galactic astronomy, including galaxy formation, it gained a wide following and made its creators famous and influential. Alas, as subsequent observations clearly demonstrate, it is quite wrong. But, so far as I know, none of its distinguished creators has been able to admit the truth. That wouldn't matter if CDM hadn't gained so many uncritical devotees that it is holding up progress in Cosmology and Cosmogony to this day. At least that's my opinion.

I confess that I've been a bit of a Pygmalion myself – over my Hidden Galaxies Theory (but then I've committed almost every one of the cardinal sins in this chapter). And it is true that if you don't love your child enough you won't fight for it as hard as you should. The fault lies not so much in loving too much but wanting to die in the right. Because a beautiful theory is superseded – as most must be in this provisional world – doesn't make its creation, or its creators, any less admirable. I believe we should admire people more for what they *try* to do and how they go about trying, not so much for their eventual success, which may well be a matter of pure luck [think of Amundsen, Shackleton and Scott]. If we did that people wouldn't fear so much being proven wrong. But don't be a Pygmalion.

(13:17) BLINKERED THINKING

is actively refusing to acknowledge that the truth may lie beyond the limited range of hypothesis one is willing to consider. The blinkered world is simpler, safer and more comforting to those of a lazy, timid or rigid disposition. Inside their blinkers they can enjoy their picnic without having to worry about the huge python wriggling towards then through the long grass – they can even prove, using heavy Weights, that no such creature exists. It appeals especially to those who would like us to revere and/or pay them for their expert certainties: priests, oracles, fanatics, pundits, psychologists, dictators, economists, financial and management experts, statisticians….. The rest of us grow up and learn to live with Uncertainty – and low Weights.

Blinkered Thinking differs only from Forgetting the Tarantula (13:16) in that it is deliberate, a fault of character rather than of forgetfulness. Examples: My God will protect me; the Nazis would never by-pass our Maginot Line: the Stalin government is much maligned; No one will ever invade us (Stalin, Scottish Nats.); Unrestricted immigration is harmless (European intellectuals); We'll beat the hell out of them gooks (US military command in Vietnam); Nobody in this game will cheat; I can calculate the Probability of that to 4 places of decimals (statistician); the new post-colonial governments will obviously look after their peoples far better than we did ……..

If I have seemed overcritical in this chapter let me call up Carl Sagan one of my heroes, for support. In a wonderful collection of essays entitled *"The Demon Haunted World"* {9 } he had these things to say:

(a) "Finding the occasional straw of truth awash in a great ocean of confusion and bamboozle requires vigilance, dedication and courage. But if we don't practice these tough habits of thought we ……risk becoming a nation of suckers, up for grabs by the next charlatan that saunters by." (p 39)

(b) "The Method of Science, as stodgy and grumpy as it may seem, is far more important than the findings of science." (p 22)

(c) "The values of science and the values of democracy are concordant, in many cases indistinguishable………." (p 38)

If there is a single lesson to be taken from this chapter it is that *the Internet is changing the balance between individuals and experts – in favor of individuals.* **In future it will pay us to make many more decisions for ourselves, particularly as we now have the brain-fuel. But that involves a sound understanding of Common Sense Thinking – together with its limitations and pitfalls.**

SUMMARY

Widespread weaknesses in thinking include:

(1) Misunderstanding the machinery of Common Sense; e.g. Popperism [13:2].

(2) Tunnel Vision: absence of the right hypothesis. [13:6], {13:15}.

(3) Underestimating alternative hypotheses, particularly chance. [13:7]

(4) Too many alternative hypotheses for the evidence to choose between them [The 'Principle of Limited Variety' or PLV]. This superabundance of choice must prevent many otherwise interesting subjects, such as Economics and Psychology, from ever becoming sciences. [13:3]

(5) Prejudice and 'Confirmation Bias'. Overweighting favored clues, underweighting disagreeable ones [13:8, 13:9 & 13:16]

(6) Taking Psychology and Psychiatry too seriously as factors in the process of decision. Neither is a science, nor can ever be so; the so called Demarcation Problem. Evolution will sift out most harmful irrational quirks anyway.[13:5]

(7) Overvaluing 'cleverness'. It doesn't seem to play much of a role in good decision taking, or in Science. [13:10]

(8) Misplaced Deference to self appointed experts and high-falutin' symbols such as 'dreaming spires' and ivied walls. [13:10]

(9) Careerism; a hierarchical habit of mind which leads to conformity. [13:11]

(10) Groupthink, which speaks for itself. [13:12]

(11) Credulity towards Statistics [11:9].

(12) Ignorance of The Principle of Animal Wisdom (10:4).

(13) Being influenced by some historical asymmetry in The Burden of Proof (13:9)

(14) Forgetting that Weights require a knowledge of $\bar{H}$ as well as H

[WIMP; (13:6) and Appendix A1]

(15) Using Analogies uncritically, especially confusing CLOSED systems for OPEN. (13:15)

(16) Falling in love with one's own 'brilliant' ideas, like Pygmalion.(13:16)

(17) Forgetting the Tarantula in the urn – the totally alien hypothesis (13:15).

THINKING FOR OURSELVES
CHAPTER 14
THE EXTRAORDINARY HISTORY OF THINKING

Draft 14/9/18 (12.8 kw)

"There are in (Optics) demonstrations which do not produce the same kind of certain Laws as Geometry – which is based on fixed and incontestable Principles. Here we first assume the Principles, make predictions from them, and look to see if those predictions are verified by experiment. There is no other way to proceed. We obtain thereby a degree of probability which very often is scarcely less than direct proof. And when there is a great number of such verified predictions, particularly if some of them are brand new, then it seems to me there ought to be a very strong confirmation of my ideas." [Christiaan Huyghens from the Preface to his 'Treatise on Light', 1690]

(14:1) YOU ARE NEVER GOING TO BELIEVE THIS.

It is the contention of this book that thinking, in the sense of deciding on the basis of evidence, is a very ancient survival mechanism shared with our forbears and cousins in the animal world. It is a form of gambling based on incomplete and imperfect information – which is all most of us can have in real life. While it may seem crude, it can obviously be damned effective otherwise we wouldn't be here to argue about it. Indeed those who were less good at it are now extinct.

Since we all desperately need this skill Nature couldn't afford to leave it to parents to teach it to their offspring. In other words it is not cultural but an organic part of our being, deeply hidden, automatic and therefore not easily unearthed by logical analysis. It is not surprising therefore that philosophers, scientists and logicians have confused themselves, and nearly everyone else, in trying to decipher how it works. Any discussion of the matter before Darwin and Wallace (1858) was in any case bound to fail (history often happens in the wrong order). And, unfortunately, later attempts have been heavily confused by those earlier misconceptions based on deduction, religion and mathematics. Because they saw thinking as the highest evocation of the human spirit many of our scholarly forbears wanted it to have a god-like aspect. Now that God appears to have retreated from the world we can look at the matter with an entirely fresh eye; serious thinking is about making decisions within our limited world, not drawing final and absolute conclusions about Eternity.

As a reader why should you bother with this history? *Because the story isn't over yet.* I suspect most contemporary educators will go to their graves before conceding that serious thinking is as natural as CST. If true then what further need do we have for many of them? Education is a colossal and ever-expanding industry with vast investments in plant, manpower and training. It is hoping and expecting to devour something like a third of your life and much of your savings. Before you surrender to its business plan shouldn't you find out about its dubious past and indeed its questionable present?

It is possible that we will never sound the bottom of Common Sense Thinking (CST), but it is certainly worth having a try. We shall take the view that combining evidence together and taking a gamble on the Odds obtained from the Detective's Equation is the trick which makes it work. It is certainly crude enough for simple animals to use – evaluating only one clue at a time. But eventually it could become highly sophisticated as a mind evolved capable of gathering, holding and weighing several clues together *simultaneously*. And if we are right about it we can now transcend the limits of our memory capacity by taking up pen and paper to combine any number of clues together in an Inference Table. That promise alone makes it worthwhile exploring the full intricacies of the DE.

Taking this point of view then, the history of Thinking becomes a search for what obscured the truth of this animal subject for so long. That is the theme of this chapter. Historians call this a 'Whig' point of view: imagining one is at the apex of civilization, then looking down and back to the crucial steps that led up towards the summit. The snag of course is that one might not be standing on the summit at all. For lack of an obvious alternative I will take this vantage point for now but, as Einstein put it, "Whoever undertakes to set himself up as a judge of Truth and Knowledge is shipwrecked by the laughter of the gods".

The literature on Thinking is vast, but I note that practicing scientists rarely take much notice of such literature today and there is little evidence of any more reverence in the past. Practitioners regard science as a hands-on trade to be picked up from one another, and from their most successful heroes. If that is so we can disregard the philosophical and sociological literature as largely irrelevant to the subject at hand. That is certainly too harsh, but it is broadly true and greatly simplifies our quest.[12] [But see **Appendix 7** '*Philosophy and our book*']

---

[12] To quote from a rare bird , both a distinguished scientist and a distinguished historian of science, George W Corner:{1} '..few scientists have attempted to analyze their own thought processes. They get on with their work and arrive at their conclusions without deeply considering the mental pathways they followed. Those who have attempted to analyze the scientists method of thinking, from Francis Bacon and John Stuart Mill to twentieth century writers, have mostly not been scientific investigators but philosophers and logicians. Working scientists have generally regarded with indifference the attempts of philosophers to categorize the steps of

The history of Thinking is quite literally staggering; for two millennia it has lurched from one side of the truth to the other like the drunkard who ' …went to Birmingham by way of Beachy Head'. The most learned men set off on a 2000 year wild-goose-chase in pursuit of Deduction. But when they finally caught the creature it turned out to have a limited knowledge of Science. Then a Dutchman called Christiaan Huyghens (1690) glimpsed the real secret hiding in a gambling den (see above) – but couldn't persuade his contemporaries to come in and embrace it. Fifty years later (1739) a Scottish philosopher, David Hume, argued that it all defied logic. However in 1761 the Reverend Thomas Bayes died in Tunbridge Wells leaving an enigmatic theorem in his posthumous papers. Thomas Price his brilliant Welsh executor claimed that ".. it would give the first clear account of analogical and inductive reasoning" and got it published in 'The Proceedings of the Royal Society'. The British ignored it, so twenty years later Price went over to Paris to try and convince the French. A brilliant Frenchman called Laplace got to hear about it, but proved, or so he thought (1812), that it wasn't practically relevant. Then Statistics got invented and universities set up Statistics departments (1900) inside Schools of Mathematics – a truly disastrous decision, if understandable. Total chaos reigned for a century because mathematicians don't believe in Induction, the heart and soul of Science. Students were either terrified into submission – or ran away. Scientists put their hands over their ears and sulked off to their work-benches: 'A plague on all of you – philosophers, mathematicians, statisticians, historians, sociologists… maybe *you* don't know how it works – but we can do it all the same, we use it every day.' And that is

---

scientific thought, for example sharply contrasting 'deduction' and 'induction', for they do not recognize these processes as distinguishable in their own work."

roughly where we are today. But the truth is out there all right; Huyghens clearly glimpsed it back in 1690. In fact it has been around for tens of millions of years. Seagulls are experts at it. It's in our very bones. I contend that we should attend more to them – and less to the hubbub of scholars.

(14:2) THE GEOMETER'S WORLD.

To our knowledge the Ancient Greeks (~ 800 to ~ 200 BC) were the first serious thinkers about Thinking. They were obsessed with Geometry. Above the gates of his Academy Plato had inscribed "Let no man ignorant of Geometry enter here". In their eyes Geometry proved that men could arrive at truths about the real (i.e. the physical) world by taking Deductive thought, and that became the paradigm for much of science for nearly two millennia. As late as 1686 Isaac Newton was writing about the Solar System in terms of 'Theorems' and 'Lemmas' as if it was an exercise in Euclidean geometry. That geometry was based on axioms (fundamental presuppositions) that were taken to be 'self-evident', so obviously true of the real world that no sane mind could doubt them. Later Christian admirers of the Greeks somehow convinced themselves that Geometry, indeed Mathematics in general, was a miraculous portal into the very mind of God:

"When the first mathematical, logical and natural uniformities, the first laws, were discovered, men were so carried away by the clearness, beauty and simplification that resulted that they believed themselves to have deciphered authentically the eternal thoughts of the Almighty. His mind also thundered and reverberated in syllogisms. He also thought in conic sections, squares and roots and ratios, and geometrized like Euclid. He made Kepler's laws for the planets to follow; he made velocity increase proportionally to the time in falling bodies, he made the law of sines for light to obey when refracted…..He thought the archetype of all things and devised their variations; and when we rediscover any one of these wondrous institutions, we seize his mind in its very literal intentions."   William James '*Pragmatism*' (1907).

Doubts began to creep in only in the 18th century whe {2/2} mathematicians themselves began to question Euclid's arguments more closely, in particular his "Parallel Axiom": 'Through a given point P not on a line L there is one and only one line in the plane of P and L which does not meet L'. Not everybody found this axiom 'self-evident'. For instance on the surface of a globe a 'straight line' [defined by Euclid to be 'the shortest distance between two points'] becomes a 'Great Circle'. And no such 'straight line' can be drawn through P which doesn't meet every other such straight line, because all Great Circles meet twice. To the objection "But real Space isn't curved like the surface of a globe" – who was to say, short of a very precise physical experiment? Perfectly logical (i.e. self-consistent) geometries were then found that didn't obey Euclid's Parallel Axiom. So the longstanding identification of Euclidean Geometry with Physical Space collapsed in ruins. Doing Geometry wasn't doing Science after all. And worse was to follow with the discovery of algebras containing quantities A and B that didn't 'commute' (i.e. BA didn't equal AB) like the real numbers in the physical world. It had taken 2000 years to recognize that Mathematics was a 'game' with plastic rules (axioms) that could be changed. Whether a particular game corresponded to some aspect of the physical world could be established only by experiment. Mathematics had lost its God-like aspect, though not its usefulness. Indeed some very bizarre mathematical creations – such as imaginary numbers and matrices – have proved so generally useful that they are taught to undergraduate engineers. They are useful because they behave like *some* aspect of the physical world; today we think of them as man-made tools – like hacksaws or screws – not messengers from Heaven.

The scientific method, according to Aristotle, was 'Hypothetico-Deductive'. You observed the world, made up hypotheses about how it worked, then deduced the consequences of those hypotheses. If those consequences agreed with further observations then well and good; if not it was time to hypothesise again and repeat the same cycle. There was little or no place for experiment, Induction or gambling. Although some advocates have tried to claim otherwise, Greek science wasn't notably successful. The Greeks were great theorisers and articulate persuaders but it is hard to think of much good science (as opposed to mathematics) that they left behind, but easy to think of much bad. Even their magnificent astronomical theories (a Global Earth and Heliocentrism) failed to survive the succeeding centuries. A network of Inductive vines has far more robust strength than any deductive temple, no matter how elegant its columns.

Even after their official divorce in the 18th century the two-thousand year union of Science with Mathematics left behind some very disturbed children. For instance there were philosophers like Karl Popper [12:2] who refused to accept the parental break {13/1}. His recently fashionable method of "Conjecture and Refutation" is nothing else than the Hypothetico-Deductive creed in other words. As far as I can understand him, which isn't far, Popper would not concede that Hume's 'Problem of Induction' can be overcome by gambling (or 'Probabilistic Reasoning'). Experiments show that it can, even if such reasoning cannot prove that its conclusions are 'absolutely true' – whatever that means.

Then there are mathematical mystics such as 'String Theorists' today who believe some 'Theory of Everything' will emerge from their equations, computer simulators who claim to find profound patterns in their digital tea leaves, and an army of deluded Statisticians who apparently believe that Inference can be done without Induction. They all hark back to an imaginary golden childhood set in Periclean Athens. In my opinion they should read their history more carefully. That Athens quickly came to a grisly and self-induced end.

(14:3) THE INSTRUMENTAL WORLD

That the world is *not* 'Self-Evident', as the geometers supposed, is arguably the greatest scientific discovery ever made. How precisely it came about, and in what order, must be the study of specialised historians, working with magnifying glasses of their own. But when at last it did come it came very quickly. Look at some dates:

1 Invention of spy-glass/telescope (about 1570).
2 Tycho's accurate observations of Mars' orbit, 1575 - 1598.
3 First modern scientific textbook; Gilbert's 'De Magnete'(Of Magnetism) appeared 1600.
4 Invention microscope ~ 1600 Holland.
5 First Scientific Society (Academei Linceii, Rome) 1602.
6 Francis Bacon's 'Advancement of Learning' proclaims that 'Knowledge is power' 1605.
7 'On the Motion of the Blood' William Harvey, 1606.
8 Galileo's discoveries with his spyglass, 1609.
10 Kepler's Laws of Planetary Motion 1609
11 Steno's 'Prodromo' ; the manifesto for Geology. 1637

12 Hooke's 'Micrographia', illustrating the microscopic world, becomes best seller. 1665. Samuel Pepys said of it 'the most ingenious book that I ever read in my life.'

13 Newton splits white light into colours and reconstitutes it; 1666.

14 van Leeuwenhoek sees his 'animalcules', 1676.

15 Newton's 'Principia' , the book that convinced us that the natural world was a machine, not a plaything of the gods. 1686

16 Huyghens explains scientific Induction (see later) 1690

The seminal role in this revolution may have been the invention of the spy-glass or telescope – because it came first and caused the greatest shock to mankind's mental state. It was probably developed in England by Leonard Digges as a natural extension of his earlier invention – the theodolite. As a surveyor he would obviously want to see further and with more precision than the naked eye would allow. His son Thomas wrote this in a publication of 1571:

"My father, by his continual painful practices, assisted with demonstrations mathematical, was able, and sundry times hath, by proportional glasses duley situate in convenient angles, not onely discovered things farre off, read letters, numbered pieces of money with the very coyne and superscription thereon cast by some of his friends upon the downes in open fields, but also seven miles off declared what hath been done in that instant in private places."

That account could hardly have been fabricated 38 years before Hans Lippershay's alternative claim to the invention in Holland. But quite possibly, like so many later inventions, it was made independently – an idea that was 'in the air' at a time when eye-lenses were becoming widely available. Poor Digges, in prison and under sentence of death by the Catholic Queen Mary Tudor, never exploited his telescope scientifically and it was left to Galileo using a spy-glass made of probably superior Venetian glass, to announce that there were mountains on the Moon; that Venus, like the Earth, orbited the Sun; that there were moons circling Jupiter; and that the Milky Way was composed of innumerable faint stars. Like Digges, Galileo was imprisoned by the Roman Church for his pains whilst poor Giardano Bruno was burned at the stake in 1600 for imagining other worlds.

Such beastly acts remind us that Thinking and Authority must often clash. After Roman soldiers killed Archimedes in 212 BC the ghastly Roman Empire – based on slavery, circuses and crucifixion, successfully stifled thinking for about 1800 years. The Romans thought of themselves as 'civilized' – literally 'living in cities' – but their town walls (or their lead pipes) appear to have entirely cut off the light of curiosity. So far as science is concerned the Dark Ages arrived not when the Romans left (~ 400 AD) but when they came. And even as their legions melted away an official Roman Church (380 AD) was set up by the Emperor Theodosius. It proved to be a mightily oppressive and long-lasting form of Thought Control. Saint Augustine of Hippo (354 – 430 AD) wrote this: "There is another form of temptation, even more fraught with danger. This is the disease of curiosity......It is this which drives us to try and discover the secrets of nature, those secrets which are beyond our understanding, which can avail us nothing and which man should not wish to learn."

The first chinks of light peeped into this dreary age in the 13th and 14th centuries AD, both in Norman Europe and Renaissance Tuscany. Why there and why then nobody seems quite certain but it may have had to do with travellers like Marco Polo (1300) bringing news of an entirely different and intriguing civilization possessed of gunpowder and printing – in China.

It was the prospect of trade with the Orient which drove first the Portuguese to develop their caravels and find their way to Asia, (Vasco da Gama via The Cape of Good Hope in 1498) and later the Spaniards via the tip of the Americas (Ferdinand Magellan in 1520). Even Columbus (1492) imagined he'd reached Cipangue (Japan). The cat was now truly out of the bag and the suppression of new knowledge coming in by sea was beyond even the torturers of The Inquisition. Finally the invention of a printing press with interchangeable type (Guthenberg 1450 Mainz Germany) ensured that the incoming light would swell and spread throughout the Christian world. Its suppression within the Muslim empire ensured their own eclipse within two centuries.

From the point of view of thinking, books brought about two separate revolutions. They by-passed the stifling power of official education. You can be sure that church and government – often in bed with one another– will control Education. Now curious men (and even women) could learn for themselves, in their own language, without having a bigoted priest instructing them in some deliberately befuddling mumbo-jumbo such as Latin. It is hard for us to comprehend what a liberation that must have been – and still is for the avid reader.

The second great effect of printing was the preservation of knowledge. Previously discoveries and ideas, even seminal ideas like the sphericity of the Earth – could be lost or suppressed for millennia. Not any longer. Now curiosity and research could move forward without any fear of slipping back. The printed pages in their millions became the ratchet of eternal unstoppable progress. Knowledge could build on knowledge. An absolutely essential part of any scientific research paper is the bibliography of references at the end – often more valuable than the author's thesis itself. Thus one scholar can trace and follow another's path through the labyrinth of learning. Evidence can be piled upon evidence, weighed and put into the scales of progress. Great stories can be told of bibliographical triumphs. We mentioned Palla Strozzi's (1400) re-discovery of Ptolemy's 'Geographica' in the library of Santa Sofia in Constantinople. Here's another. The single invention which did most to win the Second World War was the Cavity Magnetron devised by Randall and Boot at Birmingham University in 1940. Roosevelt called it 'the most precious cargo ever to reach these shores'. It was the first device capable of generating powerful short-wave radar radiation. John Randall spent his summer holiday in Aberystwyth in 1939 and one day, browsing some old books for sale in a stall on the sea-front, he came upon a dusty tome written by Heinrich Hertz the German pioneer of radio. Curious as to how Hertz could have produced short-waves back in 1877 Randall bought the book and discovered that a sort of spiral generator had been involved. That gave him the idea for the cavity magnetron and Nazi Germany was to suffer the terrible consequences as enormous bomber-streams were led right to their targets by short-wave radar [You have a cavity magnetron in your home: it drives your microwave oven].

(14:4) THE STRUGGLE FOR INDUCTION.

The three giants of 17th century Physics, Galileo, Huyghens and Newton, all wrote about the Scientific Method as they conceived it. Galileo's greatest achievement, of many, was to switch attention from the question of 'Why' (theology) to the question of 'How'. 'How does the cannonball fly?' could be answered by measurement and described by algebra – thus Galileo's obsession with Mathematics: " Philosophy (Nature) is written in that great book which ever lies before our eyes – I mean the universe — but we cannot understand it if we do not first learn the language and grasp the symbols in which is written. The book is written in the mathematical language, and the symbols are triangles, circles and other geometrical figures, without whose help it is impossible to comprehend a single word of it; without which one wanders in vain through a dark labyrinth."

Galileo also referred frequently to 'experiments' – as for example dropping a cannon-ball and a musket-ball side by side from the Leaning Tower of Pisa – to see if both fell at the same rate. Historians now believe he carried out most if not all such 'experiments' only in his head.

Newton had no such inhibitions. He thrust a bodkin behind his own eye-ball to find out how the lens worked. He was a giant in every aspect of science: in Experiment (he split and recomposed white light with prisms); in Instrumentation (the reflecting telescope – forerunner of all big telescopes); in Theory (universal gravitation); in Mathematics (Calculus) and in Method. He laid down four 'Rules for Reasoning in Philosophy' (the word 'science' wasn't used then) which included Parsimony (Ockhams Razor: Chapt.7), Uniformity, Experiment and Induction: "In experimental philosophy we are to look upon propositions collected by general induction from phenomena as accurately or very nearly true, notwithstanding any contrary hypothesis that may be imagined, till such time as other phenomena occur, by which they either be made more accurate, or liable to exceptions." It's not always easy to get at Newton's meaning – particularly as he wrote in Latin. But he appears to be elevating experiment above theory (hypothesis) without explaining how one can make the crucial step from experiment to 'proposition' (i.e. hypothesis). As far as that is concerned he elsewhere wrote "In experimental philosophy (science) particular propositions are inferred from phenomena, and afterwards rendered general by induction."

It sounds to me as if Newton thought of Induction as a process, like Deduction, capable of reaching conclusions that are certain. If so we now know he was wrong. But Newton later changed his mind and wrote in his 'Opticks'(1704) "...although the arguing from Experiments and Observations be no Demonstrations of general Conclusions, it is the best way of arguing which the Nature of Things admits of."

Even at his best Newton's writing is obscure. The net result was that his contemporaries and successors have looked upon him as a genius, even a magician. His achievements were undoubted but his methods remained hidden from sight – perhaps deliberately so. He was a solitary who didn't enjoy having to explain himself to lesser mortals and appears to have spent most of his time communing with his private god[13].

The third colossus of 17th century Physics, the Dutchman Christiaan Huyghens (1629 to 1695), came chronologically between the other two and like them had immensely wide interests: clocks, calculus, mechanics, telescopes, astronomy, navigation, but above all optics and the wave theory of light – which he invented. While spending a year in Paris as a young man he learned of the correspondence between Pierre de Fermat and Blaise Pascal in which the Theory of Probability was first developed – as an aid for gamblers. Back home at the Hague he wrote 'On Reasoning in Games of Chance' (1657) the very first treatise on Probability. He then went on to become an ace experimentalist.

In science the really big breakthroughs are often made by the first

---

[13] He had a rational basis for his belief. He reckoned that the Solar System is unstable, like a pencil balanced on its point, and wrote "The motions which the planets now have could not spring from any natural cause alone but was impressed by an intelligent agent." Modern research suggests he was right about the instability but that we are *probably* OK for the next few billion years . After that though......

man to know, and put together, two previously distinct fields – and so it was here with Probability and Experiment. As a result Huyghens could write:

"There are in (Optics) demonstrations which do not produce the same kind of certain Laws as Geometry – which is based on fixed and incontestable Principles. Here we first assume the Principles, make predictions from them, and look to see if those predictions are verified by experiment. There is no other way to proceed. We obtain thereby a degree of probability which very often is scarcely less than direct proof. And when there is a great number of such verified predictions, particularly if some of them are brand new, then it seems to me there ought to be a very strong confirmation of my ideas." [From the preface to his 'Treatise on Light', 1690]

I think the above paragraph is one of the profoundest, perhaps the profoundest in all of scientific thinking. It lays down for the first time the true recipe for scientific progress. Even now it can hardly be improved upon. It tells us that we can advance by inference, that inference depends on Induction, and that Induction relies upon the confrontation of Hypothesis with Experiment to yield a conclusion which is probable but not certain. However, by *combining* enough evidence together, we can progress towards Inductive Conviction. In my opinion Huyghens' insight is the theme tune of modern science. [The Detective's Equation is the algebraic formulation of Huyghens' theme.]

Huyghens had grasped the real secret of Induction – its ability to *compound* evidence until the Probability in favour (or against) some hypothesis amounts to virtual certainty. This argument is *fundamentally quantitative* and so makes little impact unless you see numerical

Weights multiply together to rapidly converge on convincing Odds. Huyghens did not demonstrate that so his contemporaries and successors – even Newton – missed his most crucial point. Thus Compound Induction – which might have carried the day as early as 1690 – slipped from man's grasp for another 300 years.

The great Scottish philosopher David Hume gave a profound discussion of Induction in 1739. He emphasised its irrationality using the argument ; "You can't use Deductive logic to validate the Inductive kind; nor can you use Inductive logic, because that would be to argue in a circle. There being no other type of logic Induction is indefensible." Nobody has been able to refute his argument, which has rightly had a profound effect on Philosophy, Science, Mathematics and Statistics. Scientists who need to argue from particular instances to general principles (Laws) first have to make two profound assumptions: first that there is a 'Uniformity' to Nature i.e. 'Tomorrow will be much like today' or 'The water in Peru will be much like the water in Wales'; and second 'The Principle of Limited Variety' – that there are only a limited number of hypotheses to choose from (see Chapter 13). Unless you go along with those two assumptions [which Hume did – he called them 'Custom'] it would be impossible to lead a rational life – all would be chaos*. So most of us adopt them as necessary to survival, and some of us then go on to apply Huyghens' version of Common Sense Induction to do science: i.e. 'Gather more clues, weigh them and multiply their Weights together, until you approach Inductive Conviction.'

However this argument proved to be a step too far for the innumerate – perhaps because of the notational problem which came about through ignoring Odds in favour of Probabilities. Even the highly numerate can be excused for missing the DE – when it is written out clumsily in Probabilities (14:8 below). Even scientists (who generally ignore philosophy) couldn't ignore Hume's attack and couldn't refute it either. So, after its brief sojourn in the philosophy of Francis Bacon (1605) and the science of Huyghens and Newton, Induction was forced back underground again. People used it all right – they had to every day – but generally speaking they didn't admit to it for fear of coming under philosophical fire. As one wit put it 'Induction became the glory of science but the scandal of philosophy.'

[* N.B. Even if the Principle of Uniformity applies to the natural world there is no reason to extend it to human affairs. That being so one has to be wary of applying Induction to topics such as Economics, History and Psychology.]

(14:5) THE APPEARANCE AND DISAPPEARANCE OF THE REVEREND BAYES

We next come to another highly curious episode in the already curious history of Thinking; the first appearance of Bayes' Theorem, and then its disappearance {3}. Thomas Bayes, a retiring and retired Nonconformist minister with an amateur interest in mathematics, died in Tunbridge Wells in 1761. Going through his papers his Welsh friend and executor Richard Price – of whom more anon – discovered an unpublished and incomplete manuscript entitled "An essay towards solving a Problem in the Doctrine of Chances." All but one in a million would have consigned the paper to waste. Not Price – he was by any standards a most extraordinary character. Officially Non-conformist minister for the parish of Newington Green – now part of North London, Price was a widely respected political moralist whose views were to influence both the American (1776) and French (1789) revolutions. He was also a considerable mathematician, entrusted with the vital job of working out fair premiums for the life-insurance companies then becoming popular. Price had very influential contacts both at home and abroad and if he thought something was important—then it was important, and Bayes' essay had caught his eye. He edited it, completed it and sent it off, with a long covering letter, for publication by the Royal Society in whose 'Proceedings' it appeared in 1763.

Having spent many a painful hour trying to decipher Bayes' original essay I struggle to understand the excitement. But in his covering letter Price explained "….(it) would give a clear account of the strength of analogical and inductive reasoning, concerning which at present we seem to know very little more than that it does sometimes in fact convince us; at other times does not….." Later he added: "The purpose I mean is, to shew what reason we have for believing that there are in the constitution of things fixt laws according to which things happen and that, therefore, the frame of the world must be the effect of the wisdom and power of an intelligent cause; and thus to confirm the argument taken from final causes for the existence of the Deity." God seems to have been a very busy chap in those days, particularly round North London. [Bayes and Price, being Non-Conformists, couldn't be buried in official hallowed ground. But you can find their graves in leafy Bunhill Fields near Liverpool Street Station, alongside those of other great outsiders including Daniel Defoe and William Blake.]

Many commentators have interpreted Bayes Theorem in almost as many different ways. Some think of it as 'trivial' (i.e. obvious), some think of it as revolutionary, some think of it as unsound. Let's begin with the 'trivial' version.

Obviously the Probability of propositions B & A together is the same as the Probability of A & B together so we can write, in the Probability notation:

$$P(B \& A) = P(A \& B)$$

But the second axiom of Probability Theory (10:3), the famous "Product Rule" (see Sect (14:6) for more discussion of it) is

$$P(B \& A) = P(B|A).P(A) \qquad (1)$$

So we could rewrite the first statement above

$$P(B|A).P(A) = P(A|B).P(B)$$

Or rearranging: $P(B|A) = P(A|B).P(B)/P(A)$ (2)

which is none other than Bayes Theorem in Probability notation. Who can argue with, or get very excited about that?

"Ah" the enthusiast would say "But suppose we call B some interesting hypothesis H, and A some evidence E, then we can write Bayes as:

$$P(H|E) = P(E|H) . P(H)/ P(E) \qquad (2A)$$

where P(E) is the Probability of E occurring whether H is true or not i.e. :

$$P(E) = P(E \& H) + P(E \& \bar{H})$$

Or $\qquad P(E) = P(E|H).P(H) + P(E|\bar{H}).P(\bar{H})$ (3)

if we use (1). So finally, in Probability form, Bayes' Rule is:

$$P(H|E) = \frac{P(E|H).P(H)}{P(E|H).P(H) + P(E|\bar{H}).P(\bar{H})} \qquad (4)$$

"So" the enthusiast would say "Bayes' Rule in the form (4) is offering to do the tricky problem of Induction for us. It is putting a Probability P(H|E) on some hypothesis even if there is only a single piece of evidence E bearing on it. Now we have an algebraic formula for Huyghens' Inductive scheme – albeit for only one clue."

"Yes," replies the critic "But it is useless because the so called Prior P(H) on the right hand side is completely arbitrary. You can't so easily cheat your way round the Problem of Induction."

We don't intend to re-fight the 250 year old battle over Bayes' Theorem (Rule) but one can see when and why it started out. Richard Price was clever enough to recognise its potential significance, where Bayes himself *may* not have done (it's hard to tell). Certainly he made no bold claims for it. 'Bayesians' fight under a number of different banners (later) but broadly speaking they all believe that David Hume's 'Problem of Induction' can be overcome by using Bayes Rule in the form (4), [or its Odds equivalent: $O(H|E) = W(E|H) \cdot O(H)$.].

Twenty years after trying to convince the Royal Society Price went over to Paris to convince the French. He met Condorcet, secretary of the French Academy, who in turn showed Bayes' essay to the great Laplace who was interested in applying Probability Theory to astronomical and tidal observations. Laplace had enunciated (without justification) a Prior-less version of the theorem independently [i.e. $P(H|E) \sim P(E|H)$], but he acknowledged Bayes' precedence, and the legitimacy of the Prior. [In his limited applications Laplace hoped to finesse the Prior problem by assuming that all hypotheses were equally likely, and so could be given equal Priors]. Later however Laplace seems to have realized that there was an ingenious way to dodge the contentious nature of Priors: to find Weights so decisive that the uncertainty in a Prior wouldn't seriously affect the Posterior $O(H|E)$ – which of course is what one is really interested in. Suppose one's Weight is 1,000 then

$$O(H|E) = 1000 \cdot O(H)$$

and it doesn't now matter whether one chooses the Prior $O(H)$ to be 4 or ¼ ; one is still going to reach a decision decisively in favour of H.

This may have been the main attraction of the infamous 'Central Limit Theorem' which Laplace 'proved' about that time (1810). As you will recall (11:6) it purported to demonstrate that Measurement Errors would follow a 'Normal' distribution which can indeed deliver very decisive Weights in either direction (for or against). Thus the problem of the pesky Priors was dodged for almost two centuries. The problem seemed to have become a matter of academic, but not any longer of practical interest. Bayes' Theorem retreated into the shadows almost before it had come to light. Use could be made of it in games of chance where the Weights and Priors had limited elasticity (one could for instance compute the Prior Odds that one's opponent holds that Ace) but it needed not concern practical men and scientists. And when the mathematical statisticians came along at the beginning of the twentieth century they scoffed at Bayes Theorem calling it 'The Method of Inverse Probabilities' as if that was a crushing argument.

The fundamental point is that: *'Scientific inference appears to require Induction but Induction is, logically speaking, unsound. There may however be a way round it, but only in a probabilistic sense. If, and only if, you are prepared to gamble, then you can make progress.'*

This is the Faustian (or should one say Bayesian) bargain that not everybody has been prepared to take. And even some of those who have taken it have not been willing to acknowledge their sin. [They smoked but they 'never inhaled'.]

Trying to unravel exactly what men like de Moivre, Bernoulli, Bayes, Price, Laplace and Gauss meant in modern terms has foxed some of the keenest minds in Scientific History. As I read their arguments coming down across the centuries I get a fine sense of men groping feebly and clumsily in the dark for holds which will enable them to ascend towards some sensed but imperfectly perceived light. This is how science is really done; how progress is painfully made. Very rarely does some single bold leap succeed. They made 'foolish' mistakes – or missed leads which their successors found obvious. For instance the great Laplace, and great he certainly was, made all sorts of unjustified assumptions (e.g. uniform Priors) and never reached the Normal Distribution – until Gauss – using a circular (i.e. illegitimate) argument, put it in front of his nose {4}. What I admire so much about the little Norman is not his 'brilliance' but his sheer dogged persistence. He returned again and again, over 39 years, to the problem of inferring hypotheses from data! In science, as in much else, 'It's dogged as does it'. Appx. A13 *A Sketch Map of Thinking* may be helpful here.

(14:6) SUBJECTIVE BAYESIANS

If this book is to have a hero then it must be Bruno de Finetti (1906 to 1985), first an actuary in Trieste in the 1920s, later a professor of Mathematics in Rome. De Finetti was happy to acknowledge that many Odds (or Probabilities) in real life would have to be more or less, and sometimes entirely, 'Subjective' – that is to say a matter of personal judgement. If you are losing at cards in a gambling den you have to estimate whether your opponent is a cheat – you would be a fool to say 'I have no objective means of knowing'. You have got to do the best you can, using personal experience and judgement. He began his book 'Theory of Probability' by declaring "My thesis, paradoxically, and a little provocatively, is simply this: PROBABILITY DOES NOT EXIST….the only relevant thing is uncertainty". What he meant was that it wasn't the cards, or the dice, or the electrons, or the measurements that mattered, but only the degree of personal 'uncertainty' in your own mind as to their import. He first put Probability Theory on a broad but sound footing, basing it on gambling. One's Probability for an hypothesis H was to be measured by the fraction of all one was prepared put into the pot to win a bet on H turning out to be true. He then showed that the famous Product Rule of Probability Theory i.e. $P(A\&B) = P(A|B).P(B)$ follows quite naturally from this definition whenever a bet was 'coherent' [A 'coherent' bet is one fair to both sides. If one contestant sets the Odds the other should be free to choose which side to take. Much betting is 'incoherent' i.e. rigged so that the bookie usually wins, the punter usually loses]. And once you have the Product Rule, Bayes' Rule is elementary. Thus he founded the school of 'Subjective Bayesianism' to which I, and I believe most Common Sense thinkers belong, even if they don't know it. And in his two-volume book *'Probability Theory'* {11/10} (on p 169 vol. 1) he came very close to, but didn't quite grasp the DE. His important idea of "coherence" is explained in Appx. A13.

    De Finetti's version of Uncertainty was both honest and liberating but, when they found themselves in the raffish company of gamblers, sportsmen, financiers, bankers and the like, it was too much for many scientists, most statisticians and all mathematicians. De Finetti was drawing no distinction between 'scientific thinking' and the other varieties. He responded to critics by writing; "Some authors (notably R.A. Fisher) criticise applying these ideas to problems of scientific research, regarding them as essentially economic in nature and incompatible with pure research. There do not exist two entirely different forms of valid reasoning, one suitable in a commercial context, the other for pure research. No one working in a scientific field considers it beneath him to use the same arithmetic operations, or calculating machines as are needed for commercial purposes." Well said.

Unfortunately for him, and unfortunately for progress, de Finetti never received the credit he deserved during his own life. This may have been due to a number of factors. As a young man he was a vociferous Fascist (as were many of his fellow countrymen). He wasn't an academic until later in life but had to earn his living. His writing is confusing, to say the least, and he wasn't translated into English until 1975, 50 years after his pioneering work.

(14:7) OBJECTIVE BAYESIANS.

In 1939 Harold Jeffreys, an astronomer and geophysicist at Cambridge University, gently reminded statisticians that Bayes Theorem, including a Prior, was actually right – and could be proved on the basis of rather general considerations, or 'desiderata' as he called them [the mathematicians 'proof' we gave above is too narrowly based to make the theorem widely useable in science]. The problem of the contentious Priors could be overcome, Jeffreys claimed, provided everyone stuck to 'Objective (i.e. impersonal) Priors'. If so then Bayes Theorem would be unimpeachable and could be useful – even indispensable [as in the case of the 'Cancer Scare ' (4:6) or the Sally Clark Problem (11:10)]. Finding such Objective Priors was in practice often fiendishly difficult and then one had to resort to vague or rather silly Priors such as 'All hypotheses are equally likely.' But it was a way to proceed and it gradually attracted adherents who called themselves "Objective Bayesians". They thrive today and their bible is Jeffreys' 'Probability Theory' first published in 1939. They are heartily loathed by conventional or 'Frequentist' statisticians. The Statistics pontiff R.A. Fisher sniffily said of the book in his published review: "…a logical mistake on the first page… invalidates all 395 formulae in (it)."

Sir Harold Jeffreys', as he became, *was* a scientist but his writing was obsessively punctilious and monumentally dull (he and his wife wrote a text book on Mathematical Physics so boring that after university I burned my copy in the garden as an act of revenge). Worse still he lost credibility in the 1960s with his unshakeable opposition to Continental Drift when its truth was becoming triumphantly apparent. It wasn't his opposition so much as the manifestly fallacious nature of his argument. He maintained that because Alfred Wegener's 'explanation' for Drift was incorrect then so must be the phenomenon itself. It is harder to think of a sillier argument – particularly from a be-knighted expert on 'Scientific Inference.' Even so Jeffreys founded what was to become a very influential school which appealed to those who wanted scientific thinking to be as rational, as logical and as objective as possible. Edwin Jaynes, its modern American champion, even wanted to make it so objective that CST could be carried out by a 'robot' – as logic can be built into a computer.

To my mind Objective Bayesianism is completely unsustainable because the Weight in Bayes' Rule, which of course it uses, is: $W(E|H) = P(E|H)/P(E|\bar{H})$ which requires one to define an $\bar{H}$ which, remember, stands for "All those hypotheses, apart from H, that could be responsible for E ". But how on Earth could we do that *objectively* without pretending that we know *all* the relevant hypotheses in the Universe? Such would only be possible in a CLOSED world – like a card game. It is noticeable that text-books on Objective Bayesianism shy away from discussing OPEN systems and Systematic Errors. That's ducking reality.

Although both Jeffreys' and de Finetti's adherents call themselves 'Bayesians' they are philosophically an interstellar distance apart. The 'Objective Bayesians' emphasise logic, dispassion and objectivity, the 'Subjective Bayesians' emphasise experience and Common Sense. The former cling to the ideals of mathematics, the latter to the precepts of empirical science. Decide which school you belong to yourself before you attend to articles written under the Bayesian banner otherwise you are likely to finish up very confused indeed, as I did. I wasted years imagining that I must be an Objective Bayesian, because most of the books in English were written by that school whose students are inclined to propagandize for their beliefs and ignore the alternative. As soon as I realized it was logically OK to be a Subjective Bayesian I switched to that persuasion because I regard experience mostly of value, especially in a field with which I am familiar. Indeed I regard it as generally foolish to disregard experience in the name of 'objectivity'; Nature has equipped most of us non-mathematical creatures with all manner of life-preserving instincts – for judging people, for detecting liars, for sensing danger and fraudulence, for distinguishing between big Weights and small………… and so on and so on. Certainly those instincts are not fool-proof, but Evolution will mostly have winnowed out those which are self-destructive. Thus I regard it as folly to disregard instinct and experience without good reason. It seems to me that Objective Bayesianism is a last hurrah from the misguided mathematical community who once believed that science was a branch of Euclidean Geometry. It is no accident that they are, like their equally misguided Frequentist rivals, excessively mathematical. As far as I can see their scheme will only work if they can ignore $\bar{H}$, i.e. can assume their world is CLOSED (not real); disaster. I won't try to persuade you that they are wrong. If you believe that a robot can generally be made to think as well or better than you can yourself then who am I to disagree?

Alas, more by custom than design, all the main camps stuck to the Probability notation in preference to the Odds notation and so they missed the significance, if not entirely the existence of The Detective's Equatio. It's worth reminding ourselves of the reasons why that miss was so fateful:

(a) On its own Bayes' Rule is rather pallid and not that convincing (see above). It's like a ladder with only one rung`: ingenious but neither convincing or effective.

(b) BR gives no glimpse of the *multiplicative* nature of Induction – which is most of its glory [Huyghens' point]. Once that is grasped Inductive Conviction can be seen as generally superior to the long sought but elusive Deductive variety.

(c) The DE can incorporate weak and imprecise clues which can nevertheless be decisive in alliance.

(d) It provides the essential mechanism for weighing conflicting evidence.

(e) It makes clear sense of Ockham's Razor which, history shows, is an absolutely key argument in scientific debate. And not only in science.(Ch.7) [see Sect (14:11) below].

(14:8) WHY THE DETECTIVE'S EQUATION REMAINED HIDDEN

We now come to a question of huge historical significance – absolutely huge: 'Why wasn't the Detective's Equation discovered long ago? After all it is the beating heart of all Inference, of all Common Sense Thinking.

The answer is simple: 'Thinkers used Probability rather than Odds and in terms of Probability it happens that the Detective's Equation is a ghastly tangle.' Let me write it out for you in the case of only three clues:

$$P(H|E_1,E_2,E_3,......)$$
$$= \frac{P(E_1|H) \times P(E_2|H) \times P(E_3|H) \times P(H)}{P(E_1|H) \times P(E_2|H) \times P(E_3|H) \times P(H) + P(E_1|\bar{H}) \times P(E_2|\bar{H}) \times P(E_3|\bar{H}) \times P(\bar{H})}$$

which I've had to shrink just to get it on a page. Would any animal have adopted it as a survival mechanism? Is it any wonder that it wasn't spotted or used in this form even by philosophers? And even it was occasionally spotted from time to time, as is likely, who was going to advocate it as a method for Common Sense Thinking? And if they had tried who was going to listen? [See Appx A 13]

We can see the roots of the problem when we compare Bayes' Rule written in terms of Probabilities, i.e. Eqn.(4) above:

$$P(H|E) = \frac{P(E|H).P(H)}{P(E|H).P(H) + P(E|\bar{H}).P(\bar{H})} \qquad (4)$$

with its simple Odds equivalent: $O(H|E) = W(E|H).O(H)$. While the latter leads quite naturally to the DE the latter leads to the horrendous mess which heads this section.

By denying themselves Odds serious thinkers missed the Detective's Equation, one of the great accidents of history. By comparison the Battle of Waterloo doesn't rate. We've discovered something profound and not at all obvious here; NOTATION

MATTERS. The way in which we formulate a problem – be it into words or into mathematical symbols – can dramatically affect our ability to solve it. This has become quite obvious in Mathematics – Arabic numerals versus Roman in the discussion of Zeno's Paradox being the earliest example. Indeed Mathematics is largely the effort to find better ways of formulating problems; all those arcane symbols are not merely witch-doctors' skulls. And there can be huge practical consequences. When submarine telegraph cables were first laid under the seas in the 1850s they were horrendously slow – a few words a minute – and thus impractically expensive for most people and purposes. Then in 1874 a self-taught young telegraphist called Oliver Heaviside came across James Clerk Maxwell's monumental 2-volume work on Electricity and Magnetism in his public library in London. He couldn't understand a word of it but he went home and told his parents that he was going to give up his job, live on potatoes, and devote his life to trying to find out why telegraphy was so slow. He had to invent an entirely new mathematical notation called 'Vector Calculus' to translate Clerk-Maxwell's clumsy electromagnetic equations into a much more transparent form. Once he'd done so he immediately spotted the horrendous mistake which Lord Kelvin – the great theorist of telegraphy, and later President of the Royal Society – had made in his own mathematical analysis. Several people, but not Heaviside, used Heaviside's insight to make fortunes out of speeding telegraphy up by a thousand times or more. Heaviside was also able to discover why radio waves bent around the globe; up until then a complete mystery. Yes notation does matter! One of the reasons Einstein stuck to German, even after he went to live in America, was the apprehension that his thinking ability depended on his native tongue. He might well have been right.

If there are two possible notations it really pays to learn both – fractions and decimals for instance. So thinkers ought to familiarize themselves with both Probabilities and Odds. To that end I have provided the next set of exercises:

EXERCISES (14:1) PROBABILITIES AND ODDS

and Appendix A13 *"A Sketch-Map of Thinking"*.

(14:9) BAYESIANS AND DETECTIVES

Nobody called themselves, or was called a 'Bayesian' until the second half of the 20th century – and the origin of the word itself is obscure. It can have little or nothing to do with the use of Bayes' Theorem because all parties concede that, in one form or another, it must be true; even Frequentists call it 'trivial'. Worse still there are at least two tribes of Bayesians with a yawning philosophical gulf between: simon-pure 'Objective Bayesians' generally found hanging around ivy-league universities; and down-and-dirty 'Subjective Bayesians' as happy in a gambling shop as a research lab.

What I believe distinguishes all Bayesians from other kinds of thinkers is the greater emphasis they place on Inductive as opposed to Deductive thinking. At one extreme you have the disciples of R.A. Fisher and Karl Popper who will not concede that Inference can be done with

Induction at all. At the other extreme are the followers of de Finetti who do not distinguish Scientific Inference from any other variety. I am suggesting that anyone who is prepared to bypass Hume's Problem of Induction by gambling is a Bayesian. The word perhaps started out as a euphemism; it conferred a degree of respectability, or at least disguise, on thinkers who would rather not openly defy the rage of ogres like R.A. Fisher who appointed themselves to be the righteous guardians over Scientific Thinking – when all they were were misguided mathematicians.

Why all the sound and fury you may ask. I suspect it has much to do with claiming territory and guarding job-security. If de Finetti's vision is right then what further need have we for mathematically trained statisticians to do our Inference for us? They will become as redundant as ostlers or topmastmen.

(14:10) THE DETECTIVE'S EQUATION VERSUS BAYES' THEOREM.

Some academics become agitated, even angry, if you highlight the Detective's Equation, and not Bayes' Theorem, as the centrepiece of Common Sense Thought – as I certainly do. Let me try to explain why they are right – and why they are wrong.

They are right because, historically speaking, BT became known to (some) scholars before the DE, notably Bayes himself and Richard Price in 1761, then Laplace 20 or so years later. They further point out that iterating the BT, using more clues, is equivalent to using the DE – which is also true, no argument.

They are wrong because it is precisely the multiplicative nature of the DE which makes it the powerful backbone of CST. BT on its own is not that obvious and certainly not that useful – which accounts for the historical fact that it has dropped out of sight so often and for so long{3}, and then had to be disinterred all over again. Take a ladder and saw it into sections each containing one rung. Each H-shaped section is the equivalent of BT. But it is only when you glue the sections back together that you can appreciate at once what a powerful tool you have for surmounting all manner of mental obstacles. It's the chain not the link, it's the spine not the vertebra, it's the ladder not the rung which really matters in practice. In that sense BT is merely the DE in the special case when there is only one clue: interesting, occasionally useful, but not all that exciting. But one glimpse of a ladder leaning against a wall is all one needs to understand its capabilities and its value.

The historical argument is in any case debateable. Bayes himself may not have understood the full import of his theorem and some have pointed to Laplace as its real instigator, though Laplace himself did not appreciate the Prior until it was pointed out to him by Condorcet who had been briefed about Bayes by Price {3}. In any case it was Huyghens much earlier (1690) who first recognized the vital importance of multiple clues for successful Inference, even if he didn't write down a formal equation.

My own point of view is that both the DE and BT are simple algorithms that reach far back into animal evolution, and in their Odds form (only) you can see just how obvious they are. Even the Weightings – Strong (4 or ¼), Weak (2 or ½) and Neutral (1) [as advocated here] are within reach of primitive nervous systems, and indeed safer than the more sophisticated values promoted by Statisticians and others. The single-clued (i.e. Bayesian) version of the algorithm may have developed first but it would have been dangerous to act on only one clue at a time, if it were possible to wait and act on two or more. Thus there would have been strong pressure to evolve toward a multi-clued (i.e. Detective's) version.

Given its fundamental survival value an animal's Inference Mechanism would need to be automatic (unconscious). Thus scholars might find it difficult to disinter. However from time to time they were bound to stumble over it in one or other context and in one or other form: thus Huyghens (1690), thus Bayes (1761), thus Laplace (~ 1790), thus de Finetti (1930's), thus Turing (?1940), thus Robert Schlaifer US (late 1940's){*}, thus myself (2010)….. Claiming any glory for Nature's handiwork therefore seems inappropriate and is certainly no part of my ambition. It is my duty however, as one who is trying to explain to non-academics how Common Sense Thinking may work, to do so in the very clearest way. And that is to highlight the ladder not the rungs, the chain not the links, the `Detective's Equation and not Bayes' Theorem. I make no apology whatsoever for doing that. (See Appendix A13 for the rich interplay between Odds and Probabilities.)

(14:11) OCKHAM'S RAZOR

You will have realized that so far in this potted history I haven't mentioned Ockham's Razor. That is not because it isn't important – on the contrary – but because it is an even slipperier customer than the DE. It operates deep inside our psyche, determining which hypotheses shall be allowed to enter the arena and which shall not. Most aspirants never make it into contention: OR sees to that. Suppose you tell me there's a Martian living in your bedroom – but it is shy and slides away into another dimension every time anyone enters to look at it. I immediately dismiss your story as either a fabrication, or a symptom of madness. But why? Because OR tells me that it will be a waste of time to follow up the hypothesis that your story might be true. How precisely it reaches that judgement, and reaches it fast, I cannot say. But that it is performing a valuable service one cannot doubt. It is the hidden censor that keeps us sane. Without the Razor my head would fill up with such a multitude of distracting and undecidable propositions that I could never concentrate on those few which might determine my fate. So, long before the DE comes into play, our Razor has pared down the legion of hypotheses that the DE will ever be allowed to seriously weigh. Any thinking animal must have such a Razor, so presumably it lies very deep in our being, not easily accessible to rational interrogation. All the same we will have to try because the modern world is full of the kind of hypotheses which a multi-million-year-old organ was never evolved to assess.

In any practical situation the amount of evidence we might find will be limited. Therefore no purpose is served by considering hypotheses for which one's Prior O(H) is too low [or too high]. The evidence then can't, in practice, be such as to overturn it. My Prior on your Martian is so low that it would be fruitless to weigh the evidence in your favour. Likewise with the interpretation of dreams: since the number of interpretations is unlimited the Prior in favour of any one should be vanishingly small – rendering any discussion of it quite pointless. Our internal censor must thus be constantly at work dismissing hypotheses that it considers pointless to entertain.

The situation becomes more interesting when several hypotheses are allowed into the ring for consideration. That is when Ockham urges us to pick the simplest first. Why? Because, even with strictly limited evidence, we could still reach a decision about it either way. Being simple it is rather rigid, and cannot be manipulated to fit much evidence. And if it doesn't fit it can be dismissed. But if it does fit then the Odds in favour of it fitting *by chance* must be rather low, and consequently the Odds in favour of it being actually right must be correspondingly high. In other words we can have a thoroughly constructive discussion of the hypothesis – even with limited evidence. On the other hand, as we saw in the case of the Expanding Universe [Sect 7:1], a complex hypothesis is more malleable, absorbs more evidence in its precise definition, leaving less over on which to assess its correctness. The Odds for it or against become less decisive, and the whole discussion becomes less satisfactory. If we portray the matter in terms of the DE – where the compounded Odds must be multiplied together – the malleability (complexity) of an hypothesis forces us to take out the most decisive Weights in order to define the hypothesis with sufficient exactitude. And you don't have to take out many Weights (one may do) to render the remaining compounded Odds worthless – or at least unconvincing.

Although it is commonly named after the mediaeval scholar William of Ockham (1285 – 1347) the Razor is so fundamental that it was recognized long before him and was know as 'lex parsimoniae' (the law of Parsimony) in Classical times. The problem is that some scholars thought of it as an attribute of Nature (the 'Ontological' explanation) and some merely as a maxim for productive thinking (the 'Epistemological' explanation) – a confusion that continues to this day. The greatest scientists have nearly all been big fans, but have seldom been able to explain why – not convincingly. Thus Newton highlighted it in his 'Four Principles for Reasoning in Philosophy' but explained "Nature is pleased with simplicity and affects not the pomp of superfluous causes." His rival Gottfried Leibnitz (1646 – 1716), observing that light miraculously makes its way from one place to another always by the route that takes the least possible time, took that "as support for the metaphysical principle that God governs the universe in such a way that a maximum of 'simplicity' and 'perfection' be realized." [If you don't believe in miracles then why *does* light behave so?] More recently Einstein used

Parsimony very directly to pick out from many possibilities that equation of gravitation which is mathematically the simplest, afterwards remarking : "God would not have passed up the opportunity to make nature this simple." No wonder lesser mortals have found themselves confused about parsimony.

With two explanations to choose from – the Ontological (Nature / God) and the Epistemological (as merely a maxim for efficient thought) it seems to me we can use Ockham's Razor itself to eliminate the former. As we have seen OR works efficiently so why call upon a vague God – or some mystical preference in Nature – to explain the matter? That is the very kind of unnecessary extravagance which any principle of 'parsimony' (which after all means 'meanness') would urge us to ignore.

There is additional evidence that OR is still a wise maxim for thinkers even when Nature and God are entirely out of the picture. Take military intelligence: R.V. Jones who was chief of Air Ministry Intelligence during the Second World War, and was responsible for assessing the many dangers which the Luftwaffe presented to Britain [including radio-beam-navigation for its bombers, radar, The VI and the V2] wrote a very fascinating book about his experiences 'Most Secret War' {6}. At the end (p 523) when he was summarising the outcome he waxes eloquent on Ockham's Razorgoing so far as to call it "The Cardinal Principle of Military Intelligence……for if you start allowing more complicated hypotheses than are essential to explain the facts, you can launch yourself into a realm of fantasy where your consequent actions will be become misdirected." Earlier he had said: "Time after time when I used Ockham's Razor in Intelligence it gave the right answer when others were indulging in flights of fancy leading towards panic." (P 372)

It is important to emphasis that OR will *not* always lead to the right answer. The Planets do not orbit the sun in circles, but more complicated ellipses. Einstein was forced to retreat from his simplest law of gravitation and add an extra mathematical term. All motion is not circular – as Galileo assumed on the grounds of Parsimony. The Genetic Code [for translating DNA into proteins] is not the simplest Nature could have adopted {3/13}. But looking first at the simplest hypothesis is the most constructive way to open any discussion.

In modern times, particularly with the recognition of 'Noise' and the advent of computers, much can be added on the subject of Parsimony, and I urge scientists in particular to read the material on it in in the wonderful book by Hugh Gauch Jr. whose quote heads CHAPT. 7{*}. Some of his points are surprising, even disturbing. For instance a wrong theory can sometimes make better predictions than the right one – because (being simpler) it may do a better job of ignoring the Noise. So accurate prediction isn't everything, not even in science.

So interesting is Gauch's book (he's a distinguished agricultural scientist) that I cannot end this section better than make two further quotes from it :

"When the truth is not even considered, so that all the hypotheses are false, evidence has a misleading character." (p 277)

"….those scientists who also consider the theory's parsimony, rather than only its fit to the data, are often the ones on the cutting edge of science." (p 321)

(14:12) THE OBSTACLE COURSE: HOW WE GOT HELD UP FOR TWO THOUSAND YEARS

It is time to go back and list, in approximately chronological order, the obstacles that Learning has placed in the way of CST:

(1) The confusion of Science with Mathematics (Geometry in particular) and the consequent adoration of Deduction. [From Euclid 330 BC until the 18th century AD. Alas it even lingers on today.]

(2) Asking the unanswerable question 'Why' rather than the tractable question 'How'. [Up until Galileo ~ 1600].

(3) Despising or at least neglecting experiments. Historians have attributed this to the slave nature of Greek society: 'gentlemen', i.e. philosophers, didn't soil their hands. Later the Roman church frowned heavily on experiments – which might well, and eventually did, contradict its dogma. The Protestants – who believed that a man could commune with his god without the interference of the Church – were more tolerant and it may be no coincidence that experimenters such as Digges, Brahe, Gilbert.......began to flourish almost as soon as the Roman Church lost its veto.

(4) Mistaking Thinking for a divine evocation which *homo sapiens* alone shared with God. This is a powerful myth embedded in all the Abrahamic religions. It was hard to resist before Darwin and Wallace offered a plausible alternative to a Great Designer responsible for all the manifold wonders of Nature, including Man (1859). Even today unimaginative people are inclined to dismiss the thinking powers of other creatures. They should look harder. [See below.]

(5) Permitting 'authorities' to control Education, and thus to a great extent Thinking, was and remains an extremely foolish policy. They can't be relied on to do it disinterestedly, nor, to judge from the results, to do it well. The naïve view that "Education equals Enlightenment" is a myth hardly supported by history. (More on this later).

(6) Avoiding Induction because it appeared to have no logical (deductive) basis. [Dates from a slanted reading of David Hume (1739) and thrives in Statistics departments to this day – often disguised in jargon]. If only people had attended to Huyghens (1690).

(7) Shunning Probability as a way round Induction – because it smacked of gambling.

(8) Dividing over the meaning of the 'Probability' of a proposition between a Mathematical meaning based on 'Relative Frequency in a large number of trials' and a humbler, more general meaning such as 'Degree of rational belief in'. This was the perfect recipe for centuries of confusion. The first is too narrow to be of much use outside Mathematics; the second too vague to be workable without dragging in gambling.

(9) Supposing errors of Measurement usually follow 'The Normal Curve' because the 'Central Limit Theorem' appeared to prove it so [Laplace, 1810. It is mistaking bad-simple for good-simple. Still widely prevalent; may have contributed to the financial crash of 2007].

(10) Rejecting Priors whenever they seemed elastic or based on instinct rather than logic. [Still practiced by 'Objective' Bayesians.]

(11) Misunderstanding and therefore undervaluing Ockham's Razor because it appeared to have no widely acknowledged basis. [Even Newton believed that "Nature is pleased with simplicity, and affects not the course of superfluous causes." while Einstein waffled about God.]

(12) Siting Statistics departments inside university faculties of Mathematics was an understandable mistake – but one with very grave consequences. [1900 onwards]. It put mathematics in the saddle of a subject that *has* to be empirical. It has led to an incalculable loss in lives.

(13) Supposing (in consequence) that Hypothesis Testing could be done without Induction. It can't; see Huyghens.

(14) Adopting 'Probability' notation as distinct from 'Odds' notation, when both, like Fractions and Decimals, were/are essential. [Since the beginning of Probability Theory in the 17th century.]

(15) Consequently missing The Detective's Equation – which is the beating heart of Common Sense.

(16) Neglecting weak but telling evidence for lack of a combining mechanism.*

(17) Insisting on precision where it is spurious. Believe it or not Statistics Tables are published accurate to 4 places of decimals. That alone should be warning to all who take Statistics seriously.*

(18) Struggling with conflicting evidence for lack of a resolving mechanism.*

(19) Preferring Objectivity above Experience and Judgement.*

*All these last for lack of The Detective's Equation.

I suspect we'll find it much easier to understand our own mistakes if we acknowledge that we are not all that much smarter than some of our fellow creatures on this planet {5/9}. You can best judge that for yourself by observing the wild animals in your territory. For instance I am fascinated to watch the Herring Gulls that live around me in Wales, as they navigate their way through the vicissitudes of life. We have one called Aristotle who arrives at my bedroom window at 7 am every day to remind me it is time for his breakfast. They are clever enough to have adapted their whole life-style to make the best of Man's. Instead of making a thin living on the tidal shore they can be seen following the plough far inland, breaking open garbage bags, shadowing in-coming fishing vessels, dining on road-kill, nesting on roofs, utilizing the up-draughts peeling off buildings, stealing holidaymakers picnics, pillaging bird-tables, hanging around fish-and-chip shops, shadowing cross-channel ferries, *gliding* into the eye of a strong wind (work that out!), terrifying domestic cats, even thermaling intercity in search of richer pickings. All the time, like us, they have to calculate the Odds. One finds the occasional corpse (invariably a youngster) on the roadway, but judging from their numbers they must be getting their calculations right at least 99.9 per cent of the time. To attribute such behaviour to 'instinct' is plainly ridiculous. Man's arrogance, or rather lack of imagination, sometimes has no soundings. Tests show that some birds like crows have remarkable problem-solving capabilities whilst others such as Great Tits have memories capable of recalling the exact locations of thousands of seeds which the birds have cached.

The most unexpected episodes of this whole extraordinary story are still to come. First is the discovery that for most of the last 500 years Man has been thinking *less clearly* than his animal cousins.

He got tangled up in a cultural construct called 'Probability' which blinded him to the PAW and deluded him into the belief that precise measurements and sophisticated Statistics could lead him to certainties which, in truth, are completely unattainable. He pulled through though, and far surpassed the accomplishments of his cousins because of a second cultural construct – an aid to his memory – which far outweighed in its good effects, the bad effects of Probability. When he adjusts slightly and puts all the bad effects out of mind, who knows of what he will be capable.

We are however getting ahead of our story because the two big discoveries will emerge in their own good time – in the final chapter. Those who can't bear the suspense might instead peek at Appendix 8 "Certainty, Falsifiability and Common Sense' and Appendix 9 "Categorical Inference and Animal Wisdom".

The Detective's Equation does indeed suggest a plausible evolutionary path for the animal mind. Even some zooplankton might be capable of associating one piece of evidence (e.g. light) with an hypothesis (more food) and so 'decide', with Bayes, to take the nocturnal migration upward to graze on surface-feeding algae, and the diurnal migration downward in search of safety. A mature Herring Gull probably develops dozens of sophisticated Prior-sets to cope with different situations in its life. And the longer it lives the smarter it becomes. Moreover, for flight control, birds have brains which work ten times faster than our own. Thus a 30 year old gull may have accumulated the mental experience and wisdom of a 300 year old man. [Today most of us don't develop much wisdom – life is too easy to need it or provide it. Even the word itself is dying out.]

It is fascinating to note that all the obstacles here listed have between them held up CST for something like 2,100 years between the time of Aristotle (300 BC) and today. Imagine where we might now be had we been more enlightened. By that reckoning we could have had jet engines before the crucifixion of Christ! It is an amazing thought, but it does presuppose that our ancestors would have been able in a short time to take the same sophisticated steps in *technology* that we found necessary. I'm thinking of the inventions of printing, of the telescope and microscope, and more particularly the technologies that made the ocean voyaging of Darwins and Wallaces routine: anti-scorbutics, astro-navigation and marine chronometers capable of keeping Time to a precision of one second a day (Darwin's vessel The Beagle carried 24 chronometers). That seems a big ask; Development usually takes much longer and is far more costly than Research. Judging by most non-European civilizations we might instead have had no progress at all. To speculate is entertaining but History, either as a subject or an object, does not easily lend itself to Common Sense Thinking.

SUMMARY

Thinking, in the sense of deciding on the basis of evidence, has had a tangled history. Obstacles we have placed in our own way include Authority, both imposed and too meekly accepted; rotten education; awe at the apparent order in Nature; a craving for Deductive Certainty; confusing Mathematics for Science; laziness in Experiment; arrogance as a species; an imagined affinity with God; a moral reluctance to gamble, and latterly an unjustified suspicion of Induction. How could its obvious fault-line, its irrationality, be overcome? In short by Odds, by gambling. This was not the answer expected, or welcomed, by most philosophers, mathematicians or divines; no wonder they resisted. Anyway before Darwin there was no plausible Theory of Creation which could do without a God. And afterwards Thinking was plagued by clumsy notation [no 'Odds'] – which has largely concealed the Detective's Equation – the real secret of Induction [not Bayes' Rule] until recently. Full blooded Subjective

Bayesianism wasn't enunciated until the 1930s (by Bruno di Finetti) – and is still not wholly respectable.

Then we subjected ourselves unnecessarily to serious Systematic Errors, not realizing that there is a mechanism (The Principle of Animal Wisdom PAW) for discounting them by guillotining too high or too low Weights.

Finally Parsimony, one of the two most fundamental principles of thinking (along with the DE), has long been used, but for contradictory reasons, one of which implies that it is only relevant to Nature (science). We can see today that that simply isn't true. All thinkers can and should understand Parsimony, and employ it consciously. This is particularly true in a modern world where so many hypotheses cannot be assessed intuitively.

History is a fascinating but very hard subject, particularly when one is dealing with something as elusive as 'Thinking'. It's hard enough to know what is going on in one's partner's head sometimes. How much harder to be certain what passed through the mind of someone centuries dead who lived in a society which could be ruthless and punitive. The most eminent Chinese historian was castrated for departing from the (then unwritten) party line. Much of history is therefore little more than crude propaganda. I'm very conscious that I may have missed out some vital steps here. George Orwell remarked: "To see what is in front of one's nose needs a constant struggle."

But we shouldn't be too hard on ourselves, or on our predecessors. We have been groping feebly and clumsily in the dark for holds which might enable us to ascend towards some sensed but imperfectly perceived light. Isn't that what Life has been doing for over a hundred million years?

THINKING FOR OURSELVES (Disney)
CHAPTER 15
Draft (14/9/18; 8.1 kw)
SCIENCE'S PECULIARITIES

'It is obvious that humans are unlike all animals. It is also obvious that we are a species of big mammal, down to the minutest details of our anatomy and our molecules. That contradiction is the most fascinating feature of the human species. It is familiar, but we still have difficulty in grasping how it came to be and what it means." Jared Diamond; Prologue: 'The Rise of the Third Chimpanzee'. {9}

(15:1) INTRODUCTION

When we set out to look into scientific thinking we concluded that it was mostly animal common-sense, and that led us to ask 'How do animals think?' But now that we have got somewhere with that question we have to concede that animals do not build particle accelerators or send spacecraft to Mars. So in this chapter we spotlight those practices and habits of thought peculiar to human science – 'The Scientific Method' if you like. What is there to it which definitely amounts to more than Common Sense?

Why should you bother with this chapter if you are not a scientist, and have no interest in becoming one? Because, from time to time, you may be urged to "behave more scientifically." But that can be dangerous advice if your subject doesn't map faithfully onto Natural Science. Assuming that it does can then lead to tragedy – as it did in the Communist World, as it did earlier (1798) in Britain when politicians became convinced that Malthus' "Essay on Population" was 'scientific' and therefore incontrovertible. In Section (15: 13) we list some of the main reasons to be wary of introducing scientific concepts and techniques into areas where they do not belong.

But let's start with the special peculiarities of Natural Science.

(15:2) SYSTEMATIC EXPLORATION

Curiosity is not a peculiarly human quality. Our cats for instance spend at least half an hour exploring any new suitcase, box or container brought into the house. They've been in every drawer and cupboard, including the fridge and the washing machine and if we have a craftsman in to do repairs or installations they follow his every move in fascination. Systematic Exploration though is very different. It is often directed to some purpose. For instance to look for anti-bacterial agents (which led to antibiotics), to discover all the Elements, to locate the epicentres of all major earthquakes (thus came Plate-tectonics), to map the human genome.... and so on.

P.S. There has to be some element of exploration to true science, and exploration inevitably involves risk—risk of failure, risk to reputation, or even loss of life in wasted years. Treading a well-mapped path is not exploration but settlement and if most young academics can thus tread their way to a safe and permanent job that is the way they will choose. Thus much of modern science is merely a simulacrum of such, a timid career-move, a ride on a band-wagon, a road to nowhere new. In my opinion apprentice scientists who will not strike out on their own should be eased out of the profession before they eat the seed-corn.

(15:3) DEVISING NEW INSTRUMENTS.

Often the best way to explore is to devise or adapt an instrument which can uncover what was hidden heretofore: the ocean-going caravel (the first such vessel capable of sailing to windward and thus of coming back), the telescope, the microscope, the thermometer, the barometer, the chemical balance, the seismometer, the spectrometer, chronometer, oscilloscope, interferometer, particle accelerator, cloud-chamber, geiger-counter, metal-detector, magnetometer, X-ray scatterer, polarimeter.......all led to the discovery of whole new realms of science. Almost all progress in my field of astronomy over the past century has come about, either deliberately or by accident, through the exploitation of ingenious new instruments, or the improvement of old. Thus the discovery of exo-planets, planets around stars other then the Sun, came about through the gradual improvement in optical-spectrograph design to the point where observers could measure the minute wobbling of stars as they were pulled back and forth by the tugs of their own circulating planets. And space rockets, that were developed for military purposes, lofted into Space telescopes sensitive to X-Ray and Infra-red radiations which could never get through the atmosphere but which would tell of the very hot and the very cold universes we could not otherwise sense.

As we saw in Chapter 14 new instruments may have played *the* key rule in the rise of science in the early seventeenth century [Sect 14:3]

(15:4) POSING GOOD QUESTIONS

Some have argued that posing the right questions is the key art to science. But what is meant by 'right' in this connection? Ideally such a question should be:

(a) Unavoidable: i.e. cannot be shrugged off by slight of word or mind.

(b) Fruitful: i.e. suggestive of definite hypotheses and consequent programs of research.

(c) Attractive: i.e. such as to draw fresh minds and new resources into the field. Science must anyway be exciting enough to generate the inhuman degree of doggedness often required!

(d) Timely: i.e. not requiring for its answer tools and techniques way beyond current capabilities.   Examples:

(i) 'Why have the finches on different islands in the Galapagos archipelago got differently shaped beaks?' It took Darwin 25 years to publish his answer, and even then he did so only out of fear that he would be pre-empted by Wallace. Small but insistent questions can lead to parallel questions ('Why do giraffes and elephants both have 42 vertebrae?') and so on to colossal answers.

(ii) 'What is the cause of cancer?' is an obvious question but not fruitful because it is too vague. There may not be a single cause – or even a single *type* of cause. On the other hand:

(iii) 'What is the cause of the cholera epidemic now raging in Soho?' was sufficiently local in space and time to generate obvious hypotheses and suggest obvious lines of inquiry to John Snow (3:2). It was a fruitful question which quickly led to an easily generalised and momentous answer.

(iv) 'Could we synthesize recyclable oil, that is to say a fluid capable of picking up, storing and delivering solar energy to wherever it is required?' This is an attractive question if only because its answer would solve all mankind's energy problems, avoid global warming, and generate incalculable wealth. And why shouldn't we succeed in finding an answer? If cabbages can use sunlight to provide for their energy requirements why not men with all their cunning and technology? (I've dabbled in this question myself, stimulated by the question 'How did giant pterodactyls fly?' It's easy to prove that it would have been totally impossible using conventional physiology but, if pterosaurs had blood capable of converting sunlight into energy with about 5 per cent efficiency, eminently feasible then. [See ** on our Website].

(v) 'Could we engineer the atmosphere so as to avoid global warming?' This is an unattractive question because the dangers are obvious. Mankind could throw off all restraints and rely on a technology with unforeseeable consequences.

(vi) 'Can we sequence the Human Genome?' was a question which appeared to be just answerable within a decade using existing technology. And so it was done at an eventual cost of 4 billion dollars.

(vii) 'Can we put men on Mars so as to explore it?' would cost several hundred billion dollars, even if it could be done using existing or foreseeable technology. So we are not going to try, not for now anyway.

Phrasing the right question therefore is indeed a useful art. In our own case asking 'How do animals think?' has saved us from those perils of Religion, Philosophy and Mathematics which beset our predecessors who asked 'How do humans think?'

(15:5) MEASURING AND QUANTIFYING

are vital tools for scientists as a means for discriminating between this hypothesis and that. Sometimes a fairly small improvement in precision can make it highly improbable that Theory X can be right. Thus Tycho's methodical observations of Mars eventually ruled out an Earth-centred cosmos while Newton's theory of gravitation was superceded by Einstein's on the basis of fractions of a second of arc, where one second is about one four-thousandth of a degree. Ironically, precise measurements very often require colossal instruments; I have just been using a radio telescope consisting of 27 25-meter diameter dishes spread across several miles of New Mexico because anything smaller wouldn't be precise enough to pin down Hidden Galaxies.

Instrument builders have been the true pioneers of scientific exploration although for some reason they rarely receive the kind of adulation sometimes accorded to theoreticians such as Einstein and Newton. Perhaps this has something to do with snobbery; instrument builders are after all workers with grubby hands whereas Theoreticians dream in ivory towers (universities) where they can spin one great thought about the Universe after another. Then again instrumentalists tend to be 'one-trick-ponies' because it may take a lifetime to develop and refine a single instrument. By contrast theoretical ideas, from conception to publication, may take only weeks, allowing the conceiver to move on and make a mark in other fields, suggesting 'genius' is at work.

This wouldn't matter if it didn't sometimes obscure an essential truth: instrument building, not theory building, is the core skill of science. All history proclaims it, e.g.{5/1&2}

(15:6) EXPERIMENTING

is not unique to science. As cooks, handymen, gardeners and parents we experiment all the time. But experiments play a hallowed, if sometimes exaggerated role in science – when they can be done (often they can't; never in astronomy). By careful control the experimenter hopes to eliminate all but the phenomenon (E) or $\bar{H}$ hypothesis (H) of interest so that the Probability $P(E|\bar{H})$ becomes much less than $P(E|H)$ in which case the Weight of the experimental evidence E i.e. $W(E|H) = P(E|H)/P(E|\bar{H})$ becomes potentially high, and possibly decisive –but don't forget PAW! No one piece of evidence should be decisive.

(15:7) REPEATING

observations and experiments are two of the most powerful tell-tales in science. If A can't repeat his exciting results of yesterday today then surely something must be wrong. But even if he does we still don't generally believe him until B, working in a different laboratory, can repeat A's results. And even then history reveals that a third independent experiment would be wise. Thus Pons' and Fleischman's hysterically exciting work on Cold Fusion could be repeated in some laboratories, but then not in others. It took months, and in some cases years, for the dust to settle at last and for their claims alas to be discredited. Science can be exciting and competitive, and then caution may blow out through the door. Many exciting claims in medical research which sound statistically significant turn out not to repeat. The more exciting a claim the more necessary that it does indeed repeat. As we have seen Statistics by itself is too flawed to be a reliable judge.

We discussed the problem of Demarcation; what separates science from non-science in Section (14:3). Repetition is quite definitely one hall-mark. Fortunately for us scientists the natural world perdures in a way that the Economic, Historical, Legal, Personal and even Medical worlds do not. We can play things over and over again to try out different ideas and hopefully get things right in the end. There's nothing clever about that. It is just fortunate. That must form, I fear, a large part of the response to the historian Isiah Berlin's plea in Chapter One to share in Science's secrets. Non scientific subjects certainly can and should use CST but one of the strongest lines of evidence, one of our crucial checks will be missing. In that sense Science is much easier than History Law or Economics.

(15:8) RELATING

It is unlikely that a newly discovered phenomenon, particularly an exciting one, has left no trace of itself elsewhere. Thus a wise scientist will dive into the literature in search of confirmatory or conflicting evidence. It may well turn out that someone has anticipated your exciting discovery, or looked harder than you have done and found it not to be there. Both unpleasant surprises have happened to me, as they have to most experienced scientists. Sometimes I have made a mistake, sometimes they have done, or sometimes the others missed the significance of their own research. There was a dramatic instance of this when the Cosmic Background (Big Bang) Radiation was discovered by accident in 1965. Surely such a powerful radiation field, which permeates all Space, had left traces of itself before? And sure enough it had. Back in 1941 a Canadian optical astronomer called McKellar had discovered that certain interstellar molecules behaved as if they living in a heat-bath – which he couldn't explain and so passed by. We probably all do that.

So knowing or studying 'the literature' as it is called, is a vital part of Science. One can have some very nasty surprises in the library – and some very welcome ones too. It may look quiet in there but …..

(15:9) EXPLAINING

The human mind continually seeks for explanations. It has been suggested that over-arching explanations enable us to recall a huge amount of valuable information without having to store all the messy details in our brains. If nothing else Explanation may be a useful mechanism for data-compression and retrieval. Memory freaks, who can correctly recall the exact sequences in several packs of cards, do so by constructing a 'story' (explanation) to remind them of the sequence. There is much of story-telling to science and its' good story-tellers are as admired and valued as they are in other walks of life. Sometimes indeed there's more sustained interest in a good story than a true one. Thus Alexander Fleming's story that a spore of penicillin floated in through the window of his lab at Saint Mary's Hospital in Paddington and settled in a Petrie dish with spectacular results, makes for a memorable story, though it may not be true. The spore possibly came from a lab along the corridor.

It is worth noting that there is more than one kind of explanation. Before Galileo men were in search of the answer to 'Why?' Why did this God or that order so and so? Galileo turned the question to 'How?'; 'How does the arrow fly?' – a question which could answered by observation and measurement, and described by mathematics. But behind the mathematics mechanical forces and energies were supposed to be at work. Later still, by mid twentieth century, there was little left but mathematics itself; the mantra was 'Shut up and calculate.' As long as one could use equations to predict experimental behaviour the explanations in between were regarded by many theoretical physicists as irrelevant. I found this incredibly difficult to take in as a student, naively expecting atomic physics to be explicable in terms that I would be able to understand. But it cannot be so because, on the atomic scale, matter behaves like nothing in the familiar world. Maps, analogies and homely stories make no sense down there; electrons for instance can go through two holes at once. I would have to accept that, without an explanation couched in familiar terms. All I could hope for were repeatable patterns which I could use to predict. Thus equations.

So you can't divorce an explanation from the purpose it is intended to serve. Some would regard "The Theory of Evolution" as a wonderful account of living Nature: 'Organisms evolve because of their relative "fitness to survive" '. Others have retorted that until you separately define what is fit and what is not, the theory is merely an argument in a circle with no power to predict, and is therefore meaningless.

Physics undergraduates imagine that great and enduring explanations and equations emerge out of profound and enduring thoughts about the underlying physics, because that is what they are usually taught (by physicists of course). As far as I can see this is mostly a myth. Some examples:

1. Maxwell was looking for equations to describe the outcome of earlier experimental investigations by electricians like Oersted, Faraday and Ampere. Their circulating magnetic fields led Maxwell to think of fluid vortices and thus to arrive at equations descriptive of vortices. His derivation convinced no one at the time, was mathematically clumsy, and couldn't be tested. Thirteen years later Heinrich Hertz re-assembled the same equations more convincingly and was able to demonstrate their most dramatic predictions (waves) by ingenious experiments in his laboratory at Karlsruhe. The vortices, as it turned out, were an unnecessary distraction. The equations, much simplified by Oliver Heaviside's new mathematical notation, were finally accepted because, and only because, they worked. Since he'd died in the mean time it was safe to hail Maxwell as a genius. Hertz himself said: "One cannot escape the feeling that these mathematical formulae have an independent existence and an intelligence of their own, that they are wiser than we are, wiser even than their discoverers, that we get more out of them than was originally put into them." In retrospect we can see that there were very few sensible equations to choose from there, so that sooner or later someone was going to strike lucky [Sect (14:10) & my website].

2. There was however something very disturbing about Maxwell's Equations – they contained a velocity (for waves) of proscribed size (300,000 kilometres a second). But velocity was, according to Galileo and Newton, a *relative* matter. Nothing, not even light, could have a proscribed velocity independent of who emitted it and who measured it (though Michelson and Morley were throwing doubts upon that matter [1887] by measuring a speed for light independent of the Earth's direction of motion). Lorenz and others realized that something drastic would have to be 'fixed'. Poincaré suggested that adopting Lorenz's Transformations in preference to Galileo's would do the trick *provided* that the masses of bodies were made relative (i.e. dependant on their motion with respect to the observer), a drastic step required to preserve the Conservation of Momentum. It was all algebraic bodging resting on very little physical

understanding (why *was* the velocity of light a constant?'). But it worked (1905) and somehow or other Einstein, not Poincaré, got the credit for it in the public mind.

3. Quantum Mechanics (1901) grew out of Max Planck's equation for describing the colour spectrum of radiating furnaces. It too was bodged together, with zero physical insight, to fit recent experimental measurements. Only later would it be seen as derivable from the revolutionary idea, pursued especially by Einstein (1905), that energy resides in packets – or 'Quanta'.

4. Schrodinger's Equation, the working tool of atomic physics and theoretical chemistry, was bodged together (1926) out of wave-theory, spectroscopy and energy-conservation to describe low-energy phenomena on the atomic scale. And with a few extra tweaks (e.g. Pauli's Exclusion principle to explain why certain expected spectral lines were 'missing' from atomic spectra) it does so very successfully, though it breaks down at the nuclear level, where Relativity becomes important.

Many would be happier if science rested more on profound and ineluctable principles which could be teased out by taking thought. But that is not the way it seems to work. New observations come in. Theories are cobbled together to explain them, but are modified again and again as even better observations arrive. Theories (explanations) have to be ephemeral, flexible, although they may contain ingredients – such as Conservation Laws – which cannot, for very fundamental reasons, be transgressed [For instance abandoning the Conservation of Energy would amount to admitting that natural laws could not be formulated which could resist the ravages of Time]. Thus theoretical physicists juggle with equations in a process which Frank Wilczek, one of the begetters of 'the Standard Model' describes thus: "Theory proceeds by Experimental Logic….In Experimental Logic one formulates hypotheses in equations, and experiments with those equations. That is, one tries to improve the equations from the point of view of beauty and consistency, and then checks whether the 'improved' equations elucidate some feature of nature….Experimental Logic is 'validation by fruitfulness: to validate A, assume it, and show that it leads to fruitful consequences...' "{ 3}

So the naïve idea that theories are either right or wrong has taken a battering over the years. Newton-Smith {8} has suggested several criteria by which a scientific theory should be judged including:

(1) Nesting: i.e. the successes of the old should be nested inside the new. Thus, by deliberate design, Einstein's Field Equations are very closely modelled on Newton's and so enclose its triumphs.

(2) Fertility: it gives rise to new and productive investigations.

(3) Track record: it has agreed with a diverse body of observations over a significant period.

(4) Inter-theory support: i.e. it fits well with theories covering allied fields.

(5) Internal consistency: no extra fine-tuning required to make it fit.

(6) Metaphysical compatibility [i.e. doesn't require a too drastic 'paradigm shift'].

(7) Simplicity: because it will be easier to 'compute' and therefore test. [Sounds like Ockham's Razor to me.]

(15:10) A CAUTIONARY TALE

It would be natural to suppose that if the numerical predictions of a theory agree closely with measurements then this is strong support for that particular theory. However that is not *always* so. Take the famous Maxwell-Heaviside electromagnetic equations which ushered in the modern world [Richard Feynman said of them 'From a long view of the history of mankind…..there can be little doubt that the most significant event of the 19th century will be judged as Maxwell's discovery of the laws of electro-magnetism.' But what about Darwin?]. They were originally constructed by Maxwell on the idea of an odd 'fluid' theory based on 'vortices' and 'idler-wheels' — which is why they weren't originally believed on the Continent (or anywhere else for that matter). Heinrich Hertz, unaware of Maxwell's work, reconstructed them in a more satisfying manner before going on to predict and measure their consequences.

How on Earth could a wrong theory lead to the right equations? *Basically because there is only a very limited menu of equations which 'make sense'.* If you require twice as much cause to have twice as much effect (called 'linearity'), if you require local effects to be due only to local causes (so called 'field theories'), if you require laws that are independent of your velocity, position, orientation and epoch (all such things being merely incidental), and finally if you require mathematical laws that are as simple as possible, you are left with very little choice. In fact there are only three simple ones of the right kind; Poisson's Equation, the Diffusion Equation and the Wave Equation. When new phenomena are discovered they either have to fit into these 3 equations — or various combinations of these, or disobey one of your earlier requirements (for instance Fluid Dynamics is non-linear). Thus the Maxwell-Heaviside Equations are not really new, they are a combination of two of the three fundamentals. [See our website: *'The Origin of Maxwell's Equations'*]

We now realize that nearly all physical theories are *approximations*. Maxwell's Equations are strictly true only in weakish electromagnetic fields; the fluid mechanics equations — used in designing aeroplanes and predicting the weather — approximate a fluid as a smooth continuum with no lumps (or atoms) in it. And if you take out the complexities in theories it is like blurring photographs — they all begin to look more alike. Thus The Standard Model of Particle Physics — with its recently discovered Higgs Boson — is famously successful at predicting experimental results. But many particle physicists regard it as merely an approximation to some deeper underlying theory because it is too 'arbitrary' i.e. conflicts with Ockham's Razor (it has no less than 18 Free Parameters, as bad as Big Bang Cosmology). Likewise the measured 'Lamb Shift' (an obscure atomic measurement) agrees with the predictions of the theory of Quantum Electrodynamics (QED for short) *to no less than 13 places of decimals* but few physicists nowadays take that to mean that QED is anything more than an approximation to something better.

And what is true of Physics may be true of other fields too — there being such a very limited number of 'simple' differential equations. For instance Wave Functions turn up everywhere in fields as disparate as earthquakes, storms, radio, finance, architecture, motor-mechanics, trigonometry, light, bridge-building, sound, electronics, physiology…..[π = 3.141592653589732…..is so ubiquitous because it is the half-wave-length of the simplest possible wave — it is only incidentally connected with circles. See our Website; *"Where does pi come from?"*]]. This is partly why Mathematics is so useful. But it does warn us to be careful. Agreement, even tight agreement between measurement and hypothesis can never be absolutely decisive. Never.

(15:11) PUBLISHING

Scientific work is not regarded as 'proper' science until it published in the public domain where it can be scrutinised, repeated, criticised and/or be helpful to others. *Of its nature science is finally a communal activity* even where important discoveries are first made by individuals working on their own. Darwin's observations of the beaks of finches in the Galapagos Islands weren't science until they were published twenty years later in The Proceedings of the Linnaean Society. An important part of a modern scientific paper is its bibliography at the end. That should enable the reader to see where it fits into earlier work, and to map out its wider implications. It is also a rigorous test of whether the author has done her 'homework'. You can readily pick out a crank paper because it usually lacks a satisfactory bibliography. [Though Einstein's first Relativity paper also lacked one; it wouldn't have been publishable today.]

(15:12) WHAT ABOUT MATHEMATICS?

Scientific subjects in general, but not always in particular, tend to use far more, and far more sophisticated mathematics than most other subjects. The main reason is that scientists do their best to measure things, and to use those measurements to choose between hypotheses. Have the rest of us got anything to learn from them?

So far I have been rude about mathematics and in particular rude about mathematicians. That was because some of them claimed far too much for Deduction, their rather particular, even peculiar mode of thinking. As we have seen it has limited use in CST – even in science. Now I want to change my tune completely and argue that most of us would benefit enormously from knowing and using a great deal more mathematics than we generally do today. But to make that palatable and practical the traditional way of learning the subject will have to be abandoned.

It should be crystal clear by now that most CST is of the more-or-less variety rather than the Yes/No kind. That of course implies calculating which is more and which is less, so we cannot avoid mathematics – most of it so far of the very elementary and rough and ready variety. Precision is only required when exact equality is demanded, and that is not the case in most CS – which is why animals can use it.

The point is that *some vital arguments are entirely quantitative in nature.* Nothing useful or convincing can be said outside the quantitative dimension. And if we can't understand the argument for lack of mathematics, we are as helpless as children. We will look at some examples in a moment.

The problem with mathematics is that many people find it hard to learn, indeed utterly un-digestible. They are otherwise intelligent, but proud to claim that they are 'hopeless at maths.' Can anything be done to change this situation? Yes! Yes! Yes! Humans are wired up to use Induction or CS thinking – not the deductive kind which is generally far less useful. It is less useful because it requires complete and perfect information – which is very seldom available in practice. Maths teachers however thought their mission was to teach deduction by teaching deductively – proving every result from a set of axioms. That in turn was based on the mistaken idea that the truth about the world could be reached through Deduction – in particular through Geometry. We learned in chapter 14 that this is simply not true, indeed wildly wrong. Thus the case for the Deductive teaching of mathematics – the very thing so many of us find unnatural and indigestible – has collapsed, indeed it collapsed 200 years ago. We could instead be taught Maths and learn it in the style which comes naturally to us – *inductively*. Earlier in the book I've given two examples of such Inductive teaching: in [2:2] I tried to convince you that the angles inside any triangle add up to 2 right angles, and in [9:2] to convince you that the counting error in counting a sample of Q is roughly $\pm\sqrt{Q}/2$. Pure mathematicians will point out, quite justifiably, that such Inductive demonstrations do not amount to complete and satisfactory proofs – that showing that a dozen widely different triangles all contain two right angles, doesn't prove that there is not some awkward triangle somewhere that does not conform. Our answer to them is:

"We leave the awkwards to you! We want to *use* mathematics without having to worry about its possible, but unlikely exceptions. If we weren't allowed to drive cars or use mobile phones without first understanding completely how they worked, civilisation would be paralysed. We are gamblers, we are used to taking risks; we desperately need to use mathematics and we're going to – in a rather amateurish way. We'll test our amateur maths out by trial and error, and only appeal to you professionals when things are working out badly. You are paid to sort out the messy technicalities. We amateurs, who are going to use mathematics, gladly accept the occasional risk, because there are much bigger risks and uncertainties that we have to face out there."

This last is certainly true. For instance mathematics itself can rarely be applied to the real world; that world has first to be simplified into a model world with many of the complexities left out. The mathematics of the model may then be very precise, but the model itself may make a poor fit to reality, because of what has been omitted. A disastrous early example of this was Thomas Malthus' "Essay on Population" (1798) which was based on a very simple mathematical model of births, deaths, food-production and starvation. It predicted that widespread overpopulation and consequent starvation would be the lot of all mankind. Politicians believed the model and workhouses, transportation, enclosures, clearances and much other unnecessary misery, was the tragic result. But the model was wrong – or at least vastly too simple. When real people begin to starve they will, if they are allowed to, find ingenious ways to produce more food – for instance displaced crofters went down to the coast and founded the Scottish herring industry.

We then could become amateur mathematicians – enthusiastic drivers with no idea of how to replace the crankshaft bearings. But why should we bother? In the rest of this section I'm going to try and convince you by exhibiting, largely without maths, several fascinating and consequential examples.

(A) BANKING DISASTERS: Consider the simple equation:

$$T_2 = \frac{100}{L(i-i_L)} \qquad (1)$$

which was certainly one culprit in the banking crash of 2007/08. An investment bank wants to double its money over a period of T_2 years. It has 1 million of its own money to invest and sees a promising opportunity to invest it in oil-tankers at an expected rate of return of i = 5%. According to equation (1) (if we ignore L and i_L for now) that would lead to a doubling-time of 100/5 = 20 years. Too slow – far too slow. But what if the bank was to borrow L million dollars at i_L percent and put the whole lot into tanker operating. That's called 'Leveraging': you are 'levering' the effect of your original investment by a factor of L. The doubling time would be reduced, according to equation (1), if L= 10 and i_L = 3% to:

$$T_2 = \frac{100}{10(5-3)} = 5 \text{ years}$$

which is far more satisfactory. Indeed it's money for old rope – seemingly.

Why not Leverage at 50, i.e. borrow 50 times your initial investment and get:

$$T_2 = \frac{100}{50 \times (5-3)} \approx 1 year\ !$$

It seems like magic. The bankers are borrowing other people's money to double their own every year. Let the good times roll. They must be geniuses. They feel entitled to steal huge bonuses, buy yachts and erect vast skyscrapers.

Hey, but wait a moment – there's no such thing as magic. Where's the snag? The snag is right there in equation (1). What if the investment turns nasty? What if there is a slump – or too many tankers get built – a very likely result of such exuberance ? The return i will fall below the borrowing rate i_L and then T_2 is no longer the doubling-time but the halving-time on the original investment of 1 million. Suppose i goes to 0; the halving time become 100/150 years or only a few months. And i might easily become negative, say -5%, in which case the halving time falls to 100/(50 × [-8]) or about 3 months! The situation is disastrous. Somehow the bank has to off-load all its tankers on a falling market and pay back it's enormously leveraged loan of 50 million. It can't be done. The bank crashes. The loans default, panic spreads.

The point, as a simple approximate equation (1) makes crystal clear, is that high leveraging is potentially profitable but also extremely dangerous. It is all about timescales. Had the banks been less greedy, and less ignorant, they would have had more time to adjust their affairs. Bank regulators are there to see that leveraging is held down to prevent greedy bankers from capsizing the entire system – which they will do, and have done many times in the past, if they're given the chance – after all it's *other people's money*. Alan Greenspan (US) and Gordon Brown (UK) allowed leveraging to rise from about 25 in 1995 to over 50 in 2008. Presumably neither of them understood simple algebra – which is all you need to devise and understand the leveraging equation (1). Can you afford not to understand it? Have you got any investments – or a pension? If I was you I'd pick up some amateur algebra.

(B) OUTBREAKS OF WAR : What could be more important than to find out the origins of war and so be able, in some cases, to prevent it? In 1921 Lewis Fry Richardson the famous Quaker meteorologist, investigated the origin of the First World War and concluded that it was most likely to be found in a mathematical instability in the arms-race which preceded 1914. He wrote down two differential equations describing such a race and concluded that if the ratio:

$$R = t_{disarm} / t_{arm}$$

is more than one, war becomes more or less inevitable. t_{disarm} is the time it takes to disarm (say to halve your forces) while t_{arm} is the time to increase it (say to double your forces). And R was very definitely more than one in 1914. In his opinion the ghastly war was probably an accident (on both sides) which, with understanding, might have been avoided. I find his argument far more plausible than most of the others I have read. And even if it didn't apply to the First World War in particular, it still applies to arms races in general.

At any rate, if you understand the implications of Richardson's Equation, you can do something about it. Thus a "hotline" was established between the White House and the Kremlin in 1963 after the Cuban Missile Crisis of 1962 where we came within a whisker of Armageddon. So it may be that the poor devils who died in 1914 - 18 didn't die entirely in vain. It's mathematics again. (Note a certain similarity to the mathematics of banking. In both cases catastrophe follows when there is no longer time enough to adjust. It's like drunken driving.)

(C) CIPHER BREAKING: Talking of war let us look at the maths of cipher-breaking. The decisive battle of the second world war was the Battle of the Atlantic because whichever side lost that was bound to collapse eventually. U-boats cannot find convoys out on the vasty deep without radioed intelligence. Convoys cannot plot a safe course without radioed intelligence of lurking U-boat packs. Both sides were bound to use radio but then employ encipherment to conceal their plans. The Nazis used the three or four-wheel Enigma machine, the British used the five-wheel Typex. Both sides underestimated the cipher-breaking capability of the other – with catastrophic consequences. The heavy British convoy losses in the early years (up to 800,000 tons a month, or 7 ships a day!) were largely due to B-Dienst in Berlin breaking the unsophisticated Admiralty Convoy Code (which didn't use Typex). Their survival in later years was significantly owed to Bletchley Park breaking the three-wheel and later the four-wheel U-boat ciphers. {16/9}

How can you decide whether your cipher, or theirs, is vulnerable? The outcome of the war could depend upon the answer. Only mathematics, and simple mathematics at that, can provide an answer. Once the British were in possession of an Enigma machine (and you could buy them on the open market in the 20s or 30s) it was easily possible to estimate the total number of possible machine settings (150,000,000,000,000,000,000 for the three-wheel). To break it one had to guess the plain wording (letters) of a sufficiently long string of a message, a string long enough to contain enough information to identify the rotor settings of the day. Each correctly guessed letter reduces the number of possible settings by 1/26 (26 letters in the alphabet) and so a string of N correct guesses reduces them by $(1/26)$ raised to the Nth power (the Product rule in Probability – see [11:3]). With an N of only 15 you have in fact reduced the possibilities to less than 1, and *in principle* you have all the information you need to break the system of the day. Breaking it in practice is an altogether different matter which will require ingenuity -- and machinery to speed it up. The faster you can decipher the intelligence the more useful it can be. The technical crew at Bletchly devised electro-mechanical 'bombes' and later Tommy Flowers at the Post Office built them a million times faster device, essentially the first electronic computer, called Colossus. [Flowers knew that ultra-fast electronic valves could be reliable enough – *provided* you didn't turn them off. The boffins didn't believe him so he had to pay for their first electronic computer out of his own pocket. He ought to have a statue!]. The guesses, or 'cribs' used to feed the machines would often emerge from routine cipher-traffic – weather forecasts for example. If you could use direction-finding to trace a message to its source, say a meteorological bureau, you were on your way. Thus the secret traffic was never entirely secure. It would be a battle of wits and eventually technology. But ultimately its feasibility depended on the maths of the situation. Faster computers like Colossus could deliver intelligence almost in real time. More rotor wheels in the cipher machine could slow things down – sometimes to a halt. To divert a convoy away from a waiting wolf-pack you needed several hours notice, while to locate a U-boat with sufficient precision to sink it you needed to 'be current' to within a minute or two. Thus Bletchley Park was a great defensive weapon in the U-boat war, but it couldn't win it. [That required high frequency direction finding on convoy-escort ships, and sensitive airborne radars carried by Lockheed Liberator aircraft with sufficient range to patrol the Atlantic 'air-gap'. They could put the deadly hunter-killer groups on to the scent of individual U-boats, after which their fate was sealed.]

I sometimes think that code-breaking gets over-glamorized. Authors like to attribute success to lone, and preferably flawed, genii like Alan Turing. But every side breaks codes in war, and has its own broken in turn. In the 'Venona project' the Americans even broke the supposedly unbreakable 'One-time-pad' system employed by the Russian embassy in Washington. They were horrified to find that The Manhattan Project had been betrayed in detail to the Russians – though actually the Russians Kariton and Zeldovitch had worked out all the main details in 1939 before the US had even been briefed about the bomb by physicists from Birmingham University{*}. It's very hard to keep secrets for long in science. Scientists just love talking about their work to other scientists – of whatever political persuasion. We're like children.

Get your maths wrong and you can lose a war. Thank goodness the truly brilliant Werner Heisenberg, working for the Nazis, grossly miscalculated the critical mass of Uranium needed to make an Atomic bomb. It's about 10 kilograms, not 50 tons as he estimated {See R.V. Jones book}.

(D) THE MATHEMATICS OF IMMIGRATION: All over the world people are increasingly worried by mass-immigration but are struggling to express themselves because they don't know what they mean by 'too much or too little'. In Britain politicians blithely talk in terms of 'hundreds of thousands a year' – which as we shall see, is a meaningless notion. But if you analyse the situation using simple Calculus you come up with what I call 'The History Equation' because it is so very consequential:

$$T\frac{dp}{dt} = \frac{IT}{P_0} + p\left(\frac{G}{2} - 1\right) \qquad (2)$$

The History Equation describes the growth of any population which is undergoing both immigration (I) and natural reproduction (an average of G surviving children per female) and where T is the average life expectancy of the population in years. p is the multiple by which the population has grown beyond its initial value of P_0 after t years. Thus when p = 2 the population has doubled to $2P_0$; p obviously = 1 at the outset.

Many fateful and fascinating conclusions can be deduced from this History Equation. Here are two

(i) The immigration rate I on its own makes no sense; more interesting is the immigration rate as a fraction of the population, or the death rate, or the birth rate. Take the simplistic case where births exactly replace deaths, when the last term in the Equation is zero. The equation is then easily solved and the population as a whole will double on the timescale $T_I = P_0 / I$ which, for the UK (P_0 =60M, I=0.6 m/y) is about 100 years.

(ii) But immigrant populations sometimes have much higher birthrates (G') than the indigenes (G) [If you want to know how the West was won it was won in bed.

European immigrants had G's of 10 as opposed to 2 for the indigenous North Americans.] It is then more interesting to ask how long T_G it takes for the immigrant population to double as a proportion of the indigenes, even after all immigration has stopped. The History Equation then yields:

$$T_G = \frac{1.4T}{G' - G}$$

For example in Europe today G is typically 1.5 while many of the African and Asian countries from which its immigrants come have G's such that $(G' - G)$ is more than 2.8 in which case T_G is about half T or about 40 years. Numbers like 600,000 immigrants a year into the UK are meaningless when there are about 600,000 deaths as well. A little mathematics adds a lot of common sense to demographic debates. It is surely timescales like T_I and T_G which matter. NB There is a much better discussion of the dramatic mathematics of Immigration in our website under 'My books' category **TFO&&&.**

None of the above examples depended on higher mathematics: algebra and simple Calculus, including a little knowledge of Differential Equations would do. Not so long ago you were in Britain expected to learn this material for O-level, on which pupils were examined at age 15 to 16. And if this material was taught and learned inductively, instead of in the old-fashioned deductive way, it should be within the grasp of 90% of citizens, perhaps more.

To reiterate. Some important arguments are completely numerical: which is more and which is less? Without a certain level of mathematics you won't be able to comprehend – or participate. And the price of being left out can prove tragic – just think of all the families who happily waved their menfolk off to fight in the First World War. I would go so far as to say that a society composed in the majority of citizens who don't understand Calculus, at least to the amateur level, is doomed to become third rate, doomed to unnecessary tragedy. According to the UN the wealthiest people in the world are the Japanese; they are also the most literate in Calculus. Learn amateur calculus yourself, and teach it to your children {7}. References {1,2,5,6} give excellent but readable (lay) introductions to a subject which far too many find toxic, most often because they were taught by those who disliked Maths themselves.

(15:13) WHY THE SCIENTIFIC APPROACH CANNOT ALWAYS WORK

Here are some of the main reasons why one has to be wary of introducing scientific concepts and techniques into areas where they may not belong:

(a) The Natural world perdures. You can go out and consult it again and again without any fear that it will change its story. The human, historical, social, economic, legal, psychological......... worlds do not necessarily perdure, in fact they rarely do.

(b) To use the crucial tool of Induction scientists have had to make drastic assumptions about the uniformity of the Natural world, both in Space and in Time. Namely that its entities here will be much the same as it its entities elsewhere, and that today will be much like tomorrow. It might be dangerous to extend those assumptions into other spheres like politics. If so one should be wary of importing Induction into those areas.

(c) Quantifying and measuring can be extended into non-natural spheres – but beware. For instance does GDP really grasp what you intend if it doesn't include Services? You can measure IQ – but is IQ a measure of anything significant – such as acuity, Common Sense, or sound judgment?

(d) The Natural World is massively connected – if only through the universal Laws of Nature. This can act as a powerful check on our wildest hypotheses. Other worlds, for instance Politics or Chess, may exist entirely within reflecting walls.

(e) Reductionism, that is to say reducing a complex system to its simplest constituent units, has proved very fruitful in science – for instance reducing Chemistry to atoms and molecules with easily categorized and uniform properties. Beware of extending that idea to, say, humans.

In a nutshell Science seems to be simpler and easier than most of the Humanities. That, and not the scientists' brilliance, is mainly why it has been comparatively successful. Many subjects, including Economics, Psych****y, History, Management, Medicine (?)......... can never be sciences because the worlds they deal with are too complex for science.

SUMMARY

The methods of science peculiar to science are mostly too obvious to require much discussion. The three that are not are building instruments, asking fruitful questions and constructing explanations.

Instrument building has to be embedded in a sophisticated technical culture with many disparate skills in close touch with one another: optics, electrical and electronic engineering, vacuum technology, glass-working, materials science, chemistry, detectors, cryogenics .... So Science feeds off technology, and vice versa. Where such infra-structure is lacking it is difficult if not impossible to build up or retain a scientific capability, e.g. {3/2}

Separating good questions from the rest is a key art. Clarity, fertility, excitement and timeliness are all important here.{3}

Constructing satisfying explanations is an even more delicate and elusive skill. Judging from history the rules for an acceptable explanation tend to subtly shift with time. The intrusion of a clumsy church or state into the process has generally proved disastrous to both sides [The Catholic church's persecution of the Heliocentrics, the Chinese emperor's rejection of Western technology, the Nazi's dismissal of 'Jewish Physics', and Communism's promotion of Lysenkoism immediately come to mind]. As Carl Sagan put it: "The values of science and the values of democracy are concordant, in many cases indistinguishable…"{13/9}

A close fit between measurements and the predictions of a particular theory do not *necessarily* prove that that theory is right though they do lend it support. Caution is in order because whatever the theory there may be only a limited number of possible equations that make sense in that context.

History also suggests that a science only remains alive so long as it is constantly challenged by new observations. When they dry up the dead hand of Fashionable Theory suffocates the life out of it. Particle Physics is in peril because it is becoming too expensive to pursue much further by experiment. The 'Standard Model', created in the 1970s to explain 1970's experiments, is settling like a suffocating blanket over the subject, while cults such as String Theory {4} which can apparently thrive without any need for observational oxygen, are growing like poisonous toadstools.

THINKING FOR OURSELVES (Disney)
CHAPTER 16
CONSEQUENCES AND THE ASCENT OF MAN
Draft 14/9/18; 7.9 kw.

" I have never let my schooling interfere with my education."   Mark Twain.

(16:1) SPECULATIONS

If our recipe for Common Sense Thinking really works it ought to have far-reaching consequences; so let us finally take wing. In this chapter I will argue that:

(a) The creative potential of a new piece of information *is in proportion to what one already knows.* A small amount of relevant study can thus lead to huge rewards.(16:2)

(b) The decisiveness of evidence rises exponentially (virally) with the number of clues involved. Thus doubling of evidence can easily increase the decisiveness $O(H|E)/O(H)$ by a factor of 50 or more. This will have implications for research, for breadth, for collaboration, for concentration, for networking and for aging productively. (16:3)

(c) Our capacity to think is often limited by the sheer physiological effort involved. Hard thinking requires Watts of power for days or weeks, perhaps a large fraction of all we can generate. So we avoid thinking, like we avoid running up hills. Once we understand this piece of Human Thermodynamics there are advantageous lessons to be learned. (16:4)

(d) Much of conventional education is a waste of time and money because it doesn't comprehend Common Sense and attempts to teach by lecturing whereas animals like us are programmed to learn by imitating successful practitioners. Thanks to a confusion between 'Clock-Time' and 'Human-Time' we spend half our human lives in educational institutions – a quite unnecessary extravagance. (16:5 & 9)

(e) Reaching sound decisions is generally done better by committees than powerful individuals because they can bring more evidence and wider judgement to the table. (16:6)

(f) Common Sense and Religion are not concordant; whereas the former is provisional the latter is not. With 4000 deliberately incommensurate religions to choose from, most of them can't be right. 'There is no God' may thus appeal to some, on the grounds of Ockham's Razor. (16:7)

(g) The recent and spectacular ascent of Mankind is owed to the invention of writing. Writing bypasses limited memory capacity, enabling Man to exploit the exponential potential of the Detective's Equation [see (b) above]. We calculate that modern Man's power to think effectively can easily exceed that of his illiterate forbears by factors of millions. (16:8)

(h) Dishonesty destroys our power to think straight. (16:9)

(16:2) THE VALUE OF LEARNING

I am here going to argue that the value of learning is out of all proportion to its amount: that if you double your knowledge of the field you increase your capacity to generate new ideas and reach sound decisions by a factor of between 4 and 64. This argument must have manifold and important consequences.

Our basic mental tactic is the *association of ideas*. Thus every new idea can potentially interact with all the other ideas previously existing in one's head. *Hence the creative potential of a new piece of information is in proportion to what one already knows.* This is a profound insight – and yet obvious once stated.

If, as usual, we use science as our exemplar, we know that breakthroughs are often made by the person who first knows two different things and puts them together: thus Oersted – seeing the connection between Electricity and Magnetism when he was studying storms at sea (3:4); or Huyghens first seeing the connection between experiment and Probability (14:4). Thus if you know N separate facts about a field your capacity to make innovative connections will rise in proportion to:

$$N(N-1)/2$$

because each of the N facts could be potentially associated with each of the other (N-1) (the factor 2 excludes double-counting i.e. AB *and* BA) thus when you know twice as many facts about a field you could make roughly 2×2, or 4 times as many potentially fruitful associations. Doubling our relevant knowledge has quadrupled our capacity to generate new hypotheses! I call it the Squared Knowledge Theorem' (SKT).

The truth of this revelation is more than borne out by the history of science. When one reads about the great pioneers one is immediately struck by how many of them were polymaths – curious and learned about practically everything. Two examples must suffice. Nikolaus Steno (1638 to 86) was born in Copenhagen where he studied medicine. In Paris he published a treatise on the anatomy of the brain. In Florence he originated the subject of geology by studying the rock formations of Tuscany. He identified ancient sharks teeth in sedimentary strata and was the first scholar to realise that the strata of the Earth recorded the history of the Earth and he distinguished sedimentary rocks from the igneous and metamorphic varieties. His 'Prodromo' was the first manifesto of Geology (1639). Then he returned to Denmark to become physician to the King. He converted to Catholicism, abandoned science, and took up Theology. He quickly became a bishop and key adviser to the Pope. Alas his rabid asceticism led to an early death at only 48.

Thomas Young (1773 to 1829) –'The last man who knew everything' {6} was more remarkable still. He was a prodigy who educated himself in a formidable program of self-instruction saying "… whoever would arrive at excellence must be self-taught." His learning was so great that none of his contemporaries could properly appreciate him. His achievements were to include:

(i) The discovery of how the eye accommodates (focuses).

(ii) The proof that light is wave-like.

(iii) The first description of the concept of 'Energy'.

(iv) Classifying languages into groups – which included inventing 'Indo-European' (he knew many languages).

(v) Breaking – with Champolion, into the Ancient Egyptian script – using the Rosetta Stone.

(vi) The understanding that heat was radiation and energy.

(vii) Devising the theory of three-colour vision.

He made his living by writing large portions of The Encyclopaedia Britannica'. He liked "a deep and difficult investigation because... it keeps one alive." {6}

Sociological study of Nobel prize-winning *scientists* [i.e. *not* Economists] showed that their most remarkable feature was a willingness to change fields and so combine their knowledge of very different kinds.

(16:3) DECISIVENESS GOES VIRAL

The Squared Knowledge Theorem SKT explains why broad learning can lead to the fertile generation of new hypotheses. But hypotheses are only hypotheses. Deciding whether to act upon them is a form of gambling based on such evidence as is available, evidence that will generally be incomplete and only partially reliable. This is the process of Inference which combines the Weights for the clues in an algorithm called The Detectives Equation – to yield Odds on the hypothesis – given the available evidence. Thus more knowledge of the field ought to generate more evidence, more clues to go into the process of inference, thus a more powerful decision-making capability. It turns out, as we shall presently see – because the Detective's Equation is multiplicative – that that capability will grow exponentially[14] with knowledge of the field in question. Thus a doubling of

---

[14] Exponential growth is very rapid growth of the kind we expect when a pair of flies start breeding. Each generation sees roughly a doubling of the population so that after

knowledge could increase one's decision-making power in that field by an order of magnitude (roughly by 10). This should be an even greater incentive to learning.

The argument here is not unlike the one which led to the SKT, but we now divide the observations into two categories, those belonging to the hypothesis, and those belonging to the rest of the universe .See fig (15:1):

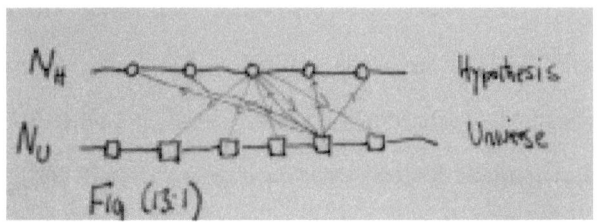

Fig (16:1) showing the possible links between the predictions of some hypothesis and observations of the real universe. Each arrowed link shows a possible clue as to the validity of the hypothesis. The number of such links, and their associated Weights, will determine the Odds on that hypothesis as determined by the Detective's Equation.

Each symbol in Fig (16:1) denotes an independent observation or fact known to you. Whether you believe the hypothesis to be true will depend on the number of sound links (clues) between the hypothesis H and the universe U, as known to you. When the links (clues) are numerous and strong (Weighty) you will decide in favour of the hypothesis; when the links are rare, weak or clash, you will decide against. It is all about the " association of ideas" once again. Strong

---

N generations it will have increased by a factor of 2^N where N is called the 'exponent'; hence the name. It is much faster than multiplicative growth. Thus if N=10 the population would have grown by the factor 2^{10} which is roughly a thousand instead of by 10, which is multiplicative growth.

consistency leads to belief; weak consistency to doubt; strong inconsistency to disbelief. Obviously the number of potential clues will increase with the product of N_H and N_U i.e. $N_H \times N_U$.

Suppose, as a result of intense study, you double your knowledge of both H and U so that N_U and N_H both double, then the number of potential clues will rise from $N_H \times N_U$ to $2N_H \times 2N_U$ i.e. it has quadrupled. So if $N_H=1$ and $N_U=3$ we have only 3 clues while after doubling there would be 2 times 6 or 12. Those extra 9 clues could potentially (below) lead to a huge increase in our 'decisiveness' i.e. our O(H|E)/O(H). It comes about because of the *multiplicative* nature of the Detective's Equation—which is most of the secret of CST [not BR on its own]. Of course having more clues won't change the Odds – if the Clues are by and large neutral. But if there is a slight preponderance one way or the other the extra clues could tip the balance: thus a mean Weight of only 1.5, when raised to the power 9 (see above), could multiply up the Odds-on by an extra factor of no less than 35!

So the Decisiveness of Evidence i.e. O(H|E)/O(H) rises (according to the DE) as $(\overline{W})^N$ where $\overline{W}$ is the mean (geometric) Weight of a clue and N is the number of clues available. The fact that N is in the exponent (index upstairs) means that Decisiveness rises *Exponentially* (i.e. dramatically) with N. Thus if a cat's brain has a maximum N of 3 while a literate human could write down a dozen then (with a mean $\overline{W}$ =2) the human's decisiveness would be $2^{12-3} = 2^9$ or a factor of 512 times higher. And in Table (10:1) to do with Hidden Galaxies we used 25 clues so the difference would be $2^{25-3} = 2^{22}$ which is a multiplying factor of over a million! I call this the Exponential Decisiveness Rule (EDR) [See Table (16:1) below for some values of Decisiveness in practice].

Some Implications:

(1) Both creative workers and decision-makers must do all they can to increase their knowledge of their immediate field and its broader surroundings. Whilst this may seem obvious, the dramatic advantage of more knowledge is not. All of us need to appreciate the Squared Knowledge Theorem and the above 'Decisiveness Rule' of CST.

(2) To that end ambitious brain-workers will miss no opportunity to read, or to attend journal-clubs, seminars, courses and conferences.

(3) Bringing new information in from outside but related fields can generate new clues which could prove decisive. Likewise outsiders with different experiences and prejudices can sometimes break deadlocks. Thus Wegener the meteorologist who overturned geology. Thus radar-engineers like Bolton and Ryle who revolutionised astronomy. Thus crystallographers like Bragg and nuclear physicists like Maurice Wilkins who drove molecular biology.{7}, {10}, {3/13}.

(4) If one can't become a polymath oneself the cheap alternative is to form a close collaboration with someone else whose knowledge complements one's own. One famous such couple were Jim Watson and Francis Crick; Watson knew the biology and biochemistry; Crick knew the physics and crystallography. With a little not altogether ethical peeking over the shoulders of Maurice Wilkins and Rosalind Franklin at Kings College London they won the race to decode the helical structure of DNA (1953).

Exactly the same argument which led to our Decisiveness Rule above can explain the effectiveness of collaborations – and not just of two people. So ubiquitous is collaboration nowadays that it is becoming hard to find scientific papers published by individuals.

Good collaborations, as opposed to lazy collaborations, are not easy to set up or maintain. Alas many collaborations are purely defensive, indulged in by those who are either unable or unwilling to do research on their own. This may significantly, and by design, reduce the amount of healthy competition in a field.

(5) Networks serve much the same purpose as collaborations – as cheap ways of extending one's reach, knowledge and experience. Experienced scientists learn that the quickest way to find something out is often to ring or e-mail an old mate. But old-mates don't grow on trees. Travelling around, working in different institutions in different countries, is an excellent way to grow one's network{*}.

(6) We have argued elsewhere that one tends to learn very fast in the process of doing active research. Thus a virtuous circle is set up in which research accelerates the acquisition of new knowledge – which in turn accelerates one's ability to do more research which…. This may go towards explaining the appearance of so called "scientific genius".

(7) Acquiring more knowledge takes time and effort. Scholars and research workers therefore need to be left in peace to get on with it. Instead nowadays success leads to promotion, and promotion then leads to manifold extra responsibilities which can steal large amounts of time, and even larger amounts of concentration, away from research. Thus brilliant young scientists become mediocre professors who in turn become busybody VIPs who have to hide their obsolescence inside large collaborations. I know famous astronomers who haven't the time to read even one scientific paper a month.

(9) Knowledge and age could and should go together, and so therefore *might* research capability. But if research stars are promoted into administration and teaching, this won't emerge. Mathematicians, it is claimed, decline with age – but then mathematicians uniquely don't use Inference so much as Deduction.

(10) It becomes easy to understand why societies and academies have proved such fertile seedbeds of scientific and other kinds of progress. They are organs for the meeting of like minds, the exchange of information and thus for increase of knowledge. Thus in the late 1700s at The Oyster Club in Edinburgh one might have seen David Hume (Philosophy), Joseph Black (Chemistry and Thermodynamics), James Hutton (Geology) and Adam Smith (Economics) supping together in the

same tavern and discussing each others work – possibly the greatest concentration of intellectual gunpowder in history. And what about The Lunar Society which met in the Birmingham area (during the full of the Moon to avoid highwayman) and which did so much to advance the Industrial Revolution.

(16:4) MENTAL LAZINESS: TO BE EXPECTED, TO BE EXPLOITED

Most of us avoid thinking hard as much as we possibly can. We prefer to get by with the shoddy variety – and with good reason; hard thinking is very draining of energy and not to be undertaken unless the potential rewards are commensurate with the effort required. Recognizing this surprising fact can be the key to deploying our mental resources in an optimal way.

Physiologically speaking humans employ a unique combination of strategies: they're warm blooded but naked; how perfectly daft you might suppose. But wait a moment; we also happen to be the most dominant species on Earth. Surely the two facts can't be un-related?

When I first began to think about this peculiarity of human physiology I realized there might be a vital connection – the Second Law of Thermodynamics [SLOT]. SLOT {5} tells us that the efficiency of any heat-engine – and we are such a one – depends on the *excess* temperature of the machine above its surroundings. Now the human body has an internal temperature of about 37 degrees centigrade which means that in the tropics (35 to 40 degrees) our Thermodynamic efficiency is virtually zero leaving us with only a Watt or less (averaged over 24 hours) to live on [A conventional light-bulb typically uses 60 to 100 Watts]. Even when we come up to the latitude of Northern Europe we have a *useful* power of only 2 to 4 Watts. I couldn't believe this tiny figure at first but it must be right – it is the inevitable outcome of our low Thermodynamic Efficiency – typically 5 per cent or less. Almost all the calories we eat and burn are given off as useless waste heat (about 100 Watts). Getting rid of that waste heat is itself a fundamental challenge for any animal, one which limits its performance – watch dogs panting in the heat. Hence the attraction of our 'naked' strategy.

Never mind the details (I've posted the whole fascinating business of Human Thermodynamics on our website. I believe it dominates human life). The fact is we have to get through life on a couple of Watts (averaged over 24 hours). We're used to the idea of our mobile phones and laptops running out of power, so what about our brains? One of the first symptoms of starvation (I've tried it) is confusion, the inability to think clearly. The brain is a very greedy consumer of calories – about 20 per cent of all so they say, thus thinking hard and long is an extravagant exercise, like running a marathon. Sensibly, most of us try to get by – thinking as little and as superficially as possible. I know I do; I try to postpone hard thinking for as long as I can in the hope that either the need for it will go away, or somebody else will do it for me – my wife, my assistant, an expert or one of my graduate students. And I don't believe I'm very different from most scientists I know – indeed I'm probably more energetic than most. I once spent 1900 hours working on a single scientific paper. It's a brilliant paper I think – though I'm decidedly *not* the best judge of that. But so far as I know nobody else has ever read it – apart from the referee. And I can understand why. It bristles with 70 numbered equations and any number of complicated diagrams and tables. If the effort to read were only one percent of the effort to write it, that still requires the full time dedication of 19 precious hours. And who's going to spend that? So progress gets held up by lack of sufficient human mental energy. I once visited an aboriginal archaeological site in Tasmania where nothing, absolutely nothing had changed over the course of 6000 years. I was horrified until I realized that Tasmanians went naked in a climate little warmer than our own. They were probably too short of energy to think much, or even dream.

This is not a council of despair; on the contrary. No matter how clever my competitors and contemporaries are they must necessarily spend much of their time switched off. Therefore, if I care to invest the mental effort, I can probably overtake them *in the specialized area which most interests me*. The same applies to all of us. I think that most encouraging.

(16:5) EDUCATION: IT'S NOT SO GOOD

If you have followed me so far in this discussion of CS you will no doubt begin to wonder whether our present educational arrangements are altogether optimal. The machinery of CST could be taught, and well taught, to humans before they are 14 years old. Thereafter it is probably best improved by the exercising of it in real-world situations where the judgment and experience vitally needed to go along with the machinery, can uniquely be picked up. One certainly has to question whether a further 10 years sitting in class-rooms learning passively – which is becoming the norm in the West – is a sensible way to go. Education as it is, with its emphasis on unfocussed information, regurgitation and deduction, can do very little to improve your powers of sensible decision-making in a particular field, while it may do untold harm – if only by stealing precious time that could be far better spent.

Let me digress on the subject of Time for a moment because I believe it is widely misunderstood. Clock-time is all very well, and we couldn't do without it on a daily basis, but it has very little to do with human-time – the frame in which we live and experience our lives. Clock-time was only invented in the 17th century, whereas human-time has been with us for hundreds of thousands of years. Confusing clock-time for [human time], hereafter in [closed brackets] to distinguish it from the other – has been a tragedy. When you are six, one year is one entire sixth of your life – and thus is very important. When you are 60 it is only a 60th, and thus is far less significant. Here-in lies a great but hidden truth: *human time is logarithmic*, not linear like clock-time. You can sense this when people talk about their lives – or write their autobiographies. They often dwell on childhood because that is when so much that really mattered to them happened. They are telling us that their time isn't clock-time, that the interval now is to be measured in terms of what has gone before, not in terms of idiot ticks on some mechanical clock. Clocks certainly measure something, but it is not something human. Using the same word 'Time' to signify two entirely different concepts was a fortuitous choice with all manner of unintended and unfortunate consequences. If you are interested more generally in this fascinating topic I have posted an essay *'Rescuing Time'* on my website. Here we concentrate only on its implications for youth and education.

Consider the following two tables, comparing a life measured in clock-years, and in [human years]:

TABLE (16:2A)

Clock	4	5	10	15	20	30	40	50	60	70	80	90
Human	[4]	[9]	[25]	[37]	[43]	[52]	[58]	[63]	[67]	[70]	[73]	[76]

TABLE (16:2B)

Human	[4]	[5]	[10]	[15]	[20]	[30]	[40]		[60]	[70]	[80]	[90]
Clock		4.2	5	6	8	12	19	29	45	70	107	166

All manner of fascinating insights emerge from studying such tables. For instance you can see why young people aged 10 to 20 often feel much more grown up [25 to 43] than the clock allows them to be, while clock-old people of 70 to 90 usually feel more youthful [70 to 76] inside. Thus a 15-year-old girl might feel inside like a mature woman [37] while an ancient of 80 would need to reach 107 to really feel his age. And they're both right. Whirring springs and vibrating crystals don't know whereof they speak.

The point I want to emphasize here is the extraordinary amount of [human time] presently devoted to education – and its alarming tendency to expand. When we leave school at 18, or University at 24, we are – in human terms – [39] and [45] years old respectively. Can we afford to spend so much of life sitting in class? I would claim not, definitely not! All those valuable young years could probably be much better spent doing, deciding and learning for ourselves in the school of hard knocks.

Let me put the matter in another way. As a research scientist I am constantly having to pick up new skills and new knowledge. The moment I stop, I drop out of the competition. So I'm constantly motivated to learn. And because I am so motivated I learn very fast. What might have taken me months or even years to learn in class will have to be learned, *will* be learned, in a matter of days or weeks. My world is moving rapidly on and I can't afford to be left behind. I need it all – and I need it now. I'm hurrying, I'm anxious, I'm determined, I'll drive myself – I have to. And that highlights the sheer inefficiency of class-room learning in general. It simply cannot have the urgency of a real-life competition. You might argue that astrophysical research isn't a good model. I believe it is: the modern world is all about change, competition, progress, adaptation. We all have to learn 'on-the-job' whether we are computer-programmers or candlestick makers. For those who won't learn, the unemployment queue awaits. This speed-up of the world, brought about by the container-ship and the internet, means that learning has become, must become, a life-long preoccupation – like it or not. The idea that you can rest on what you were taught between the ages of 5 and 20 simply isn't valid any more – if it ever was. We have to question all our educational provisions on a decadal basis, innovating as we go. We humans need to pass in and then out then back into education at our own pace, in our own direction and in our own time. Providing the mechanisms for doing so will never be easy, but they will be the measure of a sophisticated, competitive, successful and contented society. More and more years of school and college simply isn't going to hack it. By the time a course becomes a formal requirement at university – it is usually years and years out of date. My own undergraduates rebelled en-masse when I tried to teach them the elements of mobile-phone physics. It 'wasn't in the syllabus' they said and to that extent they were absolutely, but sadly for them, right.

  That brings me to my last beef on the subject of education – exams and degrees. Having spent 30 years trying very hard, and utterly failing, to set exams in astrophysics which might give *some* hint of a student's capacity to actually do astrophysical research, and watching all my colleagues fail likewise, I conclude that exams measure nothing more than – the capacity to pass exams. This perplexed me greatly, but now that we understand more about CST, I think one can see why exams are such a failure. It is almost impossible to set questions which test any of the key skills of common sense thinking: Induction, Weighting, judgment, choice of Priors, error-correction, Animal Wisdom and Parsimony... And if exams are useless, what about the degrees which are their final outcome? I would suggest that they are largely worthless. Educational establishments won't agree – because it is all many of them have to sell. Nevertheless ask yourself – whether you are a potential student, a potential employer, or a politician – if you have ever seen an exam question that is a telling measure of either common sense or wisdom. I haven't. Unfortunately Education has come to mean, more than anything else, grading people – mostly down. I suspect that grading destroys much of the self-confidence which serious thinking requires. What a disaster that could be; what a disaster.

CST is no part of a modern education. In my opinion that is a disaster. Academic scholarship, with its roots in Greek Deduction and Abrahamic religion (man has a unique relationship to God) has dismally failed to sterilize itself, as it should have done after Darwin. Having failed for 150 years perhaps it is incapable.

Having spent 50 years in the profession of Education I must confess that I'm more mystified now than when I began. About all one can be sure of is that if someone really wants to learn something they probably can. And yes there are some inspired and inspiring teachers out there – but how they do it is a mystery to me, and probably a mystery to them. So I'm rather surprised when people, especially politicians, wax dogmatic about education – which is bound to be a very subtle and above all *individual* affair. The success or otherwise of one's education could only be judged towards the end of a life anyway – and then most imperfectly. What we all really need are several wise sages and mentors to advise us full-time throughout our lives. Anything less is bound to be a trifle …………disappointing.

I believe we humans, like our animal cousins, learn by trial and error; if we are smart we learn from our own mistakes; if we are wise we learn from others' {2,3,4,8}. Here and there students do indeed learn Plausible Inference by actually doing things – in laboratories, workshops and studios, or by doing real research projects for themselves. But sitting in lecture rooms passively listening to lecturers who probably don't understand Common Sense themselves could do the average student much more harm than good – if only by robbing them of valuable time and money that could be far better spent. We animals learn above all by imitation – not instruction. Try learning to swim, or paint, of fly, or play tennis, or compose, or build a boat, or design a reactor, or cook, or write, or brush the floor, or learn French, or do Astronomy ….. from a course of lectures. It can't be done. So why do they do it – educational institutions I mean? Because it's cheap – cheap, cheap, cheap! Education is above all a business – a huge and ever expanding business: never forget that. If a university can get hundreds of bums on seats listening to a course of lectures on say astronomy, by a single lecturer they're quids in. All you need is a few slides, someone who wants to earn a living regurgitating second-hand knowledge, and an exam at the end of the course. You don't even need to heat the theatre — the students, giving off a hundred Watts each, will eventually do that for themselves. The students will leave with a degree at the end, and rather large debts – but nobody will much the wiser. In particular the students will have no idea how to do real research of any kind – because the lecturer probably couldn't do it either. If you want a real astronomer instead who *can* teach students by example, in small project teams sharing original research material brought back from far flung telescopes – it will cost. A lot. A good astronomer must be away a great deal of the time getting that data – so heshe will need an equally qualified back-up. The students will all need access to sophisticated computers, software tuition, a good library – and perhaps a laboratory as well. There will be, must be constant feedback between the astronomer and individual students – but remember, if the astronomer is to remain a competitive scientist, heshe will have to spend at least half of their own time on research, at least. Students, their parents, and those newspapers who publish league tables, are in no position to judge whether a particular course in Astronomy actually hands on anything valuable or not. Believe me, as one who has tried to pass on a hard-earned trade for thirty years, it isn't easy, and it certainly isn't cheap. And I imagine much the same is true of almost every subject taught at university. How can you learn art from someone who doesn't create good art themselves, or medicine from someone who doesn't practice on real patients? It's a self-serving fantasy, or outright dishonesty, to maintain that you can.

Enough said. If you are going to spend a small fortune on your education make sure you are going to be taught largely by imitation, and wholly by instructors who are good at the subject themselves. Otherwise you will probably be wasting both your time and your money – and you certainly won't learn how to think using Common Sense. By and large a cheap education is a bad education – and in my opinion a bad education is worse than no education at all, because it leads to delusions all round. Unfortunately a bad education is rarely recognizable as such to the poor devils upon whom it is being inflicted; at least not at the time. Later they, and their employers, will ruefully find out. T' was ever thus. Think of all those poor devils who had Greek and Latin birched into them for 500 years. Or of England, which didn't have a decent university for 600 years. Thank God for the Scots!

All I'm qualified to say is that education should be built around common sense thinking – which at present it very definitely is not. My advice to you, as an individual, is not to wait for the government, or the educational establishment, to catch up. That could take generations. But ask yourself: 'Do I want to spend more than half of my [human life] learning to pass exams – when I could be growing up and growing wise instead?'

(16:6) COMMON SENSE AND GOVERNANCE

Common sense ought to have something to say about governance, be it of a small business or a large nation-state. The Decisiveness Rule tells us that the more knowledge brought to the table the more decisive government can be. More participants can potentially double or quadruple such knowledge and so lead to a really significant – order of magnitude – increase in decisiveness.

This conclusion will surprise some, and needs qualification. Firstly the participants need to be independent – not all chosen by the chairperson to reflect their own prejudices. Second, beyond a certain number, the increase in the group knowledge with each new member will be too miniscule to offset the natural unwieldiness of a larger group. Third majority rule will have to prevail, otherwise agreement might take forever – and all members will have to support a majority decision once it is made – or resign. Fourth, the chairperson needs to know how to get the best out of their committee – but they can be trained – or rotate.

Of course dictators can be very decisive, but far less likely to reach wise decisions based on more evidence and therefore more knowledge. Their instinct to surround themselves with yes-men will quickly entomb them in a self-reflecting wall of delusion – as history so often illustrates: think of autocrats like the last Czar Nicholas, Kaiser Wilhelm the Second, Stalin, Hitler, Mao Tse-tung, or Sir Fred Goodwin who single-handedly wrecked the Bank of Scotland – the world's biggest bank.

So common sense argues strongly in favour of cabinets and committees, strongly against presidents and powerful chief executives. Unfortunately modern media, with their childish concentration on personality and style, are trying, all too successfully, to obscure this hugely important point. We need to grow up!

A provocative example is the United States with its imperial style of presidency, but its frequently gridlocked government. When you consider that the US has had every kind of natural and historical advantage over Europe, particularly since 1914, one has to question why it hasn't done so much better than Europe has for its citizens – for instance in terms of life-expectancy and per-capita wealth. Alternatively consider the per-capita wealthiest nations in the world like Japan and Switzerland. Their natural handicaps are woeful while their leaders are anonymous. Have you heard a Swiss politician ever named?

Yes this is a contentious subject – but one requiring all the common sense and objectiveness we can muster.

(16:7) RELIGION AND COMMON SENSE

The thinking mind is almost bound to include CS in its consideration of religious claims. According to Wikpedia there are about 20 'world religions' and over 4000 altogether. Since they are mutually exclusive *by design* the Prior Odds on any specific one being right must be very low. An Inference Table might be used to assess the historical evidence: for instance do the four gospels contain sufficient evidence to convince you that Jesus Christ was the son of God and that the Christian religion (in some brand) is 'the right one'? Then there is the attractive 'Argument by Design': 'Without a Great Creator how could we explain the parrot's plumage or the eagle's eye?' But then Darwin presented us with an alternative. How to choose between the two? CST offers us Ockham's Razor. Is Evolution a simpler hypothesis than God – or vice versa? (see [7:3])

Finally CST is provisional, always open to new evidence whereas, almost by definition, religion is not. Doesn't that mean that any religious system must admit that it lies outside and beyond Common Sense? I submit that it does. Whether it is to be utterly condemned on that account is another matter. I would go so far as to claim that opinions based on non-evidential arguments cannot deserve much listening to. They cannot convince me so they need not try. Above all they must not pretend to a rationality they do not possess. If you deny yourself the Right to Die, or Abortion, because of something you've heard in church, that is your affair. You have absolutely no right to deny mine, by pretending to concerns over say 'vulnerable people'. That's cheating, that's dishonest, that deserves to be condemned.

Scientists deserve and demand no special consideration in religious debate. If the great majority are atheists it is probably because of their professional respect for Ockham's Razor. They find atheism to be the simplest working hypothesis because it is so well defined – 'There is no God'. All the alternatives are far more complicated. But remember, Ockham doesn't always get it right.

(16.8) THE ASCENT OF MANKIND

Had aliens come to the Earth a mere 10,000 years ago they would have found no grounds for believing that humans would soon erect Santa Sofia, build Venice on piles in a lagoon, devise the Bayreuth Festival, organize the D-day landing, get to the Moon or launch the Hubble Space Telescope. In a blink of an eye in evolutionary terms, a marginal species has so come to dominate the Earth that we now worry that we will damage it irreparably. The zoologist Peter Medawar wrote: 'For all their intelligence and dexterity – qualities we have always attached great importance to – the higher primates (monkeys, apes and men) have not been very successful. Human beings have a history of more than 500,000 years. Only during the past 5000 or thereabouts have human beings won a reward for their special capabilities; only during the past 500 years or so have they began to be, in a biological sense, a success."{1}

What has brought about this apparently miraculous change? What made the difference? Answering that was the big challenge we set ourselves back in (7:6). My guess is that it must have been a minor but crucial improvement in our mental tactics, one that enabled us to take on far more complicated tasks, and reach far wiser decisions about them. Yes, I'm suggesting a small but critical improvement in our powers of common-sense.

Our dog Goch, as I hinted earlier, is good at common sense, and our cat Socks is even better. But neither of them can assess more than two, or at most three clues at once. That's simply not enough to devise and launch a Space Telescope. Would four do it? I doubt it. But as you will recall from (16:2) the Decisiveness of Evidence, thanks to the DE, rises exponentially:

$$O(H \mid E)/O(H) = (\overline{W})^N$$

(where $\overline{W}$ is the geometrical mean Weight of a clue and N is their number). Translated into figures the above equation yields the vitally important table:

TABLE (16:1) DECISIVENESS

$\overline{W}$	N	$(\overline{W})^N$	
4	3	2^6	= 64 (Goch)
2	6	2^6	Sherlock Holmes
2	10	2^{10}	= 1024
1.5	10	2^6	
2	15	2^{15}	= 30,000
1.5	15	2^9	= 512
1.2	15	2^4	
1.2	20	2^5	
2	20	2^{20}	a million
1.2	25	2^7	128
2	25	2^{25}	30 million

We can really see how dramatic the Exponential Decisiveness Rule is. Using writing to gather N's of more than 10, even weak ($\overline{W}$ =2) clues together yield a Decisiveness of a thousand, and thus the possibility of all manner of sophisticated planning. Even conflicting, or very weak evidence ($\overline{W}$ =1.2) can sort itself out once N advances above 20. Once we can write we can compose an Inference Table with any number of clues – as we did for Hidden Galaxies (Table 8:4 with 25 clues) and combine them together either sequentially (in any order), or simultaneously. Given the exponential nature of the Detectives Equation that implies (see Table 16:1) an improvement in our capacity to take sound decisions, by a factor of hundreds of thousands, or even millions. And those are the kind of figures we need to explain the Ascent of Man. [I am not of course suggesting that the D-day planners actually wrote out an Inference Table with 20 plus lines, but a debate with all the evidence before them laid out in papers would have amounted to much the same thing.]

It's true that most of man's greatest achievements have followed upon his invention of writing, back about 3000 BC in the Middle East (Cuneiform writing stamped on clay tablets and baked so as to become virtually indestructible). But we were 'getting going' several thousand years earlier, devising agriculture, building towns and irrigation systems. It might have been that there was a sophisticated elite who were literate even then but whose scribblings haven't survived the millennia. Indeed it seems rather probable to me given how little first-hand paper evidence we have even from the Greeks (for instance not a single first-hand word from the highly prolific Aristotle).

We started out on our quest from the radical assumption that sophisticated (e.g. scientific) thinking is no more than a slight modification of the animal variety (i.e. common sense for survival). The challenge was always going to be finding a straightforward explanation for the rocket-like Ascent of Man. Now we claim to have found it. The capacity of writing to overcome our limitation in memory, combined with the exponential nature of the very ancient animal algorithm we have called The Detective's Equation, will do the trick. Why look any further?

It's all speculation of course but experiments might be done in which different groups are challenged with difficult problems, some forced to remain dumb, others allowed to speak, yet others to both speak and write, and some even to vote. The results would be interesting!

(16:9) HONESTY – THE WINNING STRATEGY

I loathe and despise that aphorism "*Nice guys come in last.*" First because it encourages us to behave like rats – or worse, second because it is utterly wrong – as I shall now demonstrate.

In so far as life is a competition competitors must make the right decisions, and make them faster than their rivals; and nothing could handicap them more than dishonesty or self-deception. We have seen that Decisiveness is:

$$D_N \equiv \frac{O(H \mid E_1, E_2, \ldots E_N)}{O(H)} = \bar{W}^N$$

where $\bar{W}$ is the mean value of a typical clue in the Detective's Equation. Now let us calculate D_N for different values of $\bar{W}$ and N:

TABLE (16:2) OF DECISIVENESS FOR DIFFERENT $\bar{W}$

N	D_N ($\bar{W}=2$)	D_N ($\bar{W}=1.5$)
1	2	1.5
2	4	2.2
3	8	3.7
4	16	5.1
5	32	7.6
6	64	11
7	128	17
8	256	26
9	512	38
10	1024	58
11	2048	86
12	4096	130

If, in some situation, the clues have an honest Weight $\bar{W}$ of 2 say, and if, as we have suggested D_N values of between 64 and 128 to 1 are generally regarded as decisive (if we assume a Neutral Prior of O(H) for the sake of argument) then one can see that 6 or 7 clues will lead to a sound decision. But if, thanks to dishonesty, cowardice or self-deception, $\overline{W}$ drops to say 1.5, we can see that between 11 and 12 clues will be needed to reach the same certainty. In any competitive situation the more honest side would have a dramatic advantage.[15]. In simple terms : *'Dishonesty, or self-delusion, even in small quantities, is just plain stupid – an almost certain recipe for failure.*" It's not even necessary to lie; suppressing or ignoring the truth should do for any competitor in the struggle to survive, be it an organism, a business or a nation.

History is replete with instances, both positive and negative, of this comforting truth. To take but a few examples:

- `Science relies utterly on honesty and transparency. The merest tincture of corruption can spread like potato blight. For instance rotten academic psychologists have recently brought the whole psychological profession into utter disrepute (13:5).

- A large proportion of the successful industries and businesses which pioneered the Enlightenment, the Industrial Revolution and Victorian Wealth, from which almost all humankind was eventually to benefit, were instigated by Non-Conformists and Quakers who (unlike many of their Conformist or aristocratic rivals) actually practiced honesty in business, a key to their success.

- British governments, who have won virtually all of their many significant wars, have done so in considerable measure because they could borrow more money than their adversaries, thanks to a reputation for paying it back (mainly because their main creditors were their own voters), something kings and tyrants rarely did.

- Banking, insurance, stocks, bonds, pensions… all the thews of capitalism rely utterly on honesty, transparency and law. Insider-trading, monopolism, opaque financial structures, Ponzi schemes, shell companies, dark pools, unscrupulous managers and accountants, foolish Economics, deception, chicanery, lax or corrupt regulation… – from

---

[15] One can work out, using the Gamblers Formula (11:3), that a dishonest team might still win occasionally by fortuitously encountering several clues with high Weights early on, but the Odds are very much against it.

the South Sea Bubble (1720) to toxic mortgages (2008) – have time and again brought the flimsy edifice crashing down.

- The Prussian military caste, in the person of General Ludendorff, disguised from the German people that their army had been thoroughly smashed at Amiens in August 1918. Thus we all, and in particular the Germans, had to suffer another and even more ghastly smashing in 1945.

- Like most tyrants Stalin protected himself from the truth by surrounding himself with lying and cowardly toadies. Thus he was utterly unprepared for Hitler's attack on Russia in 1941, leading to tens of millions of unnecessary casualties among his own people.

- The sub continent of India fell to a few hundred Brits, and afterwards was governed by a few thousand, because they offered Indian citizens honest if not always overgenerous government. Under the previous rapacious Moghuls no ones property was safe. That's why Indians flooded into *all* the European trading enclaves – in search of honesty, security, law…... It was honesty, not brute military conquest what done it. Colonization for many colonies was preceded, and has been succeeded, by unbridled kleptocracy. In my opinion it was naive to expect otherwise.

On my first excursion deep into the Soviet Union (1986) I met a scientist from Siberia. He'd been raised in the remote Kamchatka Peninsula.

"My mother will never leave." he told me.

"Why?"

"Because, so she says, 'Only good people live there.' In a fit of paranoid rage Stalin exiled all the Moscow Jews to Kamchatka. They were thrown out into the snow with only an axe, a box of matches and a sack of potatoes each. None of those musicians and intellectuals was going to survive without a great deal of help from their fellows. Mother said all the bad ones died in the first winter, the selfish ones over the second, and the dishonest ones over the third. Only the very best people survived – because they could trust and help one another."

Personally I find all this immensely encouraging. If, as would seem to be the case both on logical and historical grounds, *'Honest Chaps Come in First'*, there is real hope for Civilization. Indeed Civilization may have succeeded, where it has done so, precisely because it was more honest than its alternatives. It was definitely my impression that the Soviet Union fell, not because it wasn't Capitalist, but because it wasn't honest. In that respect the *Nomenklatura* were no better than the Tsarists they ousted.[*Nomenklatnura* were high level Communist Party officials who enjoyed many priviliges denied to their fellow citizens e.g. travelling to the West and buying luxury Western goods in special shops set up by The Party.]

(16:10) PRIESTLINESS, HOLINESS AND CERTAINTY

It is hard to live with uncertainty, foolish to live without it. It is understandable that small children should want to know what is going to happen to them after tea, tomorrow, even next week. Adults who demand the same simply haven't grown up. They are then vulnerable to rogues and fools who are only too willing to peddle false certainties – among them priests, physicians, financial advisers, politicians, economists and educators. The up-front costs of buying such twaddle may be high but the hidden costs are unlimited. We have had to fight very hard indeed, against great Odds, to establish the Provisionality of Truth. Without provisionality we can have neither progress nor tolerance. The provisionality of truth lures us on to find out more. The certainty of it slams the door, barricades our mind against all further appeal from outside and beyond.

If we have learned one thing on our long pilgrimage together it is that truth is a journey, not a destination. Serious thinking is invariably Inductive – and therefore open to new evidence. Thus an honest conclusion is only a working truth – good enough to live by for now – but never quite final. I was tempted to describe it as "only provisional" but that might imply "inadequate" – as if there was some better, some absolute alternative waiting in the wings. Well there is not. In open worlds – as opposed to games – there can never be a way to rule out $\bar{H}$ – "the sum total of all hypotheses which could account for E, alternative to the one we have in mind" – if only because we cannot imagine them all (Appx.A8). But this is not a council of despair. The PAW – which is a key survival mechanism – urges one to go out and look for more independent evidence – which may or may not increase one's certainty in some hypothesis which one has adopted in order to live.

Provisionality is thus the passport to progress as well as the banner of tolerance. If you are confident, but not entirely certain of your beliefs, you will be happy to fly it. If you are worried that your beliefs may be entirely wrong you may, in weakness, turn away from, even persecute, those who do. When President Erdogan suppresses the teaching of Evolution in Turkish schools he is signaling that his own faith in Mohamedenism is fragile, just as dictators like Putin and Chi are signaling that they suspect that they are unworthy leaders – too fragile in self-belief to test their candidacy at the hustings.

We have been duped so often, and for so many ignoble motives, because we hoped for Certainty. In the real world it rarely ever exists.

(16:11) LAST THOUGHTS

My main reflection looking back over 24 years and 10,000 hours of constant struggle, is how haphazard and unexpected this exploration has turned out to be. I assumed that wise men somewhere knew what the Scientific Method was, and had written it down. Instead I found out that those most certain of the truth were either deluded mathematicians or academic philosophers with almost no experience of real science. That I would finish up wondering how dogs thought never occurred to me for many years. And when I did discover something that really seemed to work – the PAW version of the Detective's Equation – and showed it to my colleagues, they often reacted initially with horror. Either it was wrong, they said, or someone famous would certainly have discovered it ages ago. I can sympathize with this last point of view because I felt it myself. After all, in retrospect, it *is* pretty obvious. But then real science is so often like that. It took centuries to establish that the Earth is round; it takes only moments – in retrospect – to see that it must be so. As J M Keynes put it in a slightly different context : " The difficulty lies not in the new ideas, but in escaping from the old ones, which ramify, for those of us brought up as most of us have been, into every corner of our minds."

I have to confess, but only after a friend pointed it out, that this is actually a book about Philosophy ['the love or pursuit of wisdom']. I never meant it to be so, and feel oddly ashamed now that I know. I thought I was doing science. But my description of Common Sense Thinking is a *theory* about how we acquire knowledge and reach decisions. I cannot see how anyone could prove that animals like us actually use the PAW/DE – if indeed we do; and certainly I haven't managed to prove it here. All one can assert is that it seems to work well in practice, while it could have evolved naturally from very simple beginnings. Until someone comes up with a better algorithm it will have to do. But we must not assume that it is the end of a very long and tangled story. There may be more to come – so keep a sharp look out.[See Appendix A7 "*Philosophy and our Book*"]

Finally I am happy to report that the PAW/DE eventually worked for me and my Hidden Galaxies.. In January 2015, after 39 years of struggle, muddle and failure, Huw Lang, Juergen Ott and I used the uprated Very Large Radio Telescope Array in New Mexico to look at 17 potentially Hidden Galaxies. Two of them definitely turned out to be such while there appear to be several others accidently lurking in the wings. That certainly doesn't prove that this theory of Common Sense Thinking is right. But it is a promising start .[See Appendix A10 "*Finding Hidden Galaxies At Last*" and {8/3.]

FINAL SUMMARY (1.8 kw) (8/9/18)

Like all our other vital functions we must have inherited Thinking largely from our animal forbears. Using Science as our example – only because it is progressive, and analysing a broad range of discoveries, we found that Science advances mostly by applying Common Sense Thinking (CST) to observations – not by using Logic, Deduction or Experiment. Those are valuable modern additions, but are seemingly not the main tools of a highly sophisticated and powerful survival mechanism that must have evolved over hundreds of millions of years. Scholars have misunderstood the mechanism almost completely because it wasn't until Darwin (1859) that they became aware of our animal ancestry. They had mistaken human rationality as evidence of our special relationship to (the) god(s), and have since been reluctant to abandon unsound ideas which go back as far as the Ancient Greeks. Thus you will not be taught CST at school or university when in fact you could learn it at age 14.

So what are the tools of Common Sense Thinking, a way of reaching sound decisions on the basis of limited and only partially reliable evidence?

We are led first to Bayes' Rule in gambling. It updates our Odds $O(H)$ on some hypothesis H [e.g. 'Mammoths are dangerous'] in the light of a new clue E_1, by assigning that clue a multiplying Weight $W(E_1 | H)$ which will be used to arrive at new Odds $O(H | E_1)$ as follows :

$$O(H | E_1) = W(E_1 | H) \times O(H) \qquad (1)$$

If the Weight (which is just a number) is more than 1 our new "Odds on H given E_1" are obviously increased by the clue E_1, if less than 1 they are decreased by E_1. The Weight $W(E_1|H)$ is our estimate of the degree of association between two ideas E_1 and H.

Bayes' Rule (or 'Theorem'), which is almost self-evident in its Odds form above (1), is a crude recipe (algorithm) for incorporating a piece of new evidence (E_1) into our current thinking (H). O(H) is called our 'Prior' Odds because it is our Odds on H *prior* to considering the new clue E_1. [N.B. : the Weight $W(E_1|H)$ will usually involve guesswork or judgement which is why CST is to some extent Subjective (personal) and is why so many scholars object to it. Alas they have proposed no better alternative. Note that $W(E|H) = P(E|H)/P(E|\bar{H})$, where P(E|H) is the Probability of E if H is true, so that it also requires knowledge of $\bar{H}$ – '*all* the other hypotheses that could explain E aside from H'].

Unfortunately a single clue, a single use of Bayes' Rule, is seldom convincing, which is perhaps why it is not better known or widely taught. But what if there are several clues E_1, E_2, E_3......? If we consider them in sequence, one at a time, to update our Odds, then it is clear that:

$$O(H|E_1, E_2.....) = W(E_1|H) \times W(E_2|H) \times .... \times O(H) \quad (2)$$

where O(H) was our Odds on H (our 'Prior') *before* we considered the particular sequence of clues E_1, E_2, E_3......

Now our final conclusion $O(H|E_1, E_2...)$ on the left obviously cannot depend on the *order* in which we considered the clues E_1, E_2, E_3...... so we can, if we wish, consider them all *simultaneously* and (2), which we have dubbed 'THE DETECTIVE'S EQUATION', then becomes a quite general recipe for incorporating any amount of evidence, whether it agrees or conflicts, into our thinking. It can incorporate strong confirmatory evidence [$W(E_1 \mid H)$ high], weak evidence [$W(E_1 \mid H)$ low], neutral evidence [$W(E_1 \mid H) = 1$] and evidence *against* H [$W(E_1 \mid H)$ is then *fractional*], combining them all together [with the Prior O(H)] to reach final Odds $O(H \mid E_1, E_2 ...)$ which can form the basis for a decision.

Inference is thus a process of successively modifying the Odds on an hypothesis H as new evidence is brought to bear on it. But it has to start from an initial guess at those Odds O(H) called 'The Prior'. If that O(H) is too strong (either way) new evidence will be unable to change it much, and so Common Sense won't work. Bland or dispassionate values of O(H) can be generated by, for instance, assuming that all the possible hypotheses are equally likely (see Sect (8:5).

The Detective's Equation, which is to be thought of more as a recipe (algorithm) than a mathematical theorem, and is a natural animal extension of Bayes' Rule, has some remarkable properties – which make it ideal for Common Sense Thinking [CST]. First and foremost it is *multiplicative* so that a relatively small number of clues, if they mostly weight in the same sense, can lead to decisive Odds (one way or the other), Odds on which an animal, or a mind, can act with relative safety. And that is what we all need, be we animals, scientists or humans. Put in other words the DE can lead to *Inductive Conviction*, and thus overcome David Hume's long-standing (1737) 'Problem of Induction'. Although it is never absolutely certain Inductive Conviction is safer and more satisfying than the Deductive Conviction long sought by logicians, because it can only be overturned by a considerable burden of new and contrary evidence and not by a hairline crack in one premise (as for instance Euclidean Geometry was).

The barely known DE, which is the very backbone of CST, has many profound consequences. For instance it emphasises the importance of weak evidence because several weak clues, in multiplicative concert, can overcome one strong one. But weak clues may have rather vague (factor-of-2 ,or' binary') Weights; so vague Weights, which are anathema to most scholars, must be acceptable. But if some clues in a multiplicative chain of argument have vague Weights there is no point in searching for precise Weights for the remaining stronger clues because the precision of the product will be that of the *least* precise factor! This largely undermines the whole Statistics project with its elaborate mathematical apparatus for refining Weights. *The DE strongly suggests that the precision of evidence is less important than its variety, concordance and reliability.*

When evidence is limited, as it usually is, the Detective's Equation explains why we should always start by considering the simplest relevant hypothesis (H) first – a maxim called Ockham's Razor. A simple hypothesis will be too inflexible to fit much evidence – unless it is right. By comparison a complex hypothesis can be moulded to fit a deal of evidence and therefore becomes almost impossible to rule out – even when it is wrong. Complex hypotheses therefore lead to indecisive thinking and are not progressive [Ch. 7].

By the same token CST can only be applied to fields where the possible hypotheses are limited in number and hence where the Prior Odds are finite. This '*Principle of Limited Variety*' (PLV) means that many fields – such as Economics and Psy****y – can not ever become sciences because the finite available evidence could never supply enough weighted evidence to give any one of their unlimited possible hypotheses convincing support.

Errors in evidence can be a life-or-death matter and we identified three very different kinds of Error: Systematic Errors, Counting Errors, and Measurement Errors, the last two of concern to mankind only. Animals can discount Systematic Errors by never granting *any* clue too dramatic a Weight (more than 4 or less than 1/4) and it would seem sensible for humans to follow the same prudent maxim, after all we can't usually spot Systematic Errors either. Better to rely on a consensus of weak evidence than on one or two overwhelmingly persuasive (bully) clues – which could be systematically wrong. This is the vital PAW, the *'Principle of Animal Wisdom'*; It says "You should only use the following Weights:

$2^2 = 4$ 'Strongly in favour of the Hypothesis H'

$2^1 = 2$ ' Weakly in………'

$2^0 = 1$ "Neutral about H'

$2^{-1} = ½$ 'Weakly against the hypothesis H'

$2^{-2} = ¼$ 'Strongly against ………… ' "

where the guiding rule has been that no single strong (but unsound) clue could lead to a decisive (> 50 to 1) result in the Detective's Equation.

We found a way to estimate and deal with Counting Errors – when they are important [(9:2) to 9:4)] – and emphasised the point that Measurement Errors generally have to be *measured*, and not inferred using Mathematics [e.g. using The Central Limit Theorem] in (9:6). This, and the PAW, are the final two nails in Statistics' coffin [Chapt. 11].

Once its principles are understood the calculation of the DE can be laid out, line by line, in the form of an Inference Table (IT) like the one below, which shows three clues with their Weights, being incorporated one by one into a detective's thinking [H = 'X is guilty'] – useful for keeping account of the running Odds, useful for replacing Statistics, useful for reaching a decision and useful for persuading others:

INFERENCE TABLE (5:2): MURDER IN THE LIBRARY

Clue (E)	O(H)	W(E|H)	O(H|E)
Prior	1/6		
Motive	1/6	2	1/3
Opportunity	1/3	4	4/3
Forensic clue	4/3	1/40*	1/30

(* The PAW was ignored here and elsewhere in this table and led to a serious miscarriage of justice.)

Since inference (CST) is multiplicative, communal decisions can be reached because precise agreement as to numbers (Odds) is never required. Common Sense deals in the logic of the less/more kind, not the equal/not-equal variety. New evidence may carry a significant majority of thinkers beyond their threshold of 'Reasonable Doubt' – like a jury – into general consensus. This overcomes many philosophical worries about scientific 'consensus'.

A mind's capacity to decide will be limited by the number of clues it can hold, weight and combine simultaneously. The development of writing swept this limitation aside and almost certainly accounts for mankind's rocket-like ascent over the past 5 to 10 millennia.[16:8].

No we have not 'proved' that The Detective's Equation' is the algorithm by which the animal mind actually decides. But it could have evolved in simple steps to do so like that and eventually to do so very effectively without any mathematics [Appx.9 on 'Categorical Inference']. And combined with writing it would have taken mankind far beyond the range of his fellow creatures [Chapter 16]. So Common Sense Thinking, as we have described it, is a plausible mechanism first for survival and later for dramatic progress. In the nature of things one cannot hope to be more definitive than that.

Common Sense Thinking cannot reach absolute certainty: of its nature it is *Provisional*, always open to new evidence: moreover it involves making assumptions about $\bar{H}$ – which means "all those hypotheses, other than H, that might be responsible for E", a very tall order. Curiosity, breadth of knowledge and sound judgement are its components. It has no room or time for dogmatism, elaborate logic, advanced statistics, intolerance or "precisiveness". We claim that Provisionality has to be the secret of all progressive and tolerant civilization.

In the past our capacity to think was severely limited by our capacity to find the information we needed. We lived in the perilous world of 'experts' who did all too much of our thinking for us [Think Religion, Finance, Law, Medicine, Education….]. The Internet is thankfully overturning, capsizing that dangerous era. What will limit us in future will be our capacity to think soundly for *ourselves*. The educational establishment, with its links to classical scholarship, and its need to generate income, is ill suited to teaching us. We must and can learn CST for ourselves. It's not about 'brilliance', or passing exams, it is about finding the right tools and practicing them, and avoiding the worst snares. Welcome to the Age of The Thinking Individual.

EPILOGUE [0.6 kw]

THROTTLE UP (8/9/18)

We set out to find the secret of scientific thinking and instead stumbled over something far deeper – and ultimately more useful – the recipe for Common Sense. In our bones of course we knew it all along – we couldn't have survived without it – but it had become obscured by such encrustations of philosophy, religion and education as to be almost entirely lost to sight.

In the face of all temptations Science was mostly able to keep listening to Common Sense because of the richness of evidence out in the natural world. Atoms, rocks, finches and stars cannot be persuaded to lie, or not for long, even by the subtlest of arguments. They come in their myriads, perduring, repetitive, insistent on telling their tale. They bore such consistent witness that honest scientists sometimes had to turn their backs on philosophy and mathematics to hear Nature's tale. So in that sense Science was able to rescue Common Sense from Culture. If it has a secret, that is it.

Patient but persistent groping in the dark – that's the only hope we humans ever had. Of course we get lost in the catacombs, sometimes for hundreds, sometimes for thousands of years. As a scholar it is fascinating to look back and see how we were led astray by Greek philosophy, by Abrahamic religion, by Hume's Problem of Induction, by Laplace's Central Limit Theorem, by Statistics – but above all by our childish craving for Certainty, when Certainty is rarely available and not even necessary. And Chance of course played its fateful tricks. Had we taken a slightly different turning here and there, adopted Odds for instance instead of Probabilities, we might have invented the jet engine before the crucifixion of Christ; or conversely, still be hunting with woomerahs.

Be that as it may we don't have to relive the past. The recipe for Common Sense Thinking is straightforward, even if it must sometime take a deal of practice to get it right. Children under fifteen could be taught all the essentials without having to go on at 'school' for another ten years. Education is not always enlightenment – on the contrary. The opportunity costs of staying on in class instead of getting on with life are incalculable – but they could well be huge – both to society, and more importantly, to the individual.

This is precisely the right moment for all of us to re-acquire the skill of Common Sense Thinking. Before the Internet we were all crippled by the cost of acquiring crucial information. It was cheaper, even sensible, to leave many vital decisions to 'experts', to doctors, to lawyers, to priests, to accountants…even to governments. Not any longer. The information is out there at the press of a key. But to turn that information into sensible decisions we need to be practiced in using the tools of Common Sense: Bayes' Rule, The Principle of Animal Wisdom, Inference Tables, Ockham's Razor, and above all the Detective's Equation. You won't learn about them at school, you won't learn about them at university, indeed you are more likely to learn bad, or at least confused thinking there nowadays. Even scientists are hopeless at explaining how they think.

Despite all his confused peregrinations the miraculously fast ascent of Man from Stone-age to Space-age needs explaining. If I am right [16:8] it has to do with the multiplicative nature of Inference, of Common Sense Thinking. The invention of writing suddenly enabled humans to weight and hold more than two clues in their heads at once. And if I am right then we have only just begun. With all of us tooled up to use Common Sense properly, and to compound any number of clues together simultaneously, it will be, as they used to say on the Space Shuttle, time to 'Throttle Up'. Welcome to the Era of The Thinking Individual.

GLOSSARY THINKING FOR OURSELVES (Disney)
Version 8/9/18 (3.4 kw)

Algorithm is a procedure or recipe for carrying out some mental task.

Bayes' Rule (or '**Theorem**' is a recipe for recalculating one's Odds O(H|E) on some hypothesis H in the light of a single piece of new evidence E, and is written $O(H|E) = W(E|H) \times O(H)$. If Probabilities are used instead of Odds the rule becomes algebraically far more complicated, less natural and, what is vital, not available to animals because it requires arithmetic.[Ch.4]

The **Blackstone Ratio** means 'Odds of 10 to 1 on' – the minimum Odds enjoined on courts operating under English Common Law before they should pronounce a Guilty verdict. [Ch.8]

Categorical Inference is Common Sense Thinking entirely without mathematics. It is the version almost certainly used by animals. Instead of clues being given numerical Weights they are placed in categories such as 'strong' or 'weak'. [Appendix 9.]

Common Sense is the rather ill defined process by which common folk (and animals) are supposed to reach decisions. It is here supposed to rely on informal use of the Detective's Equation employing rough and ready Weights and Priors largely drawn from personal experience. In my opinion there is no better way to think most of the time. Deduction, beloved of logicians, and programmable into a computer, applies only to games with arbitrarily defined rules .[Ch.2, Appendices 8&9]

The **Central Limit Theorem** is an infamous mathematical result which shows that as you throw information away so any centrally peaked distribution of real data-points begins to look simpler and simpler until it resembles the bell-like "Normal Curve" of De Moivre, the simplest such curve of all – one describable by a single parameter. Well of course it bloody does! What it does *Not* show is that real data (except in rare instances) can be fairly represented by that Normal Curve. But that is precisely what many 'experts' (notably Statisticians) do assume before they practice necromancy on us. Look out for this everywhere. Whenever you hear this theorem mentioned you can be pretty certain that something suspicious is afoot. (Ch. 11)

The **Conditional Probability P(H|E)** means "The Probability of H *given* that E is true". Likewise O(H|E) means "The Odds on H *given* that E is true". Be careful of the *ordering* inside the bracket. Thus P(E|H) means "The Probability of E *given* that H is true" and is *not* simply related to P(H|E). (Ch. 4)

The **Conditional Odds O(H|E)** means "The Odds on hypothesis H *given* that evidence E is true". (Ch. 4)

Counting Errors are errors which arise from counting too small a number of items to get a representative result. Thus if we counted only 6 tosses we might get anywhere between 0 and 6 Heads instead of the representative value of 3 expected if the coin is fair. We demonstrated that if we count only N such items we expect to get results within $N/2 \pm \sqrt{N}/2$ about two thirds of the time. Thus as N increases so the 'fractional error' i.e.

$[\pm\sqrt{N}/2$ divided by $N/2 = \pm 1/\sqrt{N}]$ slowly declines. Large samples are thus more reliable than small ones – but only in the square root. But beware the cardinal sin of assuming that measurement errors are merely counting errors. In general they are NOT, and have to measured – which can be annoyingly expensive. Hence the temptation to assume otherwise. Resist it! (Ch.9)

The **Gambler's Secret** is Bernoulli's formula P(r, N) for calculating the Probability of exactly r successes in N trials where p is the probability of success in a single trial. Thus the Probability of drawing 2 Aces in a hand of 13 cards (p =1/13) is P(2, 13). Not only gamblers but everyone aged over 12 needs practice in this formula. Without it you often cannot think straight because you cannot estimate the $P(E|\bar{H})$ needed to get the Weight W(E|H) for use in the Detective's Equation. The formula for it is in Sect (11:3) and see it in action in Sect (11:7).

The **Decisiveness** of some evidence E as it bears on some hypothesis H, is defined to be O(H|E)/O(H) where O(H) is the Prior, and O(H|E) has been calculated from the Detective's Equation. E can stand in for any number of separate clues $(E_1, E_2, ... E_N)$.[Ch.16]

Deduction is the business of inferring one proposition from one or more others (called 'pre-suppositions' or 'premises') using logic alone. It is all about *consistency*. Deductive Conviction may seem preferable to the Inductive variety until one realizes that it rests entirely on the *exact* truth of the several presuppositions used. Even where they are taken as 'self-evident' they may still not be true of the real world. The further serious limitation of Deduction is that it cannot introduce new information into thinking because that new information may be inconsistent with the existing presuppositions, or lie entirely beyond them. Thus Deduction is OK for Sudoku, or Mathematics, but it is seriously inadequate for Science, or for life in general. You don't need the concept of 'Life' to do Sudoku, but you do to infer the existence of Killer Whales. Laymen imagine that Deduction plays a far more important role in Science than it actually does. (Ch. 2&3)

The **Detective's Equation** (DE) is an extension of Bayes' Rule which allows one to calculate new Odds on some hypothesis H in the light of *several* new pieces of evidence $(E_1, E_2, ...E_N)$ as

$O(H|E_1, E_2....E_N) = W(E_1|H) \times W(E_2|H) \times ... \times O(H)$. It is so vital because a single piece of evidence is seldom convincing whereas, because of its multiplicative nature, the DE can be. It is therefore more helpful to think of Bayes' Rule as a special instance of the DE when there is only one clue. (Ch.5 & 14)

The **Expected Value** $E(x_i)$ of some quantity x which can take the individual values $x_1, x_2, ...x_N$ with Probabilities $P(x_1)$, $P(x_2)$, ...$P(x_N)$ is by definition: [$x_1 \times P(x_1) + x_2 \times P(x_2) + ... + x_N \times P(x_N)$]. The Expected value is the most likely value and in the case of x is the *average* value, often written $\bar{x}$. (Appendix A3)

Free Parameters are quantities which can be adjusted so as to make some hypothesis or curve fit evidence. The more there are the 'woollier' the hypothesis becomes .[Ch.7]

Induction is the business of trying to draw general conclusions from a limited number of individual instances. In a logical sense that is impossible. However one can pile up one's Odds on some general hypothesis H by observing more and more instances, and compounding those Odds together using the Detective's Equation to eventually reach 'Inductive Conviction'. But beware of Black and White Swans. Induction is always *Provisional* – which is its glory in a world where new evidence may always come to hand. (Ch.2)

Inference or 'Plausible Inference' is the posh name for Common Sense (Thinking). See 'Common Sense' above.

The Inference Table is a simple device for displaying your working of the Detective's Equation as each new clue, and its Weight, are introduced into the argument line by line. Helps to avoid omitting crucial points (e.g. the Prior) and particularly valuable in complex arguments with many conflicting clues and Weights. Can be used to keep track of progress, to avoid confusion, to guide further investigations, to justify your thinking and to persuade (or not) others: an invaluable tool for any thinker. And once you've mastered it you can forget the equations and theorems, because they are embedded within it. (Ch. 4,5& 6)

Measurement Errors are the errors that enter any real measurement. In general they have to be *measured* (for example by repetition) and not calculated from some assumption. Reliable inferences can only be drawn from data if the true measurement-errors are first known, both in size and size-distribution. Real errors can exhibit asymmetries, outliers and 'fat tails'. The history of science is littered with disastrous conclusions based on ignorance of the true measurement errors. (Ch.9)

Minimum Consensual Odds (MCO) are the minimum Odds at which a body of judges (small j) may be willing to pass a majority verdict, for instance jurists, scientists, politicians, experts of any kind…….. It has no well defined value, indeed may be undeclared, but appears to depend on context, history or convention. Lawyers use 'at least 10 to 1 on' before passing a verdict of guilty, scientists are more picky (40 to1 on, or against, i.e. '2 sigma') perhaps because they usually have easier access to more evidence; even so they prefer 750 to 1 ('3 sigma'). [Ch.9]

The Normal Distribution (sometimes called the **Gaussian**) is a smooth bell-shaped mathematical function with thin well-defined 'tails' on either flank. In rare instances it makes a good fit to real data distributed round some mean value (as for example the number of Heads counted in samples of tosses). In fact the Normal function is no more than the crudest mathematical function which is symmetrical about a central peak. Indeed it is describable by a single parameter σ ('sigma'), its half-width about 40% per cent of the way down from its summit. Alas a variety of people took Normal to mean 'usual' here – *which it does not*. Normal distributions are no more ubiquitous in the real world than other simple mathematical constructions such as perfect straight lines, or perfect circles. Fit them if you will to real data but don't be surprised if everything then goes disastrously wrong. Statisticians, pollsters, economists, psychologists, financiers, scientists ….take special note. (Ch. 9,11 & Appendix A5)

Normal Weights. If a number departs from its expected value by t times the 'expected error' we often need to know what the Odds are that this discrepancy could be the result of sheer chance. If those Odds are high then the discrepancy is hardly remarkable. But if those Odds are low then there is probably something more than chance at work, i.e. something 'significant' and worth following up. To distinguish the significant discrepancies from the insignificant ones we assign each discrepancy (clue) a numerical Weight, which can later be used in the Detective's Equation. But to calculate those Weights we first need to know how such discrepancies are distributed about their expected values. Where that distribution is 'Normal', i.e. follows De Moivre's famous bell-curve, we can try to calculate the 'Normal Weights' [See Appendix 1 where such Weights are tabulated, together with a description of how they were arrived at and what they mean]. Alas most real data does *not* have a Normal distribution ('Normal' doesn't mean 'normal') and then one is hard put to find reliable Weights. Unscrupulous or ignorant people use Normal Probabilities all the same. Moreover Weights require $P(E|\bar{H})$ [as well as $P(E|H)$] and thus some assumptions about $\bar{H}$! Then look out! To use the table you either have to prove that your data are Normally distributed, which is generally not possible, *and* justify your chosen $\bar{H}$. If you cannot do both things, then it is better to use PAW Weights (my recommendation) for then all Weights larger than 4 or smaller than ¼ are guillotined. In effect that suppresses uncertainties, especially as to the tails of the true distribution and to $\bar{H}$. (Ch. 6 & 9, Appendix A1)

Ockham's Razor is an ancient adage urging us to prefer simpler hypotheses over more complex ones. The thinking behind it is that if a simple hypothesis fits the data well it is more likely to be right than a complex one – which can be more easily modified (fudged) to fit by altering one of its numerous details (It has nothing to do with the real universe being simple, as so many eminent scientists have wrongly assumed). An exceedingly valuable principle which helps us avoid woolly thinking. But remember, the simplest hypothesis may still not be right. Sometimes called 'Parsimony'(Ch.7 &14)

The **Odds on H, i.e. O(H)** are your betting Odds on the truth of the proposition H, where O(H) is a positive number, either more than, equal to, or less than 1 (fractional). O(H) = 5 means Odds of 5 to 1 on H, O(H) = 1/5 means Odds of 5 to 1 against H. Odds and Probabilities are simply related: O(H) is obviously equal to the Probability of H being true i.e. P(H), divided by the Probability that it isn't true i.e. P($\bar{H}$) where $\bar{H}$ means 'not H'.

The **Principle of Animal Wisdom (PAW)** advises us to confine the Weights we assign to clues to the narrow range (4, 2, 1, ½, ¼). This will help to prevent over-weighty Systematic (and other) Errors from carrying the day, and shifts the onus onto *the breadth and coherence of the evidence* as a whole. But more fundamental than that is the inability of the imagination to visualize *all* the alternatives to H that might be responsible for E. Thus it would be foolish to give too much Weight to any E. Thus the devil was blamed for causing infectious diseases before we knew about germs. Had we followed PAW we would never have burned witches. This is a *fundamentally important principle*. You can't think straight without thoroughly understanding it. (CH.10 and Appx.8&9)

The **Principle of Limited (Independent) Variety (PLV)** is an assumption needed to justify the use of Inductive Thinking in some particular context. It assumes there are only a limited number of plausible hypotheses that are to be tested against the evidence. If there are too many the Prior on any one would become too small to be overborn by the available evidence. eg: dream interpretation, Economics, Psych***y.........[Ch.13]

The **Principle of Uniformity** is an assumption needed to justify the use of Inductive Thinking in some particular context. It assumes the future will be much like the past, and that objects 'over there' will be much the same as similar appearing objects nearby. Such assumptions cannot be wholly justified, but without them it would be impossible to lead a rational life. Experience teaches us that the principle is far more likely to be valid in some contexts (e.g. astronomy) than in others (e.g. History) thus 'The Scientific Method', such as it is, cannot necessarily be translated into other disciplines such as Economics. There are subjects which claim to be 'scientific' but which are not because the P of U does not hold within them, e.g. Psychology.

The **Prior Odds O(H)** are the Odds on H that a thinker needs to declare before, i.e. *prior* to, looking at some new evidence. Why? Because the recipe used to do Induction, i.e. Bayes' Rule, is multiplicative. It has to start from somewhere. Priors have been very controversial among those many Statisticians who want their subject to be purely mathematical whereas, so they imagine, Priors carry a whiff of personal prejudice. Note that 'Prior' is not an absolute categorization: the Odds on H reached after considering some earlier evidence may become the Prior when new evidence is to be factored into the argument. Thus one can start off from the bland or neutral Prior $O(H) = 1$, and let the clues and Weights do all the talking. In my firm opinion you cannot do Science without Induction, and you cannot do Induction without Priors.[If you use Probabilities instead of Odds then you still have to have a Prior Probability P(H) on your hypothesis H.] (Ch.4)

The **Probability P(H)** is the degree of certainty one has as to the truth of the proposition H where, *by convention*, P(H) =1 amounts to complete certainty, P(H) = 0 to complete disbelief. In our philosophy this Probability is entirely subjective, i.e. entirely personal to the individual who holds it. In other schemes the Probability is supposed to be more 'objective' or impersonal, i.e. calculable from allegedly impersonal pre-suppositions. Alas such presuppositions are in many contexts often impossible to find. Our definition of Probability is called 'Subjective' and was developed in particular by Bruno di Finetti in connection with coherent bets. Even hunches are allowed to set its values. Confining Probability to the range between 0 and 1 enabled certain useful results such as $P(H)+P(\bar{H}) = 1$ to be used. On the other hand it made a pigs ear out of the Detective's Equation, which was a disaster, a complete disaster. In preference to Probabilities one can use Odds instead, and I do. However, in retrospect, we can see that it is vital to be familiar with both notations (rather like we need to understand both fractions and decimals). (Ch. 4 & 11)

The **Prosecutor's Fallacy** is a plausible argument in which the high Prior Odds on Innocence are left out. Thus the extreme unlikeliness of the accused having the incriminating DNA, which he has, must be multiplied by the high or higher Prior Odds that he is, like most people in the population, innocent. Rare events are always happening to some one: that doesn't necessarily make them guilty. Thus diagnostic tests for rare diseases can fall prey to 'false positives'. Using an Inference Table can help to avoid this fallacy (because it insists on a Prior O(H)].(Ch. 8)

The **Scientists Approximation** is when we replace the correct value of the standard deviation in a Bernoulli Probability i.e. $\pm\sqrt{Qp(1-p)}$ with $\pm\sqrt{Qp}$ [which amounts to assuming that $\sqrt{(1-p)}$ is roughly 1]. Because Qp is the average value $\bar{x}$ we have the easy-to-remember result that we expect an average value $\bar{x}$ with a standard deviation of $\pm\sqrt{\bar{x}}$, which is quite accurate enough for most practical purposes. [Section (11:4)]

Small Number Statistics are numerical arguments which are invalid because the numbers involved are too small to be significant i.e. meaningful. [Ch.9]

The **Snake Oil Delusion (SNOD)** is my name for a common and hard-to-spot fallacy in which the thinker underweights or dismisses the likelihood that any other hypothesis but the one he has in mind could be responsible for some phenomenon. It can lead to certainty where non is justified: e.g. miracles, patent-medicines, conjuring, witchcraft…… Technically called '*Affirming the Consequent*' it is common in Science. It most often arises from either lack of imagination, or ignorance of reasonable alternatives (e.g. bacteria instead of witches). (Ch.4 &13)

Statistics is the business of trying to wring reliable conclusions out of numerical data. While it uses a branch of Mathematics called Probability Theory, it goes much beyond that and wasn't developed until the end of the 19th century. There are many schools of Statistics which principally differ in their approaches to Induction, and in their definitions of Probability. Be warned that the whole field is controversial and that much statistical practice is over-ambitious, or indeed plain wrong. (Ch.11).

Statistics, the Central Fallacy of, lies in striving for single Probabilities P(E|H) of decisive value [i.e. either P(E|H) > 0.8 or P(E|H) < 0.2] because then a single 'bad' clue (Systematic Error) could lead to disastrous decisions (actions), i.e. it would conflict with The 'Principle of Animal Wisdom'. Instead one should seek for several weaker clues which might cohere in combination. But as weak clues may be imprecise no purpose is served in looking for high precision in any single Probability (e.g. the 4-figure accurate tables used by conventional Statistics).

The **Theory of Probability** is a branch of Mathematics developed in the 17th century to help gamblers calculate the Odds in games of chance where the dice or cards employed, i.e. the 'agents', are assumed to be fair. Because it is so powerful, and useful, there are temptations to introduce it into real life where the agents are not fair dice, but people, or messy properties such as weather. Then beware. At some point assumptions have to be made about the behaviour of such agents, where simple assumptions are the easiest to use, but the least likely to be true of the real world. Thus much of Statistics, which rests heavily on the Theory of Probability, is downright wrong. (Ch. 9 &11)

Systematic Errors are the 'unknown unknowns' that may be present in any thinkers brain, warping its conclusions. Very often they are plausible, or socially acceptable assumptions such as 'The Earth is Flat'. No amount of calculation can either detect or eliminate these 'Elephants in the Room'. I have suggested we discount them by adopting PAW – the Principle of Animal Wisdom, which we should use anyway.(Ch.10)

The **Weight W(E|H)** of some evidence E, as it bears on the hypothesis H, is a number, assigned to it by the thinker in question, to be used in Bayes' Rule or an Inference Table. If the Weight is more than 1 it favours H, if less than 1 it disfavours it and if equal to 1 is neutral towards it. By definition $W(E|H) \equiv P(E|H)/P(E|\bar{H})$ where $\bar{H}$ means 'not H'. (Ch.4). [The PAW strongly recommends using Weights of only one of 4, 2, 1, ½ or ¼ .]

THINKING FOR OURSELVES By M J Disney
REFERENCES BY CHAPTER WITH COMMENTARY, 20/8/21

"The most beautiful thing we can experience is the mysterious. It is the source of all true art and science." Einstein.

Much of the writing in this field, particularly from non-scientists, reads plausibly but is, by my lights, deeply flawed, but it is infernally difficult to 'see through' when you read it first. Indeed, only after I had come to a coherent view of Common Sense Thinking for myself could I look back and spot where the fallacy in a piece of once respected argument lay. So, in order not to confuse the reader likewise, I have decided to supply only a very limited set of references for each chapter, comprised of writings which I believe are basically sound – though here and there I have included unavoidable 'classics' but with health warnings attached.

Such selective referencing is of course dangerous and is rightly frowned upon by scholars – even as they employ it like silent assassins themselves. All I can plead here is 'the lesser of two evils'. Get your head into CST first, which won't take long, then go read what the philosophers and statisticians have had to say in its stead. Try the other way round and you may never get to Common Sense at all.

I would anyway expect modern readers will make their way round my biases by employing internet tools such as Google and Wikipedia to get at the truth for themselves. Good hunting.

I have resisted the modern convention of quoting dozens of very recent references to show just how learned and up-to-date I am, but haven't hesitated to quote old ones if they are, in my opinion, the best. After all this is an extremely ancient, very slowly moving subject. Unfortunately one often couldn't do that in the past because readers couldn't practically get hold of older texts. They can now, with the aid of say Google and Amazon, and that is a truly wonderful thing.

Since this book is written for the non-mathematical reader that has meant leaving out some 'classics'. That is not fatal though because higher maths has far more often obscured this field than illuminated it. Ravens don't do Calculus but they're pretty damned good at common sense. Where I have found it advantageous to include a mathematical book I have mentioned that fact most explicitly. You won't miss much if you skip through the bits containing anything more than arithmetic. In general *the more maths there is in a book about Thinking the less you should trust it.*

CHAPTER 1: CAN WE LEARN TO THINK BETTER?

{1} Medawar P.B., *Induction and Intuition in Scientific Thought*, 1969, Methuen and Co Ltd, London p 11. Medawar was the wittiest man ever to write about scientific method, if not always the wisest. Certainly he's the only one who can make me laugh out loud. He wrote two other small books in our area *The Art of the Soluble* (1967) and *Advice to a Young Scientist* (1979). His books and essays are great fun but cannot be recommended as entirely sound advice because he advocates Popperism, which I think of as complete nonsense; see (12:2) and Appx.A8.

The real heroes of my book are two naturalists Charles Darwin and Alfred Russel Wallace who, independently, first came to the idea of Evolution by Natural Selection. Here are two books which give enjoyable glimpses into their minds and into their adventurous lives:

{2} *The Voyage of the Beagle,* by Charles Darwin, 1839. Google Books.

{3} *Alfred Russel Wallace*, a life, Peter Raby, 2001, Chatto and Windus Ltd.

CHAPTER 2: DIFFERENT KINDS OF THINKING

I include a couple of references for those who wonder about the bearing of Maths on Thinking. There is very little, though mathematical types would like us to think otherwise. Cockatoos are no good at maths, but they're damn good at thinking.

{1} *Mathematics and Plausible Reasoning*, George Polya, 1968, Princeton University Press, (vol II, self contained, entitled *Patterns of Plausible Inference*). Polya knows that mathematicians think inductively most of the time but are forced to finally produce Deductive proofs. Here he is groping to find some bridge between Deduction and Induction. Very readable, but mainly for experts. Ultimately he shrinks away from Priors and Weights. Here is the schizophrenia of Mathematicians starkly displayed: they use Induction but they fear to admit it.

{2} *Mathematics: Loss of Certainty* by Morris Kline,1982, Oxford University Press. How men came to realize the shocking truth that Geometry and Science were not the same thing and more generally that Mathematics itself has uncertainties in its foundations. A masterpiece by the great historian of Mathematics, with scarcely an equation in sight.

Mathematics *can* be a lovely subject, believe me, and there are many references to it after Chapter 14. But Common Sense doesn't need it.

CHAPTER 3: HOW SCIENTISTS THINK

There are several excellent accounts of the history of science for laymen. Here are some of my personal favourites:

{1} Daniel J Boorstin, *The Discoverers*, 1985, Vintage Books, gives an enthralling account of important discoveries from ancient times until the beginning of the atomic age. You can read it from start to finish or in any order; especially strong on ancient times. I like it first because it tells the tale of geographical and scientific explorations in parallel, it explains why discoveries were made where and when, and just as importantly, why they were missed there and then.

{2} Martin Harwit, *Cosmic Discovery*, 1981, Harvester Press, UK. is the most thoughtful book I have ever read. It is ' a first attempt to collect the kind of information that might be needed to answer questions on the promise of a particular science.' Using Astronomy for his example he asks 'How was it that we first came to discover the major phenomena we now observe in the Universe? Who were the individuals responsible for the discoveries? How did they prepare for their careers? What methods led to their successes?' From the answers arrived at he goes on to ask if there are lessons here as we go on to explore the future. It seems to me that almost any exploration would benefit from this kind of analysis. For instance is a national or an international plan likely to be more or less successful than the strivings of motivated individuals? One of my desert island books.

{3} Horace Freeland Judson. *The Search for Solutions*, abridged edition 1987, Johns Hopkins Paperbacks. The author believes that science is our century's art and cherishes those special 'moments of truth', instants of exhilaration and tranquillity, which, he maintains are the fundamental attraction, which draws the individual scientist on. I agree. Poetry for the thinking mind.

{4} Richard Dawkins (Ed). *The Oxford Book of Modern Science Writing*. 2009(pbk), Oxford University Press London. This thoughtful selection covers the 20th century to the present day and so complements Boorstin's history. The 100 or so articles are gathered under four headings; 'What Scientists Study', 'Who scientists are', 'What scientists think' and 'What scientists delight in'.

{4} Edmund Blair Bolles, *Galileo's Commandment*, 1997, W H Freeman, New York. This is an anthology of 2,500 years of great science writing which makes a wonderful bedside read, though it won't send you to sleep. Who could not enjoy 'Seashells in the Mountains' by Leonardo, 'Jigsaw Continents' by Alfred Wegener or 'The Long Snowfall' by Rachel Carson?

{5} John Gribbin. *History of Western Science 1543 – 2001*. 2002, Penguin Press. I learned much from this book which is especially strong on the physical sciences. The author promises to deliver 'a sweeping narrative with plenty of stories about real people doing out-of-the-ordinary things' and he does.

{5a} Bill Bryson. *A Short History of Nearly Everything*, 2003, Broadway Books, received a number of awards and prizes, and deservedly so. With quotes like "There are three stages to scientific discovery. First, people deny that it is true, then they deny that it is important; finally they credit the wrong person." how could it fail?

{6} Richard Hamblyn, *The Art of Science*, 2012 (Paperback), Picador, London has looked between the cracks in other anthologies and found some gems: for instance how John Lind discovered the cure for scurvy, how the eye evolved and why there are ripples in the sand.

{7} Paul de Kruif, *Microbe Hunters,* 1954 (orig1926), Harcourt Brace. A trifle dated but still a very absorbing account of Leeuwenhoek, Grassi and Malaria, Pasteur, Koch and the like. Although not acknowledged, de Kruif co-authored "*Martin Arrowsmith*' by Sinclair Lewis, the best novel ever written about science . De Kruif, himself a research micro-biologist who was sacked for writing the truth about it, supplied the background, the stories and the characters and took a quarter of the author's royalties in return. Shortly afterwards Lewis won America's first Nobel for Literature.

{8} Howard Markel, 2001, '*Reflections on Sinclair Lewis' Arrowsmith, the great American novel on public health and medicine'* (see Google)

{9} Harry Collins and Trevor Pinch, *The Golem*, 1993, Cambridge University Press, presents an unusual, and to me, healthy view of science 'as fallible and untidy, a matter of craft rather than logic'. It makes its point through case studies of some notorious episodes in science including 'Cold Fusion' and the bending of eclipse evidence by Sir Arthur Eddington to 'prove' The General Theory of Relativity.

{10} Edmund Whittaker, *A History of the Theories of Aether & Electricity*, 1989 (orig 1910), Orig two volumes, now in one. Dover publications NY. This classic on the subject is plastered with higher mathematics from beginning to end. Vol 2 ch. 2 is entitled "The Relativity Theory of Poincaré and Lorenz". This surprises many non-specialists who assume that Einstein was the sole begetter of Relativity. Not so, not at all, and that's why he never got the Nobel for Special Relativity, which is at least 90 % of the whole theory. I think of Poincaré and Einstein like Columbus and Amerigo Vespucci. Poincaré got there first but it was Einstein who realized they were standing on a new continent. Einstein's unique contribution, his relativistic theory of gravitation, called General Relativity, and of interest mainly to astronomers, came out later in 1915. I've included this to supply at least some evidence for my assertion that nobody is a scientific genius, not even Einstein.

Thomas Huxley said "Try to learn something about everything and everything about something". In that spirit, as a complement to the synoptic surveys recommended above, I suggest reading one or two deep and detailed histories of particular topics and so I have chosen four masterpieces covering those most exciting developments in 20th century science: plate tectonics, nuclear physics, human origins and molecular biology:

{11} Robert Muir Wood, *The Dark Side of the Earth*, 1986 (pbk), Allen & Unwin, London. A riveting account of how the theory of Continental Drift eventually overcame its manifold opponents in a battle lasting 50 years.

{12} Richard Rhodes, *The Making of the Atomic Bomb* ,1986, Simon and Schuster. A complete history of Quantum Physics as well as the bomb. Nobody puts this down, or lends it.

{13} Horace Freeland Judson, *The Eighth Day of Creation, 1980,* Touchstone NY. How two dozen colleagues around the world cracked the structure of DNA and then the genetic code. Science as collaboration as well as competition. Very long, but an almost unique glimpse into modern minds struggling to understand biology 'profonde'. A nice foil to Jim Watson's *The Double Helix*, which I also love.

{14} Virginia Morrel, *Ancestral Passions,* Simon & Schuster 1995*,* is a gripping account of the Leakey family, in particular Louis, Mary and son Richard who all played colourful and exciting roles in the hunt for and interpretation of early humanoid remains in East Africa. It illustrates science being driven by a mixture of ambition, personality, religion, philosophy, politics and chance. Huge stories were constructed of human origins based on a handful of fossils with vast room for prior prejudices and personal ambitions to enter in. The Leakeys, at war with one another, united only against the common enemy i.e. any outsider who wanted to do some digging and some speculating themselves. Morrel is, I suspect, partial to the Leakeys' point of view so you might want to read *'Lucy's Child'* by Johannson and Shreeve (Penguin 1989) to find out what one of their competitors has to say. Anthropology reminds me of Cosmology: huge issues resting on flimsy evidence: fascinating stuff – but is it science?

CHAPTER 4: NATURAL THINKING AND BAYES' RULE.

{1} Galaxy images: NGC 7331 was taken by Vincent Peris with the Zeiss 3.5 meter Telescope at Calar Alto in Spain; NGC 4889, which is 300 million light years away, with the Hubble Space Telescope. If you are interested in seeing many spectacular images of galaxies go to websites at either eso.org or spacetelescope.org. Beware though that such images have been tarted up out of all recognition so that they look nothing like the really dim images that astronomers have to struggle with. In that sense they miss the whole point; they resemble advertisements for pizzas: they look good, but you can't actually taste them.

{2} Simon Blackburn, *Think*, 1999, Oxford University Press, is a compact account of the Big Questions in philosophy. Unusually for such books it does have a brief section within 'Reasoning' called "Chancy Stuff" which deals with `Bayes' Rule and thus with Common Sense. He says (p232) 'Science similarly contains within itself the devices for correcting the illusions of science. This is its crowning glory. When we come upon intellectual endeavours that contain no such devices, one might cite psychoanalysis, grand political theories, 'new age' science, creationist science – we need not be interested.' Hear, hear!

{3} Giulio D'Agostini, *Bayesian Reasoning in Data Analysis*, 2003, World Scientific. As a scientist I found this much the most helpful and transparent introduction to Subjective Bayesianism, the kind we discuss in this book, and the basis of Common Sense. Alas the cost of this book [£90 in hardback] is off-putting.

{4} Colin Howson and Peter Urbach, *Scientific Reasoning ;the Bayesian Approach,* 2006 (3rd edition), Open Court Publishing USA. Although written by philosophers for other philosophers it is the one such book that I find myself agreeing with at almost every turn, though I rather dislike it's imperial style. It tackles The Problem of Induction from a probabilistic point of view, demolishes the Popperites, and then goes on to mount lethal attacks on both Classical Statistics and Objective Bayesianism. We scientists and other users of Common Sense in need of protection from philosophers , mathematicians and especially statisticians, should be very grateful to these modern disciples of Bruno de Finetti. Alas they don't get as far as the Detective's Equation, tackle Error Analysis or mention Ockham's Razor. Unfortunately, in its determination to dress all its arguments up in irrefutable Logic, the book loses me in several places. Brighter people than I might be able to follow them through their intricate maze of Symbolic Logic and Calculus, and I certainly hope so because philosophers who have claimed for their own impeccable logic before (e.g. Kant, Mill and Popper) have so often turned out to be mistaken. Here is an example of their style (p9): "…we shall show that the classical methods (of Statistics) are in fact intellectually quite indefensible and do not deserve their social success. Indeed we shall argue that the ideal of total objectivity is unattainable and that classical methods, which pose as guardians of that ideal, actually violate it at every turn; virtually none of their methods can be applied without a generous helping of personal judgement and arbitrary assumption." Personally I believe they are mostly right here and elsewhere, but it is not a style of argument which we empirical scientists are comfortable with or find convincing. All the same this book has few rivals that I know of for the numerate scientist who needs to understand why Subjective Bayesianism is so much better than its rivals.

{5} Hugh Gauch Jr., *Scientific Method in Practice*, 2003, Cambridge University Press (pbk) is the best book you can buy on this general subject, not least because it is written by a distinguished agricultural scientist who has spent a life improving crop yields. It gives an unrivalled view of the history of our subject and is the authority on Parsimony (Ockham's Razor). It is however too generous to Statisticians, draws no distinction between Objective and Subjective Probability, and never mentions my hero Bruno de Finetti. But otherwise a monument of scholarship with no calculus in it. I've read my copy half a dozen times and learned something new and valuable each time.

{6} Jonathan Baron. *Thinking and Deciding*, 2008 (4th Edition), Cambridge University Press. A long long long American undergraduate textbook written by a psychologist (and I have my issues with psychologiy [13:4]). Apart from that it's not half bad, if not much fun. Chapter 5 'The Normative theory of Probability' goes into Bayes' Theorem in some depth.

{7} Samir Okasha, *Philosophy of Science: a Very Short Introduction*, 2002, Oxford University Press. For those who want a taste of this weird academic discipline, which has little if anything to do with science, this is probably the cheapest and least unpleasant way to take it.

CHAPTER 5: DECISIVE THINKING AND THE DETECTIVE'S EQUATION.

Not everybody, I realize, has the same certainty that I have from personal experience (See Appendix A11 *How this Book got Written*), that animals can think too. I strongly agree with David Hume, the great Enlightenment philosopher, who was brought up on a farm, when he wrote in 1737 *"….no truth appears to me more evident than that beasts are endow'd with thought and reason as well as men."*. Anyone who tries to claim that animals can't think should be made to realize that the burden of proof lies entirely upon them – be they scientists or hunters. They are flying in the face of all Evolution. For those who need convincing of what is an essential part of our argument – that Common Sense must have arisen through Evolution – the following selection of books might help:

{1} *Never Cry Wolf*, Farley Mowat, 1963. This is an account of 3 seasons he spent living in the wilds close to wolves by one of the greatest environmentalists of the 20th century. When I was deep in the Soviet Union back in 1986 I met a Russian scientist who hailed from Siberia. 'What's it like?' I asked. Next day he furtively handed me a copy of Landau & Lifschitz's highly mathematical '*Classical Theory of Fields'*. When I dared to open it back in my hotel room I found the centre had been hollowed out to make way for Mowat's book '*The Siberians*'. What an amazing read by one of the greatest writers and campaigners of our age.

Mowat eventually became 'President of the Arctic', representing both the native Inuit peoples and the creatures up there on both sides of the Iron Curtain. A man for the centuries. Read a later edition of his book where he has a chance to reply to his many critics. The hunting and trapping fraternities hated his book because it proved that it was they, and not the wolves, who were wiping out the caribou, as they had wiped out much else.

{2} *In the Shadow of Man*, Jane Goodall, 1970 (Amazon). Her rightly famous study of the Gombe chimps and the discovery that Humans are not the only tool users. A great and pioneering scientist, arguably the most influential one alive today. I love her remark: "We have inherited this planet from our parents – but we have stolen it from our children".

{3} *King Solomon's Ring* by Konrad Lorenz, 2002 (Orig 1950) Routledge. Lorenz was one of the pioneers of Animal Behaviour as a scientific study. His book led me to bringing up a young jackdaw which fell down our chimney. Jack taught me, beyond a shadow of a doubt, that Jackdaws could think as well, if not better than most of my acquaintances, especially my headmaster.

{4} *The Soul of the Ape* by Eugene Marais, 1989, Penguin Books. A century ago this extraordinary man went to live close to a troop of baboons in South Africa to study their extremely intelligent behaviour. His later study of Ants, written in Afrikaans, was, after Marais's suicide, stolen by Maeterlink and won him a Nobel Prize.

{5} *Almost Human*: A more modern journey into the world of baboons, Shirley C Strum, 2001, University of Chicago Press. About 'The Pumphouse Gang' which she studied for 30 years in Kenya. She found incontestable evidence of their extraordinary intelligence, complexity, planning and foresight. Alas she also discovered that baboons are far nicer beings than the academics with whom she had to mix professionally.

{6} *The Thing with Feathers* by Noah Stryker, 2014, Riverhead Books, Penguin New York. I argue that thinking is an algorithmic process – which therefore cannot depend on the underlying circuitry (neuro-physiology). The best evidence for that would be the existence of creatures with very different physiology from ours – who yet could think. And here they are – the birds. Read about Magpies who hold funerals, gulls who marry, Nutcrackers who hide 10,000 seeds over a wide area then find them nearly all again when they are buried under deep midwinter snow, crows who can solve puzzles, parrots that appreciate music, 'homers' whose capacities defeat science...........

{7} *Other Minds*, Peter Godfrey-Smith, 2017, William Collins UK. ......and octopuses, made of watery jelly and very short lived, are pretty damned smart too.

{8} Virginia Morrell, *Animal Wise*, Old Street Publishing Ltd UK, 2013, is a modern, delightful yet comprehensive account of how many creatures use their minds both to survive, and enjoy life. I loved it.

{9 Frans de Waal, *Are We Smart Enough To Know how Smart Animals Are?* , Granta Books, 2016. Our whole case assumes that human thinking can only have evolved incrementally from that of our animal forbears. To many that may seem either far-fetched or unattractive but here is a new book by an expert on apes and monkeys which makes the case for evolutionary continuity far better than I ever could. He summarizes a century of observations and experiments to show that primates (and others) plan ahead, work in teams, show empathy, imagine solutions to problems in their heads, demonstrate inference, do not live just in the present, use many tools in an intentional way, can act generously and unselfishly….. The book is also strong on history, on the way proponents of human uniqueness have misused or misunderstood concepts such as Parsimony to hold up progress for a hundred years. Their holier-than-thou attitude mirrors that of the Deductivists in the Philosophy of Science and Statistics areas who have tried to down-play Induction. Resistance to the idea that we are but another variety of ape was always going to be stubborn, but here is the almost perfect riposte. I only wish it had been more generous to Jane Goodall {2 above}. But if you read only one book in this area this must be the one!

CHAPTER 6: NUMBERS AND THINKING.

{1} Nate Silver, *The Signal and the Noise*, 2012, Penguin Books.
About the art and science of prediction, this book is a long paean of praise for Bayes' Theorem. A famously successful election (US) predictor he covers prediction in many fields including baseball, finance, politics, meteorology, poker and epidemics. Very well informed and worth reading though it is seriously marred by using the old-fashioned Probabalistic notation for the theorem which makes it seem unnecessarily opaque and arcane. But after reading our (4:2) and (14:8) you should be able to translate this book into Odds as you go along (I skipped stuff on baseball without missing the big picture) ,especially if you consult our Appendix A13 *A Sketch-map of Thinking*.

CHAPTER 7:WOOLLY THINKING AND OCKHAM'S RAZOR

{1} *Edwin Hubble: Mariner of the Nebulae* by Gale Christianson,1996, University of Chicago Press, gives the inside and outside stories of how we came to believe that the Universe is expanding. Hubble has variously been called 'The greatest astronomer since Galileo' and 'The Man who discovered the Expansion of the Universe'. The truth is 'more nuanced' as they like to put it and if this cool biography is to be believed, very different from the pop legend. Hubble appears to have been a braggart, a liar and a snob but also a very lucky man who happened to be in exactly the right place at the right time. George Ellery Hale had just completed the great "Hundred Inch" telescope in California, the only instrument then capable of dramatically improving on Vesto Slipher's pioneering work on the spectra of nebulae (galaxies) and Hubble, having won the First World War in Europe, was appointed by Hale to do that job. And over the course of the next fifteen years he very successfully did so, showing that galaxies' redshifts appeared to increase proportionately with their distances away from us. But he was too cautious or too canny to swallow the dramatic idea that this implied a universal expansion and died in 1956 still sceptical. But when the Big Bang radiation was accidently discovered in 1965 Hubble was canonised as Expansion's discoverer. It's another episode like Einstein's canonisation for

Relativity. Both men unquestionably did great science but that doesn't entitle us to assume they were either geniuses or great human beings. Einstein's work was so arcane that very few outsiders can assess it properly and so they can easily imagine that it was the achievement of a genius; Hubble's work, just as momentous in its way, is far easier to explain and comprehend and required no extraordinary talents, as you will see from Christianson's book. I rather harp on this to support my contention that the evidence for genius in science doesn't bear close examination.

{2} Charles Darwin, *The Origin of Species*, 1999 (orig 1859), Bantam Classic NY. I found this a bit of a struggle until I got to the final chapter *Recapitulation and Conclusion*. I suggest you read that bombshell first. A monument to Ockham's Razor: see Gauch's book (our {5/5}), for a thoroughgoing discussion of that topic.

CHAPTER 8 : COMMON SENSE

{1} *Sherlock Holmes*, Arthur Conan Doyle, Penguin Complete edition, Penguin Books UK, 1981, 2001.

{2} *The Hidden Universe*, Michael Disney, J.M. Dent, London,1984 : Macmillan USA 1985. Out of date now but the most detailed review of Hidden Galaxies for the layperson.

{3} *Taking a Dim View*, Adam Hadhazy, *Discover* magazine, March 2018, pp 32-39. Brings Hidden Galaxies bang up to date.

{4} *Hidden Galaxies Netted at Last, Appendix* A10, this book.

{5} *The Low Surface Brightness Universe*, International Astronomical Union Colloquium 171, Ed. by J I Davies et al, Astronomical Society of the Pacific Conference Series Vol 170, 1999. This and the next reference contain the kinds of discussions and debates that go on between professional scientists at conferences when they are struggling to understand a new and interesting topic, in this case Hidden Galaxies and Dark Matter. Not easy for the layperson to follow perhaps but here is the ring of authentic scientific debate. For instance I air my views near the start then get shot down later by Zwaan et al. Out of this discussion my perplexity at the mysterious Scientific Method grew, and hence this book 20 years later .

{6} *Dark Galaxies and Lost Baryons*, International Astronomical Union Symposium 244, Ed. by J.I Davies & M.J Disney, Cambridge Univ. Press, 2008. Brings the above discussion more up to date. There is a list of *Provocative Statements and Controversies* in the field at the end, on p 399. Out of such disagreements science marches forward. That's one reason we scientists go to conferences – they can be painful but stimulating.

{7} *Where are the Missing Galaxies?* Adrian Cho, *Science*, 2007,317, 594. Cho is a professional science journalist on the staff at *Science* who attended the above conference at Cardiff University, where I work. His report shows just how heated scientific debates can be. After all we *are* passionate about our subject – and so we need to be.

{8} *Written in the Stars,* is a quartet of novels by me about both the Hidden Galaxy story and the building and exploitation of the Hubble Space Telescope. Published 2020 on Amazon. See **mjdisney.org**.

{9}*Cosmology, the Science of the Universe* by Edward Harrison, 2nd Edition, Cambridge University Press. Nearly everyone who has studied Cosmology seems tempted to write a book about it so there are literally hundreds to choose from, most of them I have to say, highly superficial. This is a subject with few facts and masses of speculation so it is vital to understand the lines of reasoning being used, and the possible weaknesses within them. Having taught the subject myself for 20 years I found this book most helpful to serious students, without being too intimidating mathematically. It's dated now, but this subject moves slowly, despite all the razzamataz.

CHAPTER 9: ERROR ANALYSIS{1} *Chances Are*: Adventures in Probability by Michael & Ellen Kaplan. 2007, Penguin Books. Probability lies at the heart of all Common Sense and most science – and yet it is poorly understood. Here though is a very readable account of it full of lightly worn scholarship, dazzling stories, wisdom and glittering prose. Buy, read, treasure.

CHAPTER 10: SYSTEMATIC ERROR: THE ELEPHANT IN THE ROOM.

{1}*Oliver Heaviside* by Paul J Nahin, 2002, Johns Hopkins paperbacks. Although radio-waves have changed the world more than any other single invention, their history is now either forgotten or so simplified as to be a mockery. And that is typical of discoveries and inventions in general. That's a great pity because if we are going to understand and nurture science and invention we need to know exactly where it succeeded and why. So this biography of a forgotten Victorian obsessive, who revolutionised the world, yet died of malnutrition, is a healthy, and fascinating, reminder. Never mind the maths, of which there is a lot, it's a fascinating read anyway.

{2} *Sapiens* by Yuval Harari, 2014, Harvill Secker, is a sort of history of the mind which covers about 100,000 years. Very good on cultural myths and their importance for human development. They may well be the Elephants in your room.

CHAPTER 11 ; STATISTICS OR TERROR ANALYSIS.

{1} *Dicing with Death* by Stephen Senn, 2003, Cambridge University Press. A racy account of medical statistics. At the front is a lengthy discussion of the infamous Boy/Girl problem, (which I think is wrong – see Appendix A4 for my take on this fascinating problem.) But read it for yourself.

{2}*The Ecological Detective* by Hilborn and Mangel, 1997, p54 Princeton University Press, gives what I consider to be the correct answer (1/2) to the Boy/Girl problem.

{3}Muller,F.H.1940, *Tabakmissbrauch und Lungcarcinom*, Z.Krebsforch, 49, 57-85. Early Stats of Lung Cancer.(Google)

{4} *Smoking and Cancer of the Lung*, Doll, Richard & A Bradford Hill, 1950, British Medical Journal, Sept 30th, 739 – 748. (Google), the classic paper on the subject, and readable.

{5}*Mortality in relation to smoking: 50 years' observations on Male British doctors*, Doll R. et al. 2004, BMJ,doi: 10.1136 /bmj.38142.55479.AE (Google); if you're ever tempted to smoke read this first.

{6}*Bayesian Reasoning in Science* Howson C, Urbach P., 1991, *Nature*, vol 350, pp 371 - 4 .This is an absolutely key reference because it pulls the rug out from under conventional Statistics. Easy to read, calculus unnecessary. Required reading for both phobics and fans of Statistics. Can't recommend it too highly. (Google)

Since I have attempted in this chapter to undermine the whole case for conventional Statistics I feel I should, out of fairness, now reference the three alternative BIBLES on that subject:

{7}*Statistical Methods for Research Workers* by R.A. Fisher, 1925,Oliver and Boyd, Edinburgh is the most famous book ever written on Statistics by the most famous statistician who ever lived. It has gone through dozens of impressions.

Fisher's early career was spent at an agricultural research station advising innumerate crop scientists on how to conduct and analyse their experiments. His cook-book is full of recipes for people like them who would not appreciate whether the underlying mathematical reasoning was sound – which in many cases it is not. This power went to Fisher's head and the nasty little man became a tyrant who could never admit to his many mistakes. He wasn't always wrong though and did some apparently valuable work on Evolution theory. Alas he became a knighted professor at Cambridge with great power over the statistical landscape. More of a priest than a scientist, from there he did a great deal of harm by preaching completely wrong-headed ideas. For instance he thought Probability had to be Frequentist and railed against the use of Priors, particularly the common-sense (Subjective) variety. He even succeeded in turning his many personal vendettas into world-wide statistical wars.

Fisher's childish behaviour should have alerted people, far earlier than it did, to the unhealthy state of Statistics, but alas Statistics could always hide behind higher mathematics (that's why we need to be so suspicious of it). Rational people argue on the basis of evidence; gangsters like Fisher use character assassination, the old-boy network, back-stabbing, and intellectual and institutional snobbery to pursue a personal or religious agenda. If nothing else the Fisher story illustrates the folly of scientists being awarded or accepting titles and prizes. If your arguments are any good they don't need social embellishments. If they are so embellished then what is wrong with them? I don't see how social climbing and intellectual honesty can be reconciled.

Alas much of the biomedical field is still mesmerized by Fisher's 'dangerous' statistical propaganda, and I use 'dangerous' deliberately. It is impossible to estimate how many laboratory animals, or patients being used as guinea-pigs, have suffered or died unnecessarily – but we could be talking casualty lists of wartime proportions. For example Rhesus Monkeys were driven almost to extinction during the campaign to develop a polio vaccine.

Among others, biomedical scientists mustn't be allowed to get away with lazy statistical habits – for instance using cook-books like Fisher's, opaque statistical computer packages, or worst of all hiring professional statisticians. In my experience most of those have a priestly mentality, antipathetic to common sense [after all they are the intellectual grandchildren of one of Fisher, Pearson, Neyman, Jefferies or Jaynes – see below or (10:8)]. No, if we are responsible for other creature's lives, we cannot be, and should not be absolved from thinking for ourselves. If you are going to carry out medical, veterinary, social or otherwise fateful experiments, you should first look at Fishers ghastly book and ask if you would like to face a jury one day armed with nothing better than his dogmatic and often foolish theology. I wouldn't. And if I was on the jury I wouldn't let you get away with it.

{8} *Theory of Probability*, 1939 (1960). Jeffreys H., Oxford University Press, is the bible of the Objective Bayesians. Sir Harold Jeffreys was, like Fisher, another be-knighted Cambridge don with semi-religious views on Inference and Statistics. Its main virtue is that it wasn't written by Fisher and it does allow, at least in principle, for Priors. Unfortunately the Priors have to be 'Objective' – presumably as a defence against Fisher who hated Priors of any kind (except his own). To escape from Fisher all too many academics, including myself, fled into the arms of Jeffreys – and there found themselves stranded far far short of common sense. Had de Finetti's work – which preceded Jeffreys' – been translated into English 40 years earlier, we might have been spared this "Objective" and now widely fashionable aberration.(see Poincaré's objection in the next section).

The book is highly mathematical and very dry.

{9} *Probability Theory, the Logic of Science* by E.T. Jaynes, 2003, CUP. I must have spent an entire year reading, with colleagues, this infuriating, fascinating book which is as weighty as the Bible – in several senses. Jaynes was an American physicist who went to Cambridge for a year, fell under the spell of Harold Jeffreys and his disciples, and returned home to found the Church of Objective Bayesians in America. Jaynes seems to have read everything about applying logic to science, thought about it critically, and written about it trenchantly. Like his Saviour Jeffreys he's obsessed with the 'objectivity' of scientific thinking. Every Prior you use, every clue you weight, if you are to be a scientist, has to be objective. That sounds right at first – which is why he is so persuasive. It seemed worth wading through all the advanced mathematics because one is a pilgrim marching towards the only true light, where of course salvation is waiting.

Jaynes is highly knowledgeable about, and scathing of, all the other four or more rival sects of Statistics which fought so bitterly with one another throughout the twentieth century. Such bitter religious wars confused and intimidated all us outsiders [9:8]. One feels there must be a jealous God in there somewhere, but which flavour he preferred, and to which Church he belonged, wasn't clear.

Only gradually it dawned that Jaynes was wrong, very wrong too. Either he forgot about, or chose to ignore, the tarantula in the Urn. His 'religion' if you like, applies only to CLOSED systems, i.e. games, not the real world. He was such an extreme 'Objectivist' that his long term aim, in principle at least, was to construct a 'robot' to carry out 'objective scientific inference'. But as Henri Poincaré long ago pointed out, that would be foolish. If you find yourself losing heavily in a gambling den you would be silly to remain until you have 'objective evidence' that you are being cheated. Sometimes in life, very often in practice, you have got to make a best guess, and act on it. Objective Bayesianism is in my opinion a form of religious mania which is quite simply wrong, i.e. out of synch with Evolution by Natural Selection, and therefore Common Sense.

All the same I gained much from this colossal, learned and opinionated tome, principally knowledge of what was wrong with the creeds of the *other* sects. Anyone with much faith in Statistics, of whatever variety, needs to read about the ghastly religious feuds which have disgraced this subject throughout the twentieth century. Just reading through Jaynes' acidulous References is an education in itself. But remember, ultimately, he's basically wrong himself.

{10} *Probability Theory (2 Vols)* by Bruno de Finetti, 1974 (English Translation) Wiley New York, is the bible of the Subjective Bayesian sect. de Finetti provided the robust mathematical defence which Common Sense needed when it was and still is attacked by mathematicians and statisticians. Starting from a much more common-sense definition of Probability than they do, one based on what fraction of the pot you are prepared to bet on a proposition, de Finetti was able to prove The Product Rule and Bayes' Theorem and even got to within a whisker of The Detective's Equation (vol 1 p 159), only failing to take the final step and translate it into Odds. Unfortunately for history the book wasn't translated into English until 1974 though de Finetti's core ideas go back to the 1920s and 1930s. Had his common sense been appreciated earlier he might have made short work of Fisher, Jeffreys and Jaynes, but he had to earn his living as an actuary in Trieste for much of his early life before becoming a professor of mathematics at La Sapienza University in Rome. Science is such a fiercely competitive subject that the slightest handicap can bring the unlucky one down; being born into the wrong language, or even going to the wrong university, can mean the difference between fame and failure. And that's not just a tragedy for the individual; de Finetti's obscurity may have held up human progress for a generation. How many mute inglorious Miltons are there still out there? It has been suggested that on-line university courses may discover some of the great minds presently hoeing potatoes. Let's hope.

I would love to recommend these volumes to readers but alas they are highly mathematical and, to my eye, thoroughly disorganised. De Finetti was a much clearer thinker than he was a writer; he never sticks to one subject, there are few examples, and much of the book is made up of parenthetical remarks about subjects several chapters away, either earlier or later. What a great pity because de Finetti appears to have seen the truth more clearly than anyone. It's much better to read D'Agostini's {4/3} or Howson and Urbach's discussion of his work {4/4}, than De Finetti's book itself.

NB A brilliant young British thinker Frank Ramsey, even before de Finetti (1931, proved that a consistent logic of Subjective Probabilities was possible, but alas he didn't get very far with it, not even as far as Bayes'

Theorem because alas he died at the age of only 26. See Cheryl Misak's fascinating biography 'Frank Ramsey' 2021 on Amazon Kindle.

{11} *The Black Swan* by Nicholas Taleb Nassim, Random House NY 2007; Penguin UK 2008 (pbk). A wonderful polemic aimed at all those, particularly economists, who believe the real world can be understood using simplistic mathematical models such as The Normal Distribution ('Gaussian'). He comes to many of the same conclusions as we do but arrives at them from an entirely different point of view founded on the turbulence of real history (he was a refugee from his homeland) and the delusions of the financial system (he was later a NY 'Quant') which led to the 2008 financial meltdown. An extremely well read scholar whose conclusions are based on his own observations and reading, and not what he was taught at some college. He has chapters entitled : "The Bell Curve, that Great Intellectual fraud" (15) and ""Locke's Madness, or Bell Curves in the wrong places" (17). I could quote endlessly but confine myself to this, which shows he pulls no punches (p 275, Penguin revised edition 2010) " Several hundred thousand students in business schools and social science departments from Singapore to Urban- Champagne, as well as people in the business world, continue to study "scientific" methods, all grounded on the Gaussian, all embedded in the ludic fallacy…. This chapter examines disasters stemming from the application of phony mathematics to social science. The real topic might be the dangers to our society brought about by the Swedish academy that awards the Nobel Prize." In short, read this book as soon as you can. It could save you from unnecessary misfortunes; a life-belt for the mind at sea.

{12} *Ionnadis J.P.A*, 2005, *Why Most Published Research Findings are Wrong*. PLoSMed 2(8):e124 see

www.plosmedicine.org. A much discussed paper these days but I give my reasons in (11:8) for thinking it irrelevant.

{13} *The Superego, the Ego and the Id in Statistical Reasoning*, Gerd Gigeranzer, 1999, originally in. *A handbook for data analysis in the behavioural sciences*. Hillsdale, NJ: Erlbaum, 1993 (pp. 311-339) Contact author for electronic version (seb.gigerenzer@ampib-berlin.mpg.de

A trenchant attack on the use of Statistics in the Social Sciences, Psychology in particular. Journal Editors and Statistics text-book writers sensed that the different Statistics sects [in particular the Frequentist schools of Fisher, Neyman and Pearson] were inconsistent with one another, but they tried to paper it over, not understanding that they were totally irreconcilable. Thus innumerate social science students were utterly bamboozled into propagating a lot of nonsense in consequence, all dressed up in Statistical jargon-speak, including 'p-values' and 'Chi-squared significance tests'. Yet another reason to be wary of Psychology – see Sect. (13:4).

CHAPTER 12:PERSUASION
{1} *Francis Crick* by Matt Ridley, 2008, Harper Perennial, pp 66 -7 (Perutz&DNA)
CHAPTER 13: POOR THINKING
{1} *A Proof of the impossibility of inductive Probability,* Popper K & Miller D., 1983, *Nature*, 303, pp 687-8. Here is Popper claiming to show why Inductive Thinking using Probabilities is impossible. I'm sure he's wrong, but it's not easy to spot the fallacy in the argument – which is often the problem with philosophers using logic – they're so damned plausible and articulate, the chief skills required in that profession I would suppose. See if you can spot it.
{2} *The Wealth and Power of Nations,* David Landes:. Little Brown, 1998. Economics may be more of a religion than a science but it casts a long shadow across the modern world. Intelligent citizens need to understand its chequered history if only to discount those who would use it as a weapon – politicians in particular. This is a readable history of its frequent disasters and odd successes.
{3} Milton Friedman, *Capitalism and Freedom*, 1962, University of Chicago Press. 'The Economist of the Century' according to Fortune magazine, winner of the 'Sort-of-Nobel-Prize-in-Economics', expresses his opinions. How this very bad book can be remotely thought of as science defeats me. Near the beginning (p4) he writes; "...the great advances of civilization, in industry or agriculture, have never come from centralized government." This is complete and utter nonsense; as factually wrong as stating 'The Earth is flat.' [Think only of: sewerage systems, clean water, computers, space-satellites, anti-biotics, jet engines, radar, the Internet, broadcasting, machine-tools, anti-scorbutics, satellite navigation, astro-navigation, ...on and on and on.] To believe such balderdash you have to be either a fanatic, a fool, a complete ignoramus, or perhaps just an economist. And yet a lot of people followed Friedman down into his fantasy-land of 'Free-market Fundamentalism', including Mrs. Thatcher and President Reagan. If nothing else however it does illustrate the yawning gap between Science and Economics and the complete folly of confusing them, even for a second.
{4} Sigmund Freud, *The Interpretation of Dreams,* 1997 (orig 1900), Wordsworth Editions Hertfordshire, is an even worse though much more famous book than Friedman's, illustrating the ocean that separates psycho-babble from common-sense and science. Worth skim-reading for that reason alone.
{5} Jeffrey Masson, *Against Therapy,* 1992, Flamingo, is a devastating attack on Psychotherapy – not just some particular school or individual, but upon the whole notion that certain individuals can set themselves up as experts on the subject of human behaviour, can describe 'normal' behaviour and diagnose deviations from it, and then recommend or indeed carry out 'treatments' aimed at putting things right. He argues that it is at best arrogant bullshit, is always a way of making money out of other people's misery, and more often than not is harmful, often tragically so to those upon whom it is practiced. Freud is condemned as a monster out of his own private letters, Jung preyed on his female patients and would be serving a long prison sentence today, and so on and so on. As a physicist and once passionate disciple of psycho-analysis he wades into virtually all of

the modern American schools likewise and uses the evidence of controlled experiments to argue that they are equally without substance. He convinced me. If nothing else *Against Therapy* illustrates how mumbo jumbo can rule in any world which doesn't have a clear understanding of Common Sense. Don't pay a therapist a penny until you've read this first.

{6} *Irreproducibility in Psychological Research*, 2015, 'Centre for Open Science' (Nosek et al.), 'Science', http:://doc.org/68c:2015. The seminal results in 61/100 Psychology Papers proved not to be repeatable. So can one rely on anything in this subject, or anybody? Better not.

{7} *An Officer and a Spy* by Robert Harris [Hutchinson, London 2013] is a fictionalised but I hope faithful account of the notorious Dreyfus affaire in France. In 1894 Alfred Dreyfus, a junior officer in the French army, was accused of spying for Germany, court martialled and sentenced to solitary confinement in unspeakable conditions on Devil's Island – where the authorities hoped he would die. Gradually it emerged that it was a conspiracy amongst the senior ranks of the Army and Secret Service to cover up the real truth and the real villain – who had aristocratic connections. Everyone in France – and elsewhere – was induced to take sides by a biased press and flimsy evidence. An absorbing account of mass hysteria. Whose side would you have been on – and why? And could it happen here? Think of the so called 'Miners Riot' or the 'Hillsborough Disaster'. My own brief time behind the Iron Curtain made me realize how easy it is for authorities to control us through distorting information. We have to be on our guard, especially in a country like Britain where we usually trust our government.

{8}Naomi Oreskes and Eric M. Conway, *Merchants of Doubt*, 2010, Bloomsbury Press NY. See (11:10). How the nay-sayers on Global warming cooked the books. Shocking!

{9} Carl Sagan, *The Demon Haunted World,* 1996, Random House, is a wonderful collection of essays about the value of scepticism, the perils of denying it, and the concordance between scepticism, democracy and civilization. It's subtitle is "Science as a candle in the dark". Do yourself a big favour and read it. I particularly love Chapt. 12 '*The Fine Art of Baloney Detection'* which contains a list of tools for the sceptical thinker and a list of Dos and Don'ts for those who would argue honestly. It's practically a university education in itself. I could quote endlessly from it, but better you read such a wise and entertaining writer for yourself.

CHAPTER 14: THE EXTRAORDINARY HISTORY OF THINKING.

{1} George Corner, 1968; remarks taken from the Forward to Medawar's book see reference {1/1} above.

{2} *Philosophy of Science: an historical Anthology* ,Timothy McGraw et al (Editors), 2009, Wiley-Blackwell. I don't know anywhere else where you find so many important historical sources of scientific thinking between two covers. For instance without this book I might never have encountered Christiaan Huyghens' important work [12:4].

{3} *The theory that would not die* by Sharon Bertsch Mcgrayne, 2011, Yale University Press. I am so glad McGrayne has written this history of Bayes' Theorem for without it sceptics would never believe what a roller coaster ride it, and indeed Common Sense Thinking in general, has had over recent centuries. When I read it I felt like Livingstone encountering Stanley in darkest Africa. So I wasn't entirely mad after all! Precisely because it is common sense Bayes' Rule gets re-discovered time and again by different people in different contexts, and then somehow gets forgotten, which I find harder to understand. The Detective's Equation would be in the same case if only thinkers used Odds more often – see{Good}. Almost no maths in this book, which is both a plus and a minus. But a real treasure for the thinking mind, and unique.

{4} Stigler, Stephen, *The History of Statistics: The measurement of uncertainty before 1900*, 1986 Belknap Press Harvard, and Sta*tistics on the Table*, 2002, Harvard University Press. As we all know, history is written by the winners and I've certainly been tempted to pick out those threads which lead towards my own points of view. To try and avoid that sin where the contentious topics of Probability and Statistics are concerned I have generally, but not always, relied on these works by a professional historian of Statistics, which I definitely am not. I can only hope he's right.

{5} Good I.J., 1983, '*Weight of Evidence: a brief survey*,' *Bayesian Statistics 2, Conference Proceedings*, Elsevier(North Holland), pp 249-270. This is chiefly interesting for its account of Alan Turing's 'Banburyismus' technique developed at Bletchley Park in WWII to help break ciphers which *may* have had something to do with the Detective's Equation . Good later became Turing's assistant and is portrayed in that absorbing film about code-breaking '*The Imitation game*' (2014). Banburyismus was a vitally important technique for feeding clues into the 'bombes' (primitive computers) to dramatically lessen the number of cipher-wheel settings they had to try out. In divulging what went on Good appears to have been breaching national security which – rightly or wrongly – lasts for 50 years. It's intriguing to imagine that The Detective's Equation, like the design of Colossus, might have been concealed by the Official Secrets Act.

{6} *Most Secret War* by R.V Jones. 1978, Hamish Hamilton. The thrilling tale of Intelligence and counter-Intelligence between the RAF and the Luftwaffe in the Second WW; from the RAF side naturally. Full of great stories and a paean of praise for Ockham's Razor as the key to straight thinking.

CHAPTER 15: THE PECULIARITIES OF SCIENCE

{1}*It Must Be Beautiful*: great equations of modern science, Graham Farmelo (ed), 2003, Granta Publications, London. This is a wonderful account of the subject, requiring little mathematics, in which a dozen experts bring their favourite equations to life. Each essay is well worth the price of the whole book. Frank Wilczek's account of the Dirac Equation ('*A Piece of Magic)*, Igor Aleksander's account of Shannon's Information Theory (*Bit by Bit*) and Stephen Weinberg's *Afterword* I found especially captivating .

{2} Robert P Crease. *The Great Equations*, 2008, Robinson London. We don't do much maths in our book, we don't need to, so this is a welcome and very readable account of some key equations in scientific history including Maxwell's Equations of Electromagnetism and Schrodinger's Equation of Atomic Physics.

{3} *The Lightness of Being* by Frank Wilczek, 2008, Basic Books. About Mass, Aether and the Unification of Forces. I can't see anyone writing a more readable account of the foundations of theoretical physics and how it is being built nowadays. Don't believe in the aether? Then read this. The Vacuum is absolutely full of the strangest stuff.

{4} *Not Even Wrong* by Peter Woit, 2006, Jonathon Cape, is the story of String Theory, the kind of sad nonsense, pi in the sky, into which scientific theory tends to descend when it is not constantly challenged by new observational evidence.

{5} *The World of Mathematics(3 vols)* by Kasner and Newman, 2003 (orig 1940), Dover. An anthology of Mathematics and its applications throughout the ages. One of my desert-island books. What more can I say? And no you don't need calculus; but it may make you eager to learn.

{6} *The Princeton Companion to Mathematics*. Timothy Gowers (Editor), 2008, Princeton University Press. If you've ever wanted to know about mathematics this miraculous compendium has got it all, or nearly all. The world's top mathematicians have come together to distil their passions, puzzles and achievements in 1000 pages. If only I'd had this when I was a boy! You can read about half of it without calculus. The articles are mostly independent of one another. There's an e-version; useful because the original is big.

{7} *Teach Yourself Calculus* by Abbot and Wardle, 1970, Hodder, UK. Old but in my opinion still does what the title promises, and as I learned Calculus effortlessly from it myself as a boy I know it works. Lots of worked examples – which is what one needs as an autodidact. Books on the subject by pure mathematicians are often disasters for students because they are concerned with logical nit-picking

{8} *The Rationality of Science*, 1981, W H Newton-Smith, Google books, free download.

{9} *The Third Chimpanzee*, Jared Diamond. 1991, Vintage London. That's us. How the human species changed, within a short time, from just another species of big mammal to a world conqueror. Riveting, like all Diamond's books.

CHAPTER 16: CONSEQUENCES; THE ASCENT OF MAN

{1} Medawar P.B, 1972, *The Hope of Progress*, Wildwood House, London, p 126

{2} *On Human Nature*, Edward O Wilson, 1978, Harvard University Press. How far can Evolutionary theory explain human behaviour? A wise but tentative answer.

{3} *Reindeer Moon* by Elizabeth Marshal Thomas, 1987, William Collins Ltd, is an unforgettable novel in which an anthropologist expert on hunter-gatherer societies portrays what it is like to grow up in one.

{4} *Affluence without Abundance*, 2017, James Suzman, Bloomsbury Publishing, USA. An account of the disappearing African Bushmen, arguably the most successful humans to inhabit this Earth, because their way of life survived so much longer than other tribes and civilizations which, for one reason or another, disappeared so quickly.

{5} Angrist S.W. and Hepler L.G., "*Order and Chaos*", 1973, Pelican. We are, like all animals, first and foremost heat engines governed by the so called Laws of Thermodynamics. The novelist C.P Snow claimed that no one ignorant of the The Second Law of Thermodynamics could call themselves educated, and I agree. Thermodynamics is a beautiful and not arcane science which dominates every aspect and moment of our lives – including, you might be surprised to learn, Thinking. Google and Amazon, with their huge server-farms on which the computer 'Cloud' resides, certainly wouldn't disagree. Keeping all those memory discs spinning, all those processing units whirring, requires vast amounts of power for which they need entire nuclear power-stations and hydro-electric dams. It's not surprising then that the human brain, no mean processor itself, can be limited in its performance by the power you can afford to supply it. And you will be as horrified as I was to find out how little we *can* afford – a mere Watt or two most of the time. It's all about our low thermodynamic efficiency (which is nevertheless probably higher than any other animal's). Anyone who cares about thinking should therefore be very interested in Thermodynamics. For instance when you are studying should you turn the thermostat up or down? Your answer could make all the difference. This is a short, easy-to-read account of a fascinating and fateful subject. Get educated [see also our online article 'Human Thermodynamics' {website}].

{6} *The Last man who knew Everything* by Andrew Robinson, One World Publishing, 2006 UK is the biography of Thomas Young, and a very good one. A prodigy who could read before he was two, he read the whole bible (twice) before he was 4, and went on to educate himself in a formidable process of self-instruction saying " but whoever would arrive at excellence must be self-taught". He was forced to waste years at Cambridge because it had corruptly seized hold of entry into the medical profession (Young's chosen career, which he quickly gave up in disgust). This book is rewarding for many reasons, not least because it illustrates how ruthlessly corrupt entrenched professions can become, and will become if you give them half a chance: educationalists, doctors, lawyers and money-men in particular. It's an old old story, and a nasty one, and alas it doesn't get better with time.

{7} *The Origin of Virtue*, by Matt Ridley, 1996, Penguin. How complex cooperation and selfish competition evolved side by side.

{8} *The Human Story*, Robin Dunbar, 2004, Faber and Faber, draws together some of the latest research on hominid and pre-human evolution to try and explain how our minds have evolved. Fascinating stuff; but be warned: he's a psychologist.

{9} *The Code Book* by Simon Singh, 2000, Fourth Estate London. The best book on code breaking of the many I have read. A superb writer. But it is easy (not done in this book) to claim more for the effects of code-breaking than was historically the case. Code-breaking is chiefly of help to the underdogs at the time. My novel 'Strangle' (Amazon 2020), amongst other matters, explores the effects of code-breaking on the battle of the Atlantic in WW2, the most titanic, the most momentous battle in all of history. See our website **mjdisney.org**

{10} *Prize Fight* : The Race and the Rivalry to be the First in Science, by Morton Meyers, 2012, Palgrave Macmillan. If I tell my non-astronomical friends how vicious and how immoral the behaviour of some well-known astronomers can be they mostly don't believe me; they think I've become paranoid. I sometimes can't believe it myself; I would certainly prefer not to. So it's good that someone else has taken the lid off what goes on in medical science, which sounds even beastlier. This excellent book by an insider will make your blood boil. The problem is that there are no sanctions against appalling behaviour in science, and many rewards for it. How many Nobel prize–winners and other 'eminents' can sleep o'nights I don't know. Yes I do; they're basically rotters. Unfortunately some of our founding fathers, including Galileo and Newton, set rotten examples. The wonder is that so many colleagues still behave decently. Sadly some scientists leave the profession because of the behaviour of their rivals; some voluntarily, some because they are driven.

THINKING FOR OURSELVES (Disney)

ACKNOWLEDGEMENTS (1/8/18)

This has been a long long journey. I would like to thank those who accompanied me on the road – even if we later disagreed and parted company along the way. It was bound to be thus. They include: Phil Williams (University of Aberystwyth), the greatest Welshman of his day – with whom I set out, but who tragically died on the very first stage – this is for you Phil; Peter McLeod who effectively edited the entire book, making numerous wise suggestions in every chapter, most of which I heeded; Steve Gull (Cambridge University) – who kindly gave me a samisadze copy of Jaynes' provocative book long before it was published; Harry Collins (Cardiff University) who lent me Agostini's book – which set me off on a new path; my wife Nino who stoutly and intelligently resisted Frequentist Statistics even when she was an undergraduate – so leading me towards Common Sense; Colin Howson (Univ. of Toronto) who patiently put up with my e-mails, even when he so obviously disagreed; Hugh Gauch Jnr. that great scholar from Cornell University who's own book has been such an inspiration; my son Mathias Disney (UCL Geography) for sharing his capacious learning; and finally my friend Joe Romano, now at the University of Texas Brownsville, who came with me much the longest way on what would otherwise have been a very lonely road. Thank you all.

THINKING FOR OURSELVES (Disney: 15/5/18)
APPENDIX A1: NORMAL PROBABILITIES AND ODDS; AND WHAT ABOUT WEIGHTS?

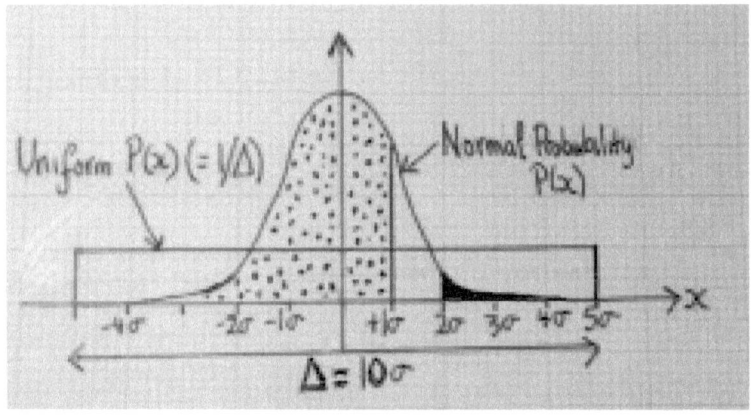

Fig (A1:1) Two common and important Probability curves, the "Normal" (bell-shaped) and the "Uniform" box. x is the discrepancy between two numbers, and the various areas under the curves correspond to the Probabilities that these discrepancies could arise by chance.

Statistics is much concerned with the discrepancies x between observed numbers N and the numbers N_H expected if some hypothesis H is true i.e. $x = (N - N_H)$. Abraham de Moivre long ago worked out the Probabilities P(x) for various x's for the case where only Counting Errors matter and came up with the famous bell-shaped curve above. That curve is described by a single parameter σ (the 'standard deviation') which is the half-width of the bell about 40 per cent of the way below its peak. It is convenient to speak of x in terms of units of σ. Thus $x = +t\sigma$ (where t is called 'the Discrepancy') refers to a point t standard deviations to the right of the origin. Since x (hence t) must have *some* value the sum of the P(x)'s, i.e. the total area under the graph, must come to 1. For all values of x (i.e. of t) de Moivre tabulated 4-figure accurate values for the dotted area (which is called 'erf(t)' standing for 'error-function'). Thus for $x = +1\sigma$ (i.e. $t = 1$) he found erf(1) = 0.8413. Thus there is an 84% Probability of x lying to the left of $t = +1$ and thus a (100-84) = 16% chance of lying to the right of $t = +1$ (this is called 'the single-tailed' Probability). Since the graph is symmetrical that means the Probability of x lying more than 1σ from the origin to *either* side is 16% + 16% or about one third (this is called the 'double-tailed' Probability). Thus the Probability of x lying within

$\pm 1\sigma$ of the origin must be two thirds. Hence the Odds are 2 to 1 that a Counting Error will be smaller than $\pm 1\sigma$ (i.e. double-tailed) while the Odds are 40 to 1 *against* a Counting Error (single-tailed) being larger than $+2\sigma$ (see fig).

Numbers can be decisive – which is their attraction to Statisticians – and others. Because they are so widely used for ruling out hypotheses I here tabulate some critical values derived from de Moivre's work. They are called (see later) "Normal Probabilities".

TABLE (A1:a) NORMAL Probabilities (single tailed) and Odds

t	$P(x< +t\sigma \,|H)$	$P(x> +t\sigma \,|H)$	$O(x> +t\sigma \,|H)$
1	.8413 (5/6)	.1587 (1/6)	5:1 against
2	.975	.025	40:1 against
3	.99865	.0013	800:1 against
4	.99998	.00002	50,000:1 against

While in the double-tailed case we get:

TABLE(A1:b) NORMAL Probabilities (double tailed) and Odds

t	$P(x<\pm t\sigma \,|H)$	$P(x>\pm t\sigma \,|H)$	$O(x>\pm t\sigma \,|H)$
1	2/3	1/3	2:1 against
2	.95	.05	20:1 against
3	.9973	.0027	400:1 against
4	.999976	.00004	30,000:1 against

Of course we need to turn these Probabilities into *Weights* before we can use them in the Detective's Equation. One's immediate instinct is to assume that the Odds in the right hand columns are the Weights (Thus the Forensic scientist and the Detective between them assumed that as $O(E|H) = 1/40$, in the Murder in the Library case, then $W(E|H) = 1/40$. But on reflection we can see this was wrong: $O(E|H) = P(E|H)/ P(\bar{E}|H)$ but $W(E|H) = P(E|H)/P(E|\bar{H})$ where $\bar{H}$ is some alternative to H as the explanation for E; (Anyway only if $P(E|H)$ is divided by something less than 1 can the Weight be more than 1.) A plausible alternative (see fig) might be that x could take any value within some range of values Δ with Uniform Probability (i.e. $P= 1/\Delta$). If we call that Uniform case the $\bar{H}$ here we can calculate (IF we arbitrarily assume $\Delta=10\sigma$):

TABLE(A1:c) UNIFORM Probabilities (double-tailed) and Odds

t	$P(x<\pm t\sigma \,|H)$	$P(x>\pm t\sigma \,|H)$	$O(x>\pm t\sigma \,|H)$
1	2/10	8/10	4:1 on
2	4/10	6/10	6:4 on
3	6/10	4/10	6:4 against

| 4 | 8/10 | 2/10 | 8:2 against |

And if we divide the Probability values in the second table by those in the last we get the:

TABLE (A1:d) NORMAL WEIGHTS (if $\bar{H}$ is UNIFORM with $\Delta = 10\sigma$)

t		W(x>±tσ |H)	
1	x *outside* ± 1σ	(1/3) ÷ (8/10)	i.e. ½
2	x *outside* ±2σ	(.05) ÷ (6/10)	i.e. 1/12
3	x *outside* ±3σ	(.0027) ÷ (4/10)	i.e. 1/150
4	x *outside* ± 4σ	(.00004) ÷ (2/10)	i.e. 1/5000
0.1*	x *inside* ±0.1σ	3.98 = 4	
0.5	x inside ± 0.5σ	3.83 = 4	
0.75	x *inside* ±0.75σ	3.64 = 4	
1.0	x *inside* ±1σ	3.41 = 3	
1.5	x *inside* ± 1.5σ	2.88 = 3	
2.0	x *inside* ± 2σ	2.38 = 2	

*The Weights at the bottom are calculated in exactly the same way as those above but the workings are left out. They could be used in graph-fitting exercises IF the errors were purely of the Counting kind. Note that they rise no higher than 4 even where the discrepancies are very small – which is right – because the Weight for narrow intervals in σ, as here, is simply the ratio of the heights of the two curves, which is never greater than 4. [For cognoscenti the $3.98 \approx 4$ arises from the $1/\sqrt{2\pi}$ in the Normal function].

All this illustrates that it is no trivial matter to find reliable Weights for Numbers because one needs to assume an $\bar{H}$ as well as an H – both usually fraught (for instance why should Δ be equal to 10σ as we assumed above for the Uniform hypothesis?). *This should lead one to use great caution before assigning Weights to Numbers.* For instance in real life most errors are NOT mere Counting Errors, in which case the beautiful bell-shaped curve with its extremely thin tails – leading to decisive Probabilities – is entirely irrelevant. To actually find the true errors and the Probabilities thereof in the tails, you would have to *measure* them – which could be very expensive – or even totally infeasible. So it's very tempting to use the so-called Normal Errors, when they are entirely invalid. Much of Statistics borrows its claimed decisiveness from using this very trick, often justified by a misunderstanding of an infamous construction called the 'Central Limit Theorem'.

Altogether then, unless you are pretty damned sure what the true errors are in your particular problem, it is better to use modest Weights of the PAW variety ('Principle of Animal Wisdom' (see Ch. 10) as follows:

TABLE (A1:e) : PAW Weights:

E	W(E|H)
x *inside* ± 0.1σ	4
x *inside* ± 0.5σ	4
x *inside* ± 1σ	2
x *outside* ± 1σ	½
x outside ± 2σ	¼
x outside ± 3σ	¼
x outside ± 4σ	¼

Using PAW Weights one gains a great deal of reliability – at the expense of decisiveness; but better to be roughly right than decisively wrong! Sect. (13:6) describes the all too plausible fallacy of confusing Probability with Weight (WIMP standing for 'Weight Isn't Mere Probability.)

TFO Disney APPENDIX A2: 'MURDER IN A UNIVERSITY TOWN'

The two trickiest issues here are the Prior and the Weight to give to the blood sample. The hypothesis I will test against the evidence is *'The accused is guilty'*.

The victim clearly disappeared not long after she was seen in the take-away queue by her friend – otherwise she would surely have contacted Georgia. Therefore she was probably lured away by someone in the immediate vicinity, most likely one of the 10,000 local residents of the student village (half of whom are male). Therefore, before considering any other evidence, the unbiased Odds on any particular village male, including the defendant, being the murderer are about 1 in 5000. So my Prior on Vilnius being the murderer is 1/5000.

As to the blood-group my Weight would be reached from the Probabilities: P(match| guilty)/ P(match| innocent) = $1/ 10^{-3} = 10^3 =$ 1,000. i.e. W(match|guilty) = 1,000 because P(match| guilty) is certain and thus equal to 1 while an innocent person will have on average only a chance, a Probability of one in a thousand of having that incriminating but rare group. As the defence barrister – whose remarks I have withheld so far – reminded the court:

"Although his blood group sounds incriminating it has to be set against the fact that there are about 5000 eligible males in that village, about 5 of whom will on average have the same incriminating group. In fact the police found 7 such altogether – 4 of whom own cars. Why aren't they in the dock instead? On its own the blood group yields Odds of 6 to 1 *against* my client being the murderer. That's very different from the one thousand-to-one *on* that my learned friend the Prosecutor tried to persuade you of earlier." [True; my O(H|bg) = W(bg|H).O(H) = 1000 times 1/5000 = 1/5.]

My own Inference Table looks like this:

Number	Clue	$P(E|H)$	$P(E|\bar{H})$	$W(E|H)$	$O(H)$
Prior					1/5000
1	Blood	1	1/1000	1000	1/5
2	Acquainted	?	?	8	~ 2
3	Motive	?	?	4	8
4	Lying alibi	1	1/8	8	64
5	No car	1/4	1/3	1/2	32
6	Can't drive	?	?	1/8	4

So I reach a combined Odds on guilt of 4 to 1 on. In my mind the accused is probably guilty but the Odds-on are not high enough (Recall the Blackstone Ratio of 10 to 1) for me to convict.

Note the importance of the Prior here, especially in connection with the blood-group evidence. Had I taken a Null Prior [i.e. $O(H) = 1$] instead of (1/1000) my final Odds would have been 20 to 1 on, and 'Guilty'. That gives one pause for thought. If only there were more evidence! That's where Science wins over other subjects such as History because, at a price, there usually is. [It's also interesting to wonder whether the police would have disclosed that there were 6 other positive blood-tests if the defence counsel hadn't been smart enough to ask for that information. Would you have done so? I probably wouldn't.]

TFO (Disney) APPENDIX A3: THE EXPECTED SCATTER (5/18)

What we are going to show is that after Q tosses we expect to get Q/2 Heads, but with scatter of Q/4, or a 'standard deviation of $\pm\sqrt{Q/4} = \pm\sqrt{Q}/2$. Thus after 100 tosses we expect 50 Heads $\pm\sqrt{100}/2 = 50 \pm 5$ Heads. This means, as we shall show later, that when we carry out 100 coin-tossing trials we expect to get between 45 and 55 Heads two thirds of the time.

The way we are going to demonstrate this very useful and general result is to use the Probabilities we calculated from the branching diagram Fig (9:1) to calculate the "Expected Value" of various useful quantities. You don't have to follow the demonstration if you will take my word for the final result, but if you won't then here it is laid out in an Inductive, or Common-sense way (as opposed to a Deductive or Mathematical way; we discuss the difference later).

The Expected value of some quantity x is *defined* to be

$$E\{x\} = \sum_{i=1}^{i=N}[x_i \times P(x_i)] \qquad (9:2)$$

where $P(x_i)$ is the probability of the value x_i turning up. Don't be put off by $\sum_{i=1}^{i=N}[........]$; its just shorthand for 'Add up all the stuff inside [....] as i rises from 1 to N.

To make sense of equation (9:2) let's do a simple example. What is the Expected value of the number of Heads after 3 tosses? Here x_i is the number of Heads, which could take any of the values $x_i = 3,2,1,0$. (i.e $x_1 = 3, x_2 = 2, x_3 = 1, x_4 = 0$). Now we can go to the vertex-numbers under 3 tosses and find, (because $N = 2^Q = 2^3 = 8$)

$P(x_1 = 3) = 1/8$; $P(x_2 = 2) = 3/8$; $P(x_3 = 1) = 3/8$; $P(x_4 = 0) = 1/8$. Now fill in Equation (9:2) as the summation sign $\sum$ indicates:

$$E\{x\} = x_1 \times P(x_1) + x_2 \times P(x_2) + x_3 \times P(x_3) + x_4 \times P(x_4)$$
$$= 3.(1/8) + 2.(3/8) + 1.(3/8) + 0.(1/8) = 12/8 = 3/2$$

which makes sense. It says the Expected value for the number of Heads after 3 tosses is 3/2 or half of 3. In other words the Expected value is the average value you might *expect* after making many trials of 3 tosses using a fair coin.

We can use the Expected Value idea to calculate many other useful quantities; for example what is the Expected value of the scatter amongst a series of numbers x_i ? By analogy with (9:2) it must be

$$E\{(x_i - \bar{x})^2\} = \sum_{i=1}^{i=N} [(x_i - \bar{x})^2 \times P(x_i).] \qquad (9:3)$$

In the coin-tossing problem we know how to calculate all the Probabilities $P(x_i)$ from the Vertex numbers and so we can, with a little bit of arithmetic, find the Expected scatter.

So now let's do what we promised: show that after Q tosses the scatter in the number of Heads will be Q/4 (the standard deviation will then be by definition the square root of this i.e. $\pm\sqrt{Q}/2$).

Start with the simple case of 1 toss i.e. Q=1. Now the expected number of Heads, i.e. the average $\bar{x}$ will always be Q/2 so $\bar{x} = \frac{1}{2}$ here. Looking at the path diagram there are only two cases: 1 Head or none. From the vertex-numbers as a fraction of N underneath P(1Head)= ½ and P(0Heads)= ½ so, by Equation (9:3):

$$E\{(x_i - \bar{x})^2\} = (1-0.5)^2.(1/2) + (0-0.5)^2.(1/2) = 1/8 + 1/8 = 1/4 = Q/4$$

So far so good: with Q=1 the expected scatter = Q/4.

Now do 2 tosses. From Fig (9:1) you can have 2, 1 or 0 Heads with Probabilities ¼, 2/4, ¼ respectively while $\bar{x} = Q/2 = 2/2 = 1$. So (9:3):

$$E\{scatter\} = (2-1)^2.(1/4) + (1-1)^2.(2/4) + (0-1)^2.(1/4)$$
$$= 1/4 + 0 + 1/4 = 2/4 = Q/4 \text{ again.}$$

After 3 tosses (Q=3) one can have 3,2,1,0 Heads with Probabilities [see branching diag.] 1/8, 3/8, 3/8 and 1/8 respectively while $\bar{x}$ =Q/2=3/2=1.5 thus

Equation (9:3):

$$E\{scatter\} = (3-1.5)^2.(1/8) + (2-1.5)^2.(3/8)$$
$$+ (1-1.5)^2.(3/8) + (0-1.5)^2.(1/8)$$
$$= (1.5)^2.(1/8) + (0.5)^2.(3/8)$$
$$+ (-0.5)^2.(3/8) + (-1.5)^2.(1/8)$$
$$= (9/32) + (3/32) + (3/32) + (9/32) = 24/32$$
$$= 3/4 = Q/4 \text{ once again.}$$

So a pattern has emerged. Three times in a row for values of Q= 1,2,3 the Expectation value of the scatter is Q/4. I am prepared to bet that that pattern will continue indefinitely for all values of Q. If you are not likewise convinced then do:

EXERCISE (9:5) EXPECTED SCATTER

If indeed the scatter is Q/4 then, in the usual way the standard deviation $\sigma = \pm\sqrt{Q/4} = \pm\sqrt{Q}/2$. After 100 tosses $\sigma = \pm 10/2 = \pm 5$ Heads. So indeed we expect to find 50 ± 5 Heads most of the time.

So we've reached a general and very useful result: when we count Q things we expect the actual counts to be scattered about the mean value by about 1 standard deviation $\sigma = \pm\sqrt{Q/4} = \pm\sqrt{Q}/2$.

P.S. Note the difference between Inductive and Deductive thinking. We didn't *prove* that $\sigma = \sqrt{Q}/2$ *always*, we only worked it out for a small number of specific cases (i.e. with Q = 1, 2 or 3), perceived an obvious pattern, and assumed that that pattern would apply in general, i.e. for any Q, however large. That's Induction. Real mathematicians would throw up their hands in horror. In Chapt. 10 I've hinted at how the rule can be *deduced* if you find that necessary. However the truth is that, even in Mathematics, Inductive thinking is generally the rule. Indeed creative mathematicians mostly get their exciting results by first doing specific examples (i.e. by Induction) and only afterwards look for deductive proofs of their conjectures. That's their messy job; they are like lawyers picking through the clauses in a legal contract, looking for flaws. We don't have to do it for them.

PSS. If this arithmetic has bewildered you a little, remember it all stems from that branching diagram Fig (9:1) where it is obviously clear that vertex- numbers tend to bunch up round the average number of Heads (= Q/2) simply because there are more alternate pathways for getting to such locations, far more ways, than there are of getting to the extreme values at top and bottom. That is the cardinal point.

TFO (Disney) APPENDIX A4: THE BOY/ GIRL PROBLEM

A family contains 2 children; one is a boy. What is the Probability that the other is a girl?

Let the hypothesis H = BG for the siblings (no order implied) and evidence E = B (one boy known).

(A) The 'Mathematicians Answer'.

There are only 3 possible 2-sibling families: BB, BG, GG. Now according to Bayes' Rule our answer is P(BG|B). Let's work in Odds, remembering that Odds Od and Probabilities P are related by:

$$P = \frac{Od}{1+Od} \qquad (A1)$$

Now Bayes: $O(BG|B) = W(B|BG) \times O(BG)$ \qquad (A2)

And $W(B|BG) \equiv \dfrac{P(B|BG)}{P(B|\,notBG)} = \dfrac{1/2}{1/2} = 1$

While Prior: $O(BG) = \dfrac{P(BG)}{P(notBG)} = \dfrac{BG}{BB+GG} = 1/2$ \quad (A3)

So (A2) : O(BG|B)= 1 × ½ = ½

And (A1): $P(BG|B) = \dfrac{1/2}{1+1/2} = \dfrac{1}{3}$ \qquad QED (A)

(B) The 'Layman's Answer'.

There are 4 possible 2-sibling families (by order):
$B_1B_2,\ B_1G_2,\ G_1B_2,\ G_1G_2$

Now Bayes: $O(BG|B) = W(B|BG) \times O(BG)$

While $W(B|BG) = \dfrac{P(B|BG)}{P(B|notBG)} = \dfrac{1/2}{1/2} = 1$ as before.

While Prior $O(BG) = \dfrac{P(BG)}{P(notBG)} = \dfrac{2/4}{2/4} = 1$ \hfill (A4)

So Bayes: $\quad O(BG \mid B) = 1 \times 1 = 1 \quad$ i.e. evens

Which makes it obvious [see(A1)]:

$$P(BG) = \dfrac{Od}{1+Od} = \dfrac{1}{1+1} = 1/2 \qquad \text{QED(B)}$$

The two arguments differ only in their Priors (3) and (4). Which is right?

In Odds form we can clearly see that (4) is right while (3) is wrong: (3) forgets that BG (in no order`) is twice as common as either BB or GG because BG contains two possibilities $B_1 G_2$ and $B_2 G_1$. Thus one should write:

$$O(BG) = \dfrac{BG+GB}{BB+GG} = \dfrac{2}{2} = 1$$

and not as we did in (A3).

Does it matter? Only when Mathematical types claim their answer to be evidence of superior insight, which they often do. Sorry chaps. (PS: In my experience this tricky problem, and it is very tricky, is better discussed in Odds, as here, than in Probabilities. In particular it allows one to spot the fallacy in the Mathematicians argument, which is otherwise difficult.)

TFO (Disney) APPENDIX A5: WHERE THE 'NORMAL DISTRIBUTION' REALLY COMES FROM (15/5/18)

Suppose we have a centrally distributed curve like Fig (9:1) but centred on x = a and with a half-width σ. How could we find an equation to describe such a curve given only that we are looking for the *simplest* possible equation i.e. the one containing the smallest number of free parameters to describe it? We use only elementary calculus; dy/dx is the notation for the slope of the graph, defined as positive when the slope is upward right. Looking at the graph:

    We want dy/dx = 0 at x = a.

    We want dy/dx = positive when x is less than a.

    We want dy/dx = negative when x is more than a.

    We want dy/dx to tend to zero as y tends to zero.

All of which can be arranged if:

$$\frac{dy}{dx} \propto -(x-a) \times y \equiv -(x-a) \times y / \sigma^2 \quad (1)$$

as you can check by fitting in appropriate x's and y's to meet the conditions above.

σ (sigma) is some so-far unspecified constant introduced to make sure both sides have the same dimensions (i.e. as y/x) and the same size.

Now (1) is a separable differential equation. We are taught in elementary calculus to solve it by getting all the y's on one side and all the x's on the other:

$$\frac{dy}{y} = -\frac{(x-a)}{\sigma} . dx$$

and we are also taught that the integral of the LHS is a natural logarithm so that when we integrate:

$$\ln y = -\frac{(x-a)^2}{2\sigma^2} + C$$

where C is as usual the undefined constant of integration. Thus

$$y = A\exp\left[-\frac{(x-a)^2}{2\sigma^2}\right] \qquad (2)$$

where A is such that ln A = C. We can make A take any convenient value we desire depending on what we want to use the equation for. For instance if we wanted the equation to represent a curve of Probabilities we would want the total area under the curve to be equal to 1 (see Fig in Appx 1)) and that would set some particular value for A. But when we are using Odds we are using the *ratio* of two Probabilities so the A's cancel out and then we might as well set A =1 for convenience. We can also get rid of the little 'a' by moving the origin of the graph to x = a; we can always do that because the origin is a matter of convenience for the purposes of plotting, not anything fundamental or real: for instance it cannot change the shape of the curve. Thus we are finally left with:

$$y = \exp\left[\frac{-x^2}{2\sigma^2}\right] \qquad (3)$$

which indeed has only one free parameter σ. If you plot it σ turns out to be the value of x when y has about 60 per cent of its peak value which, as you can see from the figure, could be used as a rough measure of the half-width of the curve, as suggested.

So, as promised, (3) is indeed the simplest curve with a central distribution – because it has only one free parameter σ. Mathematicians unfortunately label it 'The Normal Curve' and it is indeed convenient to use mathematically, though it is no more normal than a perfect straight line, or a perfect circle or ....

Note how we obtained the equation: from purely mathematical considerations with no relation to Probability, to Physics, to measurements of any kind, *indeed to anything in the real world.* So if somebody wants to use it for inferring anything about some feature of that real world *they had better first establish a fool-proof connection between that real feature and this simple algebra.* De Moivre did manage to establish such a connection for the simple case of Counting-Errors (only) such as are relevant to card and dice games. You can thus obtain the Normal Odds Table [Appendix A1] by plotting the curve on graph paper and then measuring and then ratio-ing the areas on either side of vertical lines at various positions(t-values) of $x = t \times \sigma$; ratio-ing because Odds remember is the ratio of two Probabilities (Chapt 4).

What one must *not* do, and yet what is so often done, especially by Statisticians because it is convenient, is use the Normal curve without first *proving* that it is appropriate to the real-world situation which one is investigating. Then it can lead to ridiculous and indeed dangerous results, especially where Probabilities and Odds are concerned. You can only arrive at high (or low) legitimate Weights if you know the exact distribution of your errors out in the wings, something you can generally establish only by *experiment.* That was the fatal mistake perpetrated by our forensic scientist; his Odds of 40 to 1 against the leading lady's guilt were complete nonsense and led to an innocent's 20 years hard labour.

TFO (Disney) APPENDIX A6 : THE SALLY CLARK CASE

See Chapter 11. I progress through a series of arguments:

(A)   Prob. of 1 SID = 10^{-3} and therefore of 2 in the same family = 10^{-3} times 10^{-3} (by the Product Rule) = 10^{-6} or 1 in a million. There being 650,000 live-births a year in the UK the expected number of innocent double SIDs is 650,000 $\times 10^{-6}$ or about one a year. So SC *could* be innocent.

(B)   Ah yes but SC is not a typical mother. She's university-educated, non-smoking and a successful solicitor. The figures show there's only a (1/8,300) chance of a mother like her having an innocent SID. The chance of her having two is therefore by the Product Rule (1/8,300) squared or 1 in 73 Million. Thus with 650,000 births p.a. the chance of hers being innocent are (0.65)/73 or about one in a hundred. By the Blackstone ratio she should be found guilty. [This was the Meadow testimony on which the jury convicted by a majority of 10 to 2 in 1999].

(C)   Yes but the implied assumption that the Odds of a second SID in the same family are the same as for the first is highly questionable. There could be environmental or medical factors which increase the likelihood of a second SID. Some surveys suggest that a second could be 10 times more likely than a first. Therefore the Odds against 2 SIDs occurring to the same woman by chance are not 73 Million but 7.3 Million to 1. Thus the Odds against SC's innocence are only 7.3/.65 or about 10 to 1 — at the margins according to the Blackstone Ratio. This was the obvious argument of the Royal Statistical Society, and others, in 2001, but it didn't move the Appeal Court. What did move them was the discovery, by an independent solicitor suspicious of the verdict, that the pathologist Dr Alan Williams had concealed from the defence-team that the second infant was infected with *staphylococcus aurea*, a possible cause of death. Thus SC was released, although the Odds still appeared rather against her real innocence.

(D)   The court was taken in by 'The Prosecutors Fallacy' i.e. by the assumption that because the Probability of innocence is low the Probability of guilt must be high. This amounts to ignoring the Prior odds on innocence, which might be high. In truth $O(I|E) \neq O(E|I)$ but $= O(E|I).O(I)$ according to Bayes. Since only 30 mothers a year are known to kill their own children out of 650,000 births, O(I) is about 20,000 to 1 on. So even with O(E| I) = 100,000 to 1 against, i.e.1/(100,000), the Odds on her innocence are (1/100,000).(20,000) or 5 to 1 against, not good but not quite bad enough to convict.

(E) If a mother is before a court because, and only because, she has 2 SIDs (the Evidence) then the odds on her innocence are, by Bayes:

$$O(I|E) = \frac{P(E|I)}{P(E|G)} \times O(I)$$

where I = Innocent, G = guilty.

Obviously P(E |G) =1 BUT, AND THIS IS THE POINT, P(E| I)=1 ALSO! She *does* have 2 dead babies, *otherwise she wouldn't be in court.*

Therefore $O(I|E) = 1 \times O(I)$

i.e. she deserves the same presumption of innocence as any other mother. With only 30 a year killing their children O(I) is extremely high, about 20,000 to 1 on innocence, so the woman is almost certainly innocent.

This argument was never brought up either at the trial or afterwards. It came from my father Maurice Disney (dec 2006) himself a distinguished childrens' doctor who had been involved in many SID cases and who was horrified by the SC case and the testimony of Meadow in particular. The iron rule of his profession, he said, was: 'You cannot convict on statistics alone. Look for other medical evidence, it's often there, broken bones for instance from previous abuse'.

What do you think? Dad wasn't a numerate man but if there is a possibility of innocence — which there is if you ignore the Meadow testimony — there will be $\sim 10^{-5} \times 650,000$ = 6 or 7 innocent double-SIDs a year — then the question before the court *should be "Is the woman before us one of the innocents, or a murderer?*", ` a question that cannot be answered without evidence BEYOND that she has 2 SIDs!

If that is the question then of course $P(E|I) = 1$ and $O(I|E) = O(I)$; i.e. lacking further evidence she must be regarded as innocent. That is 'Dad's Argument'.

(F) I claim my name is 'Mike Disney' But, from the statistics of names, the Odds that I am a Mike are about 1 in a hundred, that I am a rare Disney about 1 in a hundred-thousand, so the Odds that I am indeed 'Mike Disney' are, by the Product Rule, about 1 in 10 Million. So even if the Prior Odds on my dishonesty are rather small (say 1 in 10 thousand) the Odds that I am an imposter are about 1000 to 1 on.

Now you can sense there is something wrong with this argument because it could apply to anyone, irrespective of their name. What is wrong? You might like to think about it now.

LONG PAUSE FOR......

I claim one is asking the wrong question. We are asking the question 'What is the chance that a *random person* would claim that his name is 'Mike Disney'?' And indeed the chances of that *are* very low — as we discovered. But the question we needed to ask, and should have asked is 'What is the chance that a man *claiming to be Mike Disney* is actually telling the truth?' And that would be pretty high, unless the population at large is generally mendacious. Mike Disney deserves the same presumption of honesty as anyone else.

After thinking about the Sally Clark case for literally hundreds of hours spread over a dozen years I have come to believe that this weird but subtle analogy is the key to the SC problem. Other 'experts' think it is nonsense but have given no convincing reason as to why. The question facing the court should NOT have been the same as the one facing the police who were, in my opinion, quite justified in suggesting her for trial; the circumstances *were* very suspicious. I don't want to find fault with anyone here because it has proved so bloody hard for us all. But if there is a point to note for future reference The Crown Prosecution Service must recognize that any such trial without additional medical evidence will almost certainly fail on appeal.

I'm going to leave it there; convinced, but not *absolutely* convinced, that Dad was right.

I hope you are convinced that thinking can sometimes be very hard.

PS There were several after-shocks following this sad case, aside from SC's tragic death. Meadow was struck off by The General Medical Council for "Serious Professional Misconduct" but reinstated on appeal because they couldn't prove the 'Serious' bit. But if I am right his dodgy statistics, and dodgy they were, had little to do with it. I think the court asked him the wrong question, in which case the GMC got it wrong too. The pathologist was struck off as well, but not reinstated. Nor does 'The Prosecutor's Fallacy' cut the mustard; it still leaves SC under suspicion. And you can make up your own mind about the Royal Statistical Society's 2001 lucubrations by downloading them from the web at (rss@rss.org.uk). They never mentioned the dread word 'Prior' and I found their final remarks self-serving: "Although many scientists have some familiarity with statistical methods, statistics remains a specialised area. The Society urges the Courts to ensure that statistical evidence is presented only by appropriately qualified statistical experts, as would be the case for any other form of expert evidence." Um. I think not; specialised knowledge cannot compensate for lack of Common Sense.

TFO(Disney) APPENDIX A7: PHILOSOPHY AND OUR BOOK

When I sent a resumé of the book to an eminent colleague of mine [Prof. George Ellis from the University of Cape Town – formerly scientific adviser to President Mandela] he commented, in no derogatory sense I am sure, that it was 'Philosophy'. Having had a low opinion of that subject since boyhood I was shocked by the allegation, but was bound to consider it seriously, coming from George. So here I consider the charge, and refute it if I can.

What is Philosophy – which translates literally into 'Love of Wisdom'? I can do no better than quote this definition from Paul Horwich, a philosopher at New York University: "Philosophy is respected, even exalted, for its promise to provide fundamental insights into the human condition and the ultimate character of the universe, leading to vital conclusions about how we are to arrange our lives. It is taken for granted that there is deep understanding to be obtained of the nature of consciousness, of how knowledge of the external world is possible, of whether our decisions can be truly free, of the structure of a just society, and so on – and that Philosophy's job is to supply such understanding."[16]

Fair enough as to 'promise', but what about performance? Ask a hundred educated people what Philosophy has done for them: only five will attempt an answer, and only one will sound convincing. In search of 'the good life' I read much Philosophy as a youth, but was disappointed to find that most philosophers seemed hardly better than secular priests engaged in debating the number of angels that could dance on the head of a pin, with David Hume – the Scottish Enlightenment philosopher – proving the honourable exception. However, with bated talk about 'The Relativity of Space and Time' they did encourage me, as a curious 14 year old, to go and read Einstein – and I shall

---

[16] "Was Wittgenstein Right?" Paul Horwich, New York Times, March 3, 2013

always be grateful for that. After a titanic struggle with Einstein's 'popular' book on Relativity I grasped enough at least to realize that the philosophers I had been reading had no conception of that subject – which is extraordinarily difficult.

The real charge though is that Science progresses while Philosophy does not. Philosophers are still interested in what Parmenides had to say in 500 BC; scientists are barely interested in what was said by their predecessors 20 years ago – the subject moves so rapidly on. Then scientists solve problems whereas philosophers do not; even their central issues remain mired in controversy 2 millennia on. No wonder I was upset by George's charge when I felt that I was doing hard science from beginning to end. But was he right? Here is my account of the events which led to the gradual assembly of the point of view expressed in this book.

I started out trying to decipher the so called 'Scientific Method' with no settled opinions – on the contrary I was looking for hints toward solving my own problems in galaxy research. To that end I studied the history of science and compiled a list of observations – that is to say accounts of scientific discoveries and controversies from the past. That took a couple of years [Chapt. 3] but it did enable me to pose relevant questions – about for instance the importance of Deduction, Experiment and Genius –with unanticipated answers.

Then I read the 'experts' – of which there are so many. I was tugged first this way and then that, finding it much easier to agree with them by turn than to spot where they must be going wrong – as surely some of them had to be— seeing they couldn't agree. I waded through bogs of Symbolic Logic, went up to my neck in Probability Theory and nearly drowned in Advanced Statistics. What eventually became clear though was that real scientists had resigned from the field to get on with their research, leaving tribes of pugnacious philosophers and statisticians to fight over the carcase .

But I ploughed on – needing help as I did with my galaxies. The big break came through studying gambling, horse-racing in particular [see Chapt.4] –

which used Odds instead of the Probabilities which were the universal coinage amongst serious scholars. And notation in symbolic thinking can be vitally important (see Zeno's Paradox). Though I was by no means the first to translate Bayes' Rule into its Odds form I was, I believe, through years of dogged pursuit, the first to catch its full possibilities. Bayes' Rule itself can rarely generate Odds of sufficient size to be decisive – which is why it had drifted in and out of the general consciousness for centuries. But when I tried to employ it in the graph-fitting problem [Chapt.5] the Detective's Equation (DE) hove into sight.

Now, at last, I had a hypothesis of my own with which to work: *"The Scientific Method is based on Common Sense Thinking (CST) – which itself relies on the DE."*

Starting from there I did what any scientist would do: I explored the implications of my hypothesis and compared them with the real world. The first satisfying explanation for Ockham's Razor (at least to me) immediately emerged – which I felt was a triumph [Chapt. 7]. Moreover the scheme seemed to work very well on a variety of toy examples, less well on my real problem of Hidden Galaxies. It was good at dealing with conflicting evidence, weak at suppressing Systematic Errors. So I modified it – by introducing The Principle of Animal Wisdom (PAW) [Chapt. 10]. Now it worked excellently well, like a surgeon's head-light, illuminating any number of dark but critical cavities. Indeed it worked so well that I found I could dispense entirely with Statistics, and all its vicious squabbles, substituting my version of Common Sense instead [e.g Sect. (11:2) on smoking].

At last I sensed I was homing in on wisdom – but was it Science, or merely Philosophy? I felt sure it was Science – and feel so now – for I had unearthed an instrument, like a microscope, which could be used by anyone to shine a light on difficult problems. But could it be falsified – the sharp demarcation fence erected by Popper to distinguish between scientific and non-scientific ideas? Ignoring for the moment my serious reservations about Popper's thesis, it does seem that my scheme could be falsified in at least one of two feasible ways: an Artificial Intelligence system might be developed that can outperform CST – but do so in some mysterious way we cannot explain; or an experimental psychologist might demonstrate that we possess some valuable mental trait which we could not possibly have inherited from our animal forbears [e.g. 'Mind' – whatever that is]. I have used as a guiding principle that '*All the mental powers we have could only have evolved incrementally by Natural Selection.*'

The other serious charge against Philosophy is that it makes generalisations beyond its field of competence. But I hope we haven't done that. Such claims as we make are of the negative or sceptical kind, for example:

- Thinking can never lead to certainties about the real world.
- All conclusions based on evidence must be provisional.
- We neither need nor possess mental powers which our animal cousins do not share to some degree. [Although we use simple maths to do Inference (the Detective's Equation), which they do not, it turns out (Appx. 9 on Animal Wisdom) that the maths is merely a convenient but dispensable model of the underlying Categorical Inference which they can use, and which we probably use too.]
- Natural Science has no secrets from which other disciplines might learn. It is simply that the natural world largely obeys The Principle of Uniformity, and there is a practically unlimited supply of evidence – unlike in say History, or Law.

And so on.

George Ellis's charge is understandable in one sense. Trying to comprehend how we might come by sound knowledge of the external world – which is what we are all about – is called 'Epistemology' and has traditionally been assigned to Philosophy. But then Science itself was called 'Natural

Philosophy' once upon a time. But I believe we have cleared ourselves first of the charge that we have used any non-scientific means to reach our conclusions, and second of making grand generalisations that are unwarranted by our research.

Although much of Philosophy is unproductive, and often pretentious, it is fair to say that once a branch of Philosophy becomes positively fruitful it is invariably re-christened as something else – 'Science' itself being a case in point. And all subjects need constant criticism – especially from outside. Scientists themselves are plagued by narrowness of vision, illusions of grandeur, busyness and Group-think – not to mention ignorance of our methodology. Thus we should welcome criticism – wherever it comes from—and why not from Philosophy. I would like in this connection to mention Colin Howson and Peter Urbach, philosophers from the London School of Economics, who have rendered Science *two* invaluable services. They helped revive Bruno di Finetti's seminal work on Subjective Bayesianism from obscurity, and they have seen through Statistics' claims to Objectivity. We need such scholars, people read so widely that they can see the big picture as it has unfolded in Space and Time, and critics who are not over-awed by scientists' pretensions. Without such spurs Science is in constant danger of taking a wrong turn. At the moment for instance, I personally see us as being over-managed by administrators, in thrall to fashion and money, driven by a need for popularity, obsessed by issues of gender and political correctness, and far too willing to become the tools of government policy. Judging from history real science is a very tender plant indeed, one which can sicken and die as quickly and mysteriously as it sometimes springs to life. Even if Philosophy is an all too convenient priest-hole for charlatans it probably has a role to play – if only as a critic. But if it wants to be listened to, clarity should be its by-word; obscurity stinks of the witch-doctor. Scientists have very little spare time to read what philosophers think, so effective criticism needs to be crisply stated – in journals which scientists read – not only in self-regarding journals of professional Philosophy. Philosophers need to talk to us – not merely to each other.

TFO(Disney) APPENDIX A8 : CERTAINTY, FALSIFIABILITY AND COMMON SENSE.(22/2/19)

Climbers know that when you reach the top you can often look back and see a much easier rout than the one up which you have just struggled. So it is here. One can now see that the folly that has held us all up for centuries has been the search for Certainty in thinking.

So often we think by analogy. Bewitched by the certainties of ancient Greek geometry with its incontestable proofs, scholars sought for the same kind of certainties in our thinking about the real world – which is why, even today, they are dismissive of Common Sense as a sort of fumbling folk substitute which can never hope to attain the kind of certainties they are bent on seeking. Well the scholars were wrong. They were bedazzled by an imperfect analogy. Euclidean geometry is a closed system with precisely defined rules. The real world is an open system with rules that have to be unearthed by observation and experiment. In such an open world it turns out, as we shall prove, that thinking can never lead to certainty, nor even categorical falsification.

Need the follies of the past concern us now? Yes they must because they still cast blinding shadows across the present. So long as we continue to despise Common Sense Thinking (CST) as a fumbling substitute for the proper thing we will never learn to do it optimally, or teach it to our young. Worse, society will continue to regard provisionality – which is all CS can reach , as an imperfection to be overcome – rather than the heart and secret of progress itself. It is the very provisionality of a conclusion which beckons us on, while the finality of certainty is a dungeon door. In my opinion craving for certainty is a form of religious mania which is inexcusable in adults. History reveals that dogmatic certainty leads to backwardness, barbarity, persecution, even to war.

573

The aim of this appendix is to demonstrate, as categorically as I can, that thinking can never attain to certainty about the real world, not even to certainty of falsification. This truth has some very remarkable implications.

CERTAINTY IS UNATTAINABLE

Bayes' Rule [Chapt.4] is:

$$O(H \mid E) = O(H) \times W(E \mid H) = O(H) \times \frac{P(E \mid H)}{P(E \mid \bar{H})} \quad (1)$$

where H is the hypothesis under test and E is the evidence to hand. Certainty amounts to infinitely high Odds O(H|E) which, according to (1), can only be reached if $P(E \mid \bar{H})$ is zero. But because $\bar{H}$ (i.e. not-H) can be infinitely various in real life, after all it consists of *all* conceivable (and inconceivable?) hypotheses barring H, the possibility that $P(E \mid \bar{H})$ is zero can never be established. 'QED' as they used to say.

Likewise falsity [i.e. O(H|E) = 0] can never be established categorically because that would require – see (1) – that:

P(E|H) = 0   (2)

which one feels intuitively could never be established either – except in cases where H is of the kind 'All A's are B' (all Swans are white) – which are themselves claims for certainty, which we have ruled out already.

Intuition aside, what about a categorical proof? Philosophers of the Karl Popper school have not only claimed that falsification is possible but have erected a whole system of philosophy upon that assumption. But such claims are wrong because it is not possible to define an H in such a way as to render it undeniably falsifiable. That would require an infinite number of qualifications to rule out every conceivable variant of H, one of which could be consistent with the evidence E in question. To see what we mean consider some examples:

(a) Naïve H: the Earth orbits the Sun.

Evidence E: But no parallax more than 30 arc sec observed. [Parallax is the foreground stars apparently moving against the background ones because the Earth is in motion.]

Qualified H': Earth orbits Sun but stars are more than a thirtieth of a parsec away.

Evidence E': But still no parallax observed to a level of 10 arc sec.

Refined H'': Earth orbits Sun but stars are more than a tenth of a parsec away.

And so we could go on and on, never able to decisively falsify the infinite variants of naïve H. Aristarchus had pointed this out in 300 BC but others felt the enormous stellar distances implied were 'implausible'. However Aristarchus' hypothesis was eventually accepted (1500 AD) mainly on the grounds of Parsimony [Ch. 7].

(b) Naïve H: Evolution by Natural Selection.

E: But very similar large animals found on continents oceans apart. Therefore they must have been independently created.

H': Evolution in a world where continents split and drift away from one another.

E': But the cooling of the Earth reveals it is only 20 million years old, far too young for sufficient drift, or indeed for Evolution.

H'': Evolution in a world where unknown processes offset cooling.

And so on. In fact Evolution was accepted by many (1859) – mainly on the grounds of Parsimony – before radioactivity (1910) revealed that the Earth is billions, not millions of years old.

(c) H: Spiral galaxies are opaque to most of their own radiation.

E: Falsified by the observation that galaxies do not significantly change colour when they are seen from different angles.

Now this widely accepted falsification was found to require that the dark smoke responsible for the clamed opacity had to have a very particular (and rather implausible) configuration. Alter that configuration in a number of ways and the falsification collapses.

Later observations demonstrated that indeed spirals absorb half or more of their stellar radiation.

So categorical falsification is as unattainable as Certainty. William Quine long ago (1950s) pointed out that Falsification must often be unreliable because it may involve a chain of reasoning, any link of which could be at fault, not necessarily the hypothesis under test (e.g. see galaxy opacity). But it is much worse than that; it is simply impossible to define any hypothesis (beyond a claim for certainty) so completely that it can be categorically ruled out by any conceivable evidence. Hypotheses are slippery devils; look how Ancient Greek medicine and chemistry, both nonsensical, managed to slip through a wide net of evidence for 2000 years.

HISTORY AGAIN

The great disaster with Probability is historical in nature. As it started out in games of chance every possible hand of cards, or combination of dice, could be enumerated. This enabled one to calculate Probabilities over the whole range between zero and one with exactness and certainty – as we did with coin-tosses in Fig (9:1). But the moment you apply Probability to the real world you cannot, even in principle, enumerate all possible variants of H or $\bar{H}$. Exactitude, precision and certainty are all now wholly out of reach, and we are in a different world entirely, not in a lighted gaming-room but outside in the dark with only a canopy of cold stars stretching to infinity. But in my experience at least this transformation is not acknowledged as forcibly as it needs to be. I have taught it so myself, starting with dice and coins, then gliding almost imperceptibly into inferences about the real world. In so doing I must have misled hundreds of students. Now I would go almost as far as to recommend that Probability never be used in connection with Inference – with Odds or Categories replacing it. Not only is Probability misleading when it comes to the real world – it can be incredibly clumsy – as we demonstrate explicitly in Appendix 9. In retrospect the huge rows that have poisoned the use of Probability and Statistics [Ch. 11] can be seen as the unfortunate consequence of an inept or at least inappropriate concept.

MEASUREMENTS

Seem to play a hallowed role in some people's thinking. But no more than any other evidence can they categorically prove, or disprove, any hypothesis. No matter how close they are to the hypothetical value they could be there by chance, or because the hypothesis is a close approximation to some better theory (e.g. Newtonian gravity to Einsteinian). And no matter how far away they are from the hypothetical value they could be explained either by errors in the measurement itself, or by mistaken assumptions in its interpretation (e.g. 'transparent galaxies' above). Remember that Error Distributions have to be *measured* – which is often impractical – while using hypothetical Error Distributions instead can be treacherous in the extreme [Chapt. 11]. Anyway it is not the precision of a measurement which enters CST but the Weight, and the PAW alone forbids us from giving any clue – measurement or not – a Weight above 4, or less than ¼. So measurements are no magic passageways to either Certainty or falsification. There is no escaping the gathering of many clues, and gambling to reach the betting Odds of Common Sense Thinking.

IMPLICATIONS

(a)   Animal Wisdom – the unanswerable case.

We introduced the Principle of Animal Wisdom (i.e. restricting Weights to one of only 4,2,1, ½ or ¼) or PAW, as a crude remedy for disarming the Systematic Errors that have plagued Science – indeed thinking in general [Chapt. 10]. Moreover we suspect that animals couldn't categorize, or at least store for future combination with other evidence, Weights with any higher value or precision. But in retrospect we can recognize that the PAW acknowledges the profound truth embedded in Equations (1) and (2) above – that no agency, not even a superhuman one, could define *all* the alternatives to H (i.e. $\bar{H}$) that could generate E, nor *all* the variants of H that could accommodate themselves to E – no matter how implausible they look initially (e.g. radioactivity). Very high (> 4) Weights or very low Weights (< ¼) can never be justified, while very precise Weights serve no purpose when they will almost certainly have to be multiplied together with others of much lower precision, in the Detective's Equation. The PAW might be animal, but it is wisdom nonetheless, and profound wisdom at that. (I suspect anyway that had there been a survival advantage to heavier or more precise Weights Nature would have found a mechanism to exploit them).

Scientific colleagues generally find the hardest part of this scheme to swallow is the apparently crude and arbitrary nature of the PAW. They have been led to believe that an altogether more precise and more objective form of assigning Weights either has been devised (by Statistics say) or should be achievable with better measurements or higher mathematics; but I am sorry to say they are deluding themselves, or are being deluded. No amount of mathematics can ever make up for our inability to define $\bar{H}$ or H precisely, for these are insurmountable failures of the *imagination*. Even a super-being would be incapable of imagining *all* the members of an infinite set! Once that is accepted one must make do with the PAW, but compensate for that disappointment by looking for a wider portfolio of evidence, and the high Odds of which that is capable.

(b)   **You can't beat Common Sense Thinking**.

It is hard to see that there can ever be, except in detail, a more effective way of thinking than Common Sense. It searches widely for evidence, it weights the clues to the utmost that our perception of ultimate uncertainty will allow, and it multiplies those Weights together with some Prior, searching for Odds on which it might be prudent to act. What is there to be hoped for, or imagined, beyond that? Enterprises which claim to do better than CS, such as Statistics with it's 4-figure tables, were a wild- goose chase from the start. Only breadth and coherence among several clues can lead to wisdom – the kind of Odds one needs to act.

(c) **Cleverness – forget it**.

If serious thinking is as I have described it then there is no place in it for the kind of cleverness which is good at exams, or IQ tests, or crossword puzzles. And yet our whole educational system is geared to applauding and promoting that very kind of common-senseless froth. Curiosity, knowledge, being observant, breadth, doggedness, experience, imagination, judgement, caution, flexibility…. are what go to make up wisdom – not high IQ. I hope you find that truth as much a relief as I do. We can all become wise – if we try – even if we can't all become 'clever'. This conclusion is strengthened even further when we come upon Categorical Inference in Appendix 9.

It has often seemed to me, first as a pupil, then as a teacher, that the chief outcome of Education is to convince most of us that we are stupider than we are, or at least stupider than many other people. Because we don't get as many marks as others for geometry or spelling, or in some IQ test, we must be mentally deficient to a degree, and therefore unsuited to some desirable profession or other. Now it would appear that that is utter nonsense. Certainly it explains many otherwise extremely puzzling facts. For instance why is there almost no correlation between a student's ability to pass exams with very high marks (in say Astrophysics) and their ability to do research? And why, if you listen to the coffee table conversation of university professors (who are undoubtedly very clever), do you hear so much foolish nonsense? I wouldn't allow many of my colleagues to let off the family fireworks – or promote them to lance-corporal (the lowest rank in the British army above private soldier).

I don't understand how we got into this silliness about cleverness, but the quicker we grow out of it the better. Meanwhile my very strong tip is *"Don't ever allow your scholastic experiences, however bruising, to deter you from aiming high or becoming wise."* Follow Mark Twain : "I never let my schooling interfere with my education." It is worth reminding ourselves that any long-lived species like us must be born with a damned effective capacity for reckoning the Odds. As for having a high IQ – what useful purpose could it possibly serve? And look what a miserable record it has got.

(d) **Provisionality – the high secret of Civilization.**

The fundamental truth we all need to understand, be we dictators, professors, arch-bishops, children or judges, is that *all serious arguments about the real world can only reach provisional conclusions.* The best any of us can hope for are good betting odds deriving from a wide variety of evidence. Once we recognize that single truth we might be less willing to despise, imprison, even murder folk for not believing the same things as we do. Provisionality is the only soil in which any progressive civilization can germinate or flourish.

And why withhold that tolerance from our fellow animals on this planet? Now we understand that our thinking can hardly be superior to theirs, and that only literacy allows us to handle more evidence than they can, should we not be far more considerate of them too – at the very least leaving room for them to continue sharing the Earth with us, living in their manifold, free and ingenious ways? If we become extinct, as most species quickly do, animal intelligence might then evolve a second time into something more tolerant, more admirable, and perhaps more durable than our own.

This is such a vital section of the book that it must seem odd to put it in an appendix. The truth is I didn't come to fully appreciate these matters until I was 80, with no time to rewrite yet again. Anyway it is not obvious where else to put it. Logically it ought to be in early on where – because it requires some appreciation of Error Analysis (Chap. 9) it wouldn't be persuasive. And if at the end readers might feel cheated. So here it is where it can be referred to from any part of the book, with the hope of at least partial appreciation. The French have a delightful phrase 'L'esprit d'escalier' – going upstairs to bed imagining all the great things one should have said during the day but didn't. Bien sur.

P.S. Surely it is illogical (inconsistent) to claim, as I just have, that "It is certainly true that Certainty is unattainable." Certainty *is* attainable within a closed deductive system, subject always to the proviso that the axioms of that system are accepted without question. So in making a dogmatic claim that any conclusion reached by means of Induction must remain provisional I must be making a deductive statement based on some axiom I have not stated and cannot prove. The unstated axiom I am relying on is that: 'Inductive conclusions can only be reached by employing Bayes' Rule (or Theorem)'. If that is not true then my dogmatic claim is invalid. But, so far as I am aware, no one has come up with a way of reaching inductive inferences which does not rely, explicitly or implicitly, on Bayes. If somebody does so then my insistence on the provisionality of all Inductive arguments will have to be re-examined.

TFO (Disney): APPENDIX A9 "ANIMAL WISDOM OR CATEGORICAL INFERENCE (15/5/18)

The guiding principle of this thesis is that serious thinking, like any other survival mechanism, must have evolved, bit by tiny bit, from our animal forbears. Have we then succeeded in demonstrating that they could think in almost the same way as we appear to do – i.e. by using the Detective's Equation? No we haven't because they don't appear to use numbers – or multiply them together. So in this section we unearth what was, at least to me, a most astonishing discovery. You don't need numbers, or mathematics, to do Inference. The Detective's Equation is merely a convenient mathematical model for manipulating the real laws of Inference – which are categorical, not numerical.

As far as I can see the Laws of Inference, in their most fundamental form are three:

LAW I: THE CATEGORY LAW:

"As they bear on some hypothesis H all Clues can be placed in one of only 5 categories:

$$\begin{aligned}
&\text{Strongly For} &&(s) \\
&\text{Weakly For} &&(w) \\
&\text{Neutral} &&(n) \\
&\text{Weakly Against} &&(\underline{w}) \text{ i.e. underlined} \\
&\text{Strongly Against} &&(\underline{s}) \text{ i.e. underlined}
\end{aligned}$$

LAW II: THE COMBINATION LAW:

"Clues can be combined ($\star$) as follows:

$$w \star n = w$$

$$w \star w = s$$

$$w \bigstar \underline{w} = n$$

$$s \bigstar n = s \text{ ''}$$

LAW III: THE DECISION LAW:

"A decision for or against H will be reached only when the net accumulated evidence has reached:

$$s \bigstar s \bigstar s = sss \quad \text{(For)}$$

$$\text{or} \quad \underline{s} \bigstar \underline{s} \bigstar \underline{s} = \underline{s}\,\underline{s}\,\underline{s} \quad \text{(Against)}"$$

Remember we introduced this "Principle of Animal Wisdom" (PAW) to prevent a single strong (s or $\underline{s}$) but bad clue (a Systematic Error) from forcing an imprudent decision.

Amazingly these three Laws are *all* we need to do Inference (CST). To see them in operation observe the following sequence of clues:

CATEGORY Col. (1) New Clue	CATEGORY Col. (2) Accumulator	NUMBER Col. (3) Weight	NUMBER Col.(4) Accumulator
Prior	n	1	1 (2^0)
s	s★n = s	4	4×1= 4 (2^2)
w	w★s = ws	2	2×4= 8 (2^3)
w	w★ws = ss	2	2×8=16 (2^4)
$\underline{s}$	$\underline{s}$★ss = ns = s	¼	¼ ×16 = 4 (2^2)
s	s★s = ss	4	4×4 =16 (2^4)

W	w★ss = wss	2	2×2=32 (2^5)
W	w★wss = sss Decision For	2	2×32= 64 (2^6) Decision For

Note that two schemes, the CATEGORICAL and the NUMERICAL lead to *identical* decisions. The Numerical (Odds) form to the right is merely a convenient arithmetical (Odds) representation (model) of the Categorical form to the left. Now the point is that higher animals at least could surely categorize clues into weak and strong, and combine (★) two clues together along the lines of Law II. Thus an educated homo using The Detective's Equation is doing no more than a categorizing animal could when it comes to Inference, while the PAW is surely the outcome of Evolution – the extinction of forbears who allowed a single strong Systematic Error (i.e. an s or $\underline{s}$) to lure them into imprudent and fatal action.

So the mathematics of Odds is entirely irrelevant to the decision-making process! Replacing s by 4, w by 2, ★ by × and $\underline{s}$ by ¼ and so on is merely a convenient mathematical representation (model) of Categorical Inference – but nothing more. It will always reach exactly the same decisions on the basis of the same clues IF they are identically categorised.

We might have guessed this earlier when we introduced the binary system to replace the real numbers: i.e. 4 by (2^2) 8 by (2^3), ½ by 2^{-1} and so on, because then all we did to accumulate the evidence was add the indices , i.e.

$$4 \times 8 = 2^2 \times 2^3 = 2^{2+3} = 2^5 = 32$$

So we didn't even have to multiply, which means we could replace ★ by 'add' i.e. by '+', provided the neutral value n is now set to zero. Thus for the same sequence of clues as before:

Category of Clue	Number	Combination	Accumulator
Prior = n	0		
S	2	2+0	2
W	1	1+2	3
W	1	1+3	4
s	-2	-2+4	2
S	+2	2+2	4
W	1	1+4	5
W	1	1+5	6 (Decide For)

this last because 2^6 exceeds 50 to 1, which is the minimum deciding Odds. This shows that an additive model of Categorical Inference works just as well as a multiplicative one.

The net implication is that using Mathematics in Inference (CST) is merely a convenient symbolic model of a much more basic categorization process. We can do so if it is convenient, but it is in no sense a necessary part of the argument. It is the Rules I, II and III of Categorical Inference which are fundamental, and animal in nature. Having spent 65 years of my life mathematizing Inference in terms of Probabilities first, and Odds later, I am rather dazed by this discovery.

Note that we haven't had to introduce the idea of Probability, or Bayes' Theorem, not even the Detective's Equation. They are merely reflections, in mathematical form, of obvious inferential principles. The most important law is III – what we have called the Principle of Animal Wisdom (PAW) hitherto. Where does it come from? It recognizes that some strong clues may in fact be wrong, may be Systematic Errors, or the rogue elephants in the room. No animal can take the risk of acting on the basis of only 2 strong clues one of which might be wrong. Hence 3 strong clues would seem to be the minimum accumulation of evidence upon which either animals or men can afford to act. Once that is accepted all manner of other arguments follow. For example $s = 4$ in the Odds ($\times$) notation. Why? Well we tend to act when the Odds on an hypothesis exceed about 50 to 1 on and $s \times s \times s = 4 \times 4 \times 4 = 64$ to 1 on whereas $s \times s = 16$ which is comfortably below 50. We could instead have set $s = 5$ because $5 \times 5 = 25$ is also comfortably below 50, but then $w \star w = s = 5$ would have meant using $w = \sqrt{5} = 2.236$ and $\underline{s} = 1/\sqrt{5} = 0.4472$ which is much clumsier than using $s = 4$ and $w = 2$.

We are free to carry out Inference, that is to say Common Sense Thinking (CST), using Categories (s, w, 1, $\underline{s}, \underline{w}$), Odds (4, 2, 1, 1/2 ,1/4), Counts (3, 2, 1, 0, -1, -2, -3) or even Probabilities (see below). Once the clues have been categorized (Weighted) all four alternatives should lead to identical conclusions. Does that mean we have been wasting our time using Odds – when we could have been using Categories, or Counts, instead? After testing them all out on a number of real examples I do not believe so. Perhaps because we are familiar with symbols such as $\times$ and ½, and not with $\star$ and $\underline{s}$, it is actually easier to use Odds and the Detective's Equation, and so we stick to that notation. (You can test out the alternatives for yourself and make a different choice if you prefer)

What about the crucial importance of writing which, in the Odds context, we identified as the crucial secret of homo, the development which has enabled us to build space-ships whilst our equally smart cousins are still cracking nuts in the jungle (16:8)? If categorization is all that is needed to make sound inferences, why do we still need to write? Because writing extends not only the dynamic range but the *permanence* of memory. Once it is written down we have a long lasting – even inter-generational – record of the accumulated evidence, while a score of Categories – say sw, stored only in memory, can fade over the course of time, or even be entirely overlayed by the manifold distractions of later life. Big projects, as I know from my 30 year stint on the Hubble Space Telescope, require, above all, *documentation* – unbelievable amounts of it. Individuals die or move on, people forget; interconnections – possibly vital interconnections – can then be lost. For instance the main camera WFC-1 relied entirely on its Texas Instruments detectors. But when the first batch got ruined and NASA asked the firm to build replacements they couldn't because the lady who had done the vital thinning had left to have a baby – taking her forgotten skills with her. Had they been documented…..

To make these vital points we re-cast the Hidden Galaxy Inference Table below, because it is the largest, most complicated, *real* example we know of. We do it in both Odds (as we did before) and Categories so they can be compared. And we do something else: we explore what would happen if memory-leakage and distraction were allowed to affect the results. Crude models of both Leakage and Distraction suffice to illustrate the crucial importance of keeping an indelible record of the accumulating results (evidence). For the same reason parliaments and courts, which are decision bodies, always record their proceedings carefully and completely.

TABLE (A9/1) : ODDS, CATEGORIES, LEAKAGE AND DISTRACTION

1	2	3	4	5	6	7*	8**
Clue#	Date	W(E|H)	O(H|E)	Categ	Accum.Cat	Leakage	Distraction
Prior			2^{-5}	$\underline{w}$	$\underline{w}\,\underline{s}\,\underline{s}$	$\underline{w}$	$\underline{w}$
1	1975	2^{2}	2^{-3}	s	$\underline{w}\,\underline{s}$	$\underline{w}$	$\underline{w}$
2	1978	2^{-2}	2^{-5}	$\underline{s}$	$\underline{w}\,\underline{s}\,\underline{s}$	($\underline{s}\,\underline{s}$)	[$\underline{w}\,\underline{s}$]
3	1983	2^{2}	2^{-3}	s	$\underline{w}\,\underline{s}$	$\underline{s}$	$\underline{w}$
4	1984	2^{2}	2^{-1}	s	$\underline{w}$	1	[1]
5	1985	2^{2}	2	s	w	s	s
6	1987	2	2^{2}	w	s	(s)	[w]
7	1987	2^{2}	2^{4}	s	ss	ss	ws
8	1990	2^{2}	2^{6}	s	sss	(wss)	[ws]
9	1993	2^{2}	2^{8}	s	ssss	wss	ws
10	1994	2^{2}	2^{10}	s	sssss	(sss)	[ws]
11	1995	2	2^{11}	w	wsssss	wsss	ss
12*	1995	2^{-2}	2^{9}	$\underline{s}$	wssss	(ss)	[1]
13	1997	2	2^{10}	w	sssss	(wss)	w
14*	1997	2^{-2}	2^{8}	$\underline{s}$	ssss	(s)	[1]
15	1998	2^{2}	2^{10}	s	sssss	ss	s
16	1999	2^{-1}	2^{9}	$\underline{w}$	wssss	(s)	[1]
17	2002	2^{-1}	2^{8}	$\underline{w}$	ssss	w	$\underline{w}$
18	2002	2^{-1}	2^{7}	$\underline{w}$	wsss	1	[1]
19*	2005	2^{-2}	2^{5}	$\underline{s}$	wss	(1)	$\underline{s}$
20	2005	2^{2}	2^{7}	s	wsss	1	1

21	2007	2^2	2^9	s	wssss	s	s
22	2009	2	2^{10}	w	sssss	(s)	[w]
23	2012	2^2	2^{12}	s	ssssss	ss	ws
24**	2013	2^{2+2+2}	2^{18}	sss	ssssssss	(wssss)	[wsss]
25	2013	2^2	2^{20}	s	sssssssss	wssss	wssss

* In the Leakage Model, after every other clue has been accumulated, the score drops back towards the neutral value by one point (w or $\underline{w}$) depending on the value. The accumulated value *after* the leak is shown in brackets ( ).

** In the Distraction Model, on the grounds that distraction will increase the time intervals between accumulating new clues, after every other clue the model takes a 2-point drop towards the neutral value (i.e. by s or $\underline{s}$). The accumulated value *after* the distraction is shown in square brackets [ ].

Study Table (A9/1) for a few minutes. It starts off as a direct copy of Table (10:1) which kept account of the Hidden Galaxy debate as it developed over 38 years. Compare column 4 Accumulated Odds with col. 6 of Accumulated Categories. Personally I find it easier to remember 32 to 1 than wss so that is why I prefer to use Odds to Categories. But if one had a different education it might be otherwise.

Now compare columns (7) and (8) with column (6). Leakage (7) weakens the whole argument – which after all continued for 38 years, while distraction (8) destroys it altogether. And that reflects my personal recollections [family matters, other projects such as the Space Telescope, changing assistants and students, competitors, building new equipment, finding funds and time, teaching…….]. My personal memory – which often failed – and three dozen fairly disorganized research notebooks –

were the only records which persisted – and my rarely consulted diary. But they proved vital, particularly after they were indexed.

So I hope Table (A9/1) makes the case for writing, or rather indelible recording, even stronger than it was already. The real difference between Man and Apes is that we are holding a pen. Rodin's Thinker ought to have had a pen, and a notebook on his knee.

NB   Incidentally it is interesting to look at column 6 where the categorical values accumulate. By 1990 they had reached sss, by 1994 sssss. With so much certainty we should have been far more sceptical of the three subsequently unfavorable but strong clues (12), (14) and (19) – which indeed all turned out to be wrong. So Categorical Inference could have saved 15 years of unnecessary effort, doubt, and pain – had we known about it then, which we did not.

SUMMARY

Categorical Inference, with its three simple rules I, II and III, contains all that can be said about Inductive Thinking. All the mathematical superstructure erected upon it, including Probability Theory, Bayes' Rule, and the Detective's Equation, is superfluous (though it might be convenient); Categorical Inference will do the job on its own. Any mind, be it animal or human, need know only two things about a clue: is it favorable or unfavorable, strong or weak. The rules for combining clues are straightforward. Thereafter the only room for evolutionary improvement lay in increasing curiosity (to search for new and more diverse evidence) and in the more reliable accumulation of that evidence. Language would have helped because then experiences could have been shared and thus to some extent memorized collectively. But writing would have enabled a dramatic step forward in compounding the evidence needed to tackle immensely more complicated tasks.

POSTSCRIPT ON PROBABILITY. (For experts) We haven't discussed Probability in this appendix but we ought to do so because that is the notation, the coinage the 'experts' use. Let's look at the mess.

Probability Theory was developed to deal with cards and dice – in which context it was fine because the rules were defined and the agents perfectly known. Thus the Probability of getting a fifth ace was exactly zero. It even made sense to 'Normalize' the Probabilities (i.e. force them to lie entirely within the bounds of zero and one, which is pure convention.) because, as it turns out, that makes it easier than Odds say, to deal with calculations of the kind Prob(A *or* B). But there was a huge price to pay in the form of Bayes' Theorem which now has to be written:

$$P(H|E) = \frac{P(E|H) \times P(H)}{P(E|H) \times P(H) + P(E|\bar{H}) \times P(\bar{H})}$$

with its clumsy and unnatural denominator which arises from the artificially imposed need for Normalization. Compare that with the Odds form: $O(H|E) = W(E|H) \times O(H)$

or the Categorical form of Bayes':

New Outlook = Clue Category ★ Old Outlook.

And what about the PAW which, as we have seen, is central to Inference? Well it can be introduced in terms of Probability as follows: First derive:

$$P(H|E_1,E_2) = \frac{P(E_1|H).P(E_2|H).P(H)}{P(E_1|H).P(E_2|H).P(H) + P(E_1|\bar{H}).P(E_2|\bar{H}).P(\bar{H})}$$

Then insist that:

$$0.02 \leq P(H \mid E_1, E_2) \leq 0.98$$

otherwise only two clues, one of which might be a systematic error, could tip the balance towards fatal action (i.e. beyond Odds of 50 to 1). Now call $P(E_1 \mid H) = P(E_2 \mid H) \equiv p$ and set P(H) at the neutral value of ½ and you will find that;

$$P(H \mid E_1, E_2) = \frac{p^2}{2p^2 - 2p + 1}$$

which, inserted in the above inequality, leads to:

$$0.2 \leq p \leq 0.8$$

which is consistent with the PAW in Odds form, namely:

$$1/4 \leq W(E \mid H) \leq 4$$

[Incidentally the fundamental relation $W(E \mid H) = P(E \mid H) / P(E \mid \bar{H})$ can be seen as a necessary requirement for the Odds and Probabilistic forms of Bayes' Rule to be consistent, which of course they must be.]

So, in principle, Probabilists could have reached the PAW eventually, but in fact they didn't, and it's now obvious why: the truth was obscured behind a blizzard of normalizing algebra. That has had all sorts of repercussions, the worst of which was to delude themselves into the belief that very high or low Probabilities, or very precise Probabilities, were attainable – the central fallacy of Statistics. That is, above all, a failure of imagination, a failure to recognize that all manner of $\bar{H}$s and variants of H are possible [see Appendix 8].

TFO (Disney 15/5/18)

APPENDIX A10: HIDDEN GALAXIES; NETTED AT LAST.

Hidden Galaxies (HGs) proved to be far more elusive than we had anticipated. The main reasons for our failure to find them until recently were: (1) The miraculous new electro-optic cameras (like the one in your phone), while 50 times more sensitive than the old photographic plates, are 10,000 times smaller in area. Therefore we were never able to survey with a big telescope a large enough area of sky to find any. We (I) should have realised that – but we were too intoxicated by their super-sensitivity to realize their crippling limitations as survey instruments. (2) We hoped to find more or less optically invisible galaxies by the radio signals emitted by their neutral Hydrogen gas at a wavelength of 21 centimetres. Giant single-dish radio telescopes seemed ideal for this purpose. But it turns out they have neither the sensitivity to pick up diffuse Hydrogen in short observations, nor the angular precision to defeat the clustering problem (below). (3) But the real heartbreaker is the tendency of all galaxies to huddle ('cluster') intimately together. If therefore you do find an optically invisible galaxy (in the radio) you can bet there will be a bright one close by – and at the same distance. So one might naturally mis-identify that as the source of the radio signal and conclude, as many too hastily did, that there *are* no invisible galaxies. It turns out that the only remedy is a radio-interferometer telescope with the combination of angular precision, sensitivity and band-width (to cover enough frequency-channels). This is a very tall order technically because of the immense computing power required. But finally in 2014 the Very Large Array in New Mexico (www.vla.nrao.edu) was fitted with enough grunt for us to do a very small pilot survey. Juergen Ott (US NRAO) and I then looked at 17 Hydrogen radio sources picked out by my colleague Huw Lang (Cardiff University) as having "suspicious" identifications with optically bright galaxies. Sure enough at least 2 were not the bright galaxies identified in our earlier single-dish survey but were invisible on existing optical surveys. However in

2016 we went to the Canaries and took extremely deep optical images with The William Herschel Telescope on which we can just see vestiges of light associated with each (see Fig A10:1 below). No explanation for them makes much sense within the existing paradigm on galaxies. It is all very mysterious and intriguing.

From the point of view of the Scientific Method in general, progress was held up for so long by the following factors: (a) Lacking the Detective's Equation we had no means to balance conflicting evidence (thus a single strongly negative clue was seen as paramount); (b) The Burden of Proof issue allowed naïve 'optical identifiers' to get away with murder; (c) Because they had failed to find HGs themselves too many people thought of themselves as 'experts' on the matter; (d) For ignorance of PAW three major Systematic Errors dominated the scene; (e) Fatal in-breeding among Hydrogen radio-astronomers led to some appalling, even in some cases dishonest, refereeing; (f) Arrogance towards outsiders from beyond 'the guild'; (g) Careerism led to youngsters, and others, making premature, dogmatic and naïve claims; (h) Mad Big-Telescope Disease ('My telescope is big enough to see anything I look for'; a fatal complaint in my profession); (i) Compartmentalisation of knowledge into tiny pigeon-holes; (j) Ignorance and lack of self-confidence on my part. In other words we all behaved like very human beings instead of the wise natural philosophers we are supposed to be.

In retrospect one can see that a huge challenge was to master the very wide range of technical skills necessary to get an overall view of an immense landscape. Clearly that was my responsibility; I tried hard, but I wasn't quite up to the job.[One wonders how many other fields are stalled for the same reason; and what we might do about it.] My most crippling lack was a proper understanding of Common Sense as it applies to the scientific enterprise. But when The Detective's Equation turned up, and with it an Inference Table (10:1) for the whole campaign, it was as if the bright Sun had emerged from behind an eclipsing Moon.

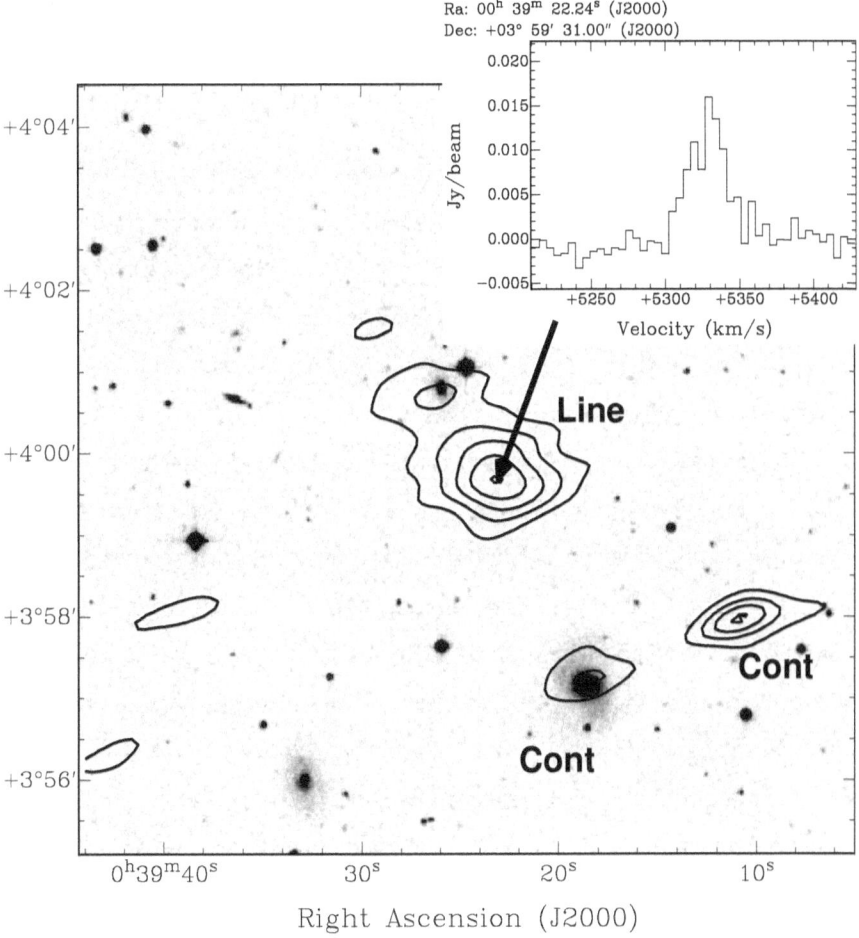

Fig (A10:1) An image of the first truly Hidden Galaxy – its called "Parkes 0039+03". The contours show more neutral Hydrogen gas than exists in our own giant Milky Way galaxy – mapped by its radio emission – while the inset shows its spectrum as a function of its cosmic recession speed. Those contours are plotted on top of a negative of the same piece of sky seen in the optical. When originally detected with a big radio dish the Hydrogen was assumed to belong to the bright optical galaxy below it, and somewhat right (the fuzzy blob). But as you can see in this much higher resolution interferometer image it has nothing to do with that galaxy, nor with the fainter one on its upper margin (which, from its speed, lies far beyond in the distance). On a far deeper optical image we can just see immensely dim specks of light associated with the Hydrogen. It appears as if the galaxy has just switched on its first stars – but that would be far too much of a coincidence – given how old the cosmos is. So for now it, and our other HG, are intriguing mysteries.

P.S. You might feel that a mere 2 HGs after all that effort isn't much to show. But now we know how to do it I am confident that our successors will find thousands. That is invariably the way in astronomy.

TFO (Disney 15/5/18)
APPENDIX A11   HOW THIS BOOK GOT WRITTEN

I set out to write this book for my own benefit back in 2001 in order to try and fight my way out of a swamp into which my own research project was then sinking. When I started I thought I knew more or less what I was going to first find out and then write. I would distil the wisdom of The Scientific Method, sprinkle a little of History, Statistics and Probability-Theory onto Common Sense, and there would be The Truth Revealed.

I couldn't have been more wrong! It took at least 9,000 hours of puzzling, reading, arguing, writing, re-writing and re-re-writing to realize that the experts disagreed violently about the Scientific Method, while no one seemed to know, or indeed to care, how Common Sense Thinking worked. It was supposed to be a low animal-process, beneath the consideration of the philosophers and statisticians who felt themselves to be the pundits in this field, while scientists, oddly, refused to talk about such matters, preferring to get on with their research rather than squabbling with the pundits who, they felt in their bones, were probably wrong.

My truth, as it eventually emerged, proved very different from my expectations. The key was Common Sense itself, that despised and neglected child of Nature. Only when that was decomposed could one see into the business of Science and into the misconceptions of it held by philosophers and statisticians. We had all been misled by history, which sometimes happens in the wrong order. The Theory of Evolution had come too late (1859) in the day to save us from the confusions of philosophy and religion. We were already enthralled by the idea that Deduction on the one hand, and a hot-line to God on the other, were the wellsprings of Rational thought, and of Scientific thought in particular. Only after Darwin and Wallace had demonstrated that we were one with the other animals did it make sense to go back and ask 'How do animals think?' because if we can think it must be because they could do so beforehand. Only when one asks that question does one get enlightening answers; Bayes' Rule, The Detective's Equation, Ockham's Razor and The Principle of Animal Wisdom (PAW) then fall naturally out of the inquiry. CST is then seen to be an algorithm (recipe) which is dependant neither on deductive logic, nor on psychology nor on neuro-science, because it could be made to work using a variety of switching systems, from bird-brains to electronics.

Writing this book has largely been an exercise in unlearning what I had either been taught or picked up from famous text-books. By definition animal thinking is not original – it is tens of millions of years old. If there is anything creditable here it lies in excavating the truth from under a heavy overburden of confused historical scholarship – mainly Greek Philosophy, Abrahamic religion, Mathematics and especially Statistics. Yes I have given an original name 'The Detective's Equation' to its key tool because it needs to be recognized very distinctly, but the tool itself is as old as the hills (Chapter 5). It has been glimpsed by others before – but then they allowed the super-incumbent confusion to fall back and re-bury it (Chapter 14). The excavation has been a dirty and exhausting business what with fruitless trenches, and caved in tunnels to say nothing of my gullibility, blundering and stupidity. I only persisted because of certain accidents in my upbringing and profession. It helped enormously that I am a professional astronomer because astronomers cannot do clarifying experiments, and have to rely, more than most other scientists, on vague and conflicting evidence. It helped that I started off as a boy wanting to become a philosopher (I naively thought it would lead to the good life) and lived in a house full of animals, including three enormous pigs who used to lie in front of the lounge fire, blocking off any heat from us kids, and a half-wild jackdaw who could count up to fourteen. We never had any doubt about animal intelligence! And it has certainly helped that in my 50-year career I have been many kinds of an astronomer – a theoretician, an applied mathematician, a computer-modeller, an optical observer, a statistical type, a computer-systems designer, an instrument-builder, a Space-scientist, a teacher and a radio-astronomer. Such feckless wanderings wouldn't have been possible before my time, and are no longer tolerated today. So I have been very fortunate in that, as I have been very lucky in living on five continents and working with some of the brightest people on this planet, particularly so while working for over 35 years within the Hubble Space Telescope community. I am not being mealy-mouthed when I say it would probably have been impossible to get within reach of CST without all that varied, incoherent experience (which included being a factory-worker for a year and a soldier for two). In my opinion Philosophers and Statisticians, no matter how clever, wouldn't have stood a chance. Cutting edge science is a challenge to all one's misconceptions. Yesterday's brilliant insight will finish in tomorrow's waste-paper-basket. Oh yes and then there's Failure. Failure is a sovereign teacher, at whose feet I have sat for forty years, trying and failing to convince my colleagues of the existence of Hidden Galaxies (Chapter 4 and after). But wait! (See Appendix 10)

TFO (Disney)

APPENDIX A12: CAN MACHINES BE TRAINED TO THINK?

is a question of some moment when Deep Blue can beat the world champion Garry Kasparov at Chess and Deep Mind can beat anyone at the much more complex game of GO.

The first point to note is that games like Chess and GO are CLOSED – that is to say all their possible configurations can be mechanically envisaged and analysed – whereas nearly all questions in the real world are OPEN – where such is not the case. For every Clue E there may exist hypothetical explanations $\bar{H}$ beyond the reckoning of machine or man, in which situation the straightforward setting of Weights becomes problematical – which is not the case in a game [Appx. A8]. Thus the unquestionable skill of computer programs at playing complex games is not relevant to the question of whether they will ever be able to think seriously.

To approach that question we list the 8 steps needed to tackle, for instance, a serious scientific problem (see Ch.3 on 'How do Scientists think?') :

A) Recognize the problem as interesting.
B) Formulate some hypothesis H_1 about it.
C) Set a Prior value on the Odds $O(H_1)$.
D) Gather clues E_i bearing on H_1.
E) Weight or categorize those clues.
F) Combine those clues (e.g. using the Detective's Equation).
G) Come to a decision for or against H_1.
H) If against, go back to B, formulate a new H_2 and start again.

One can imagine that a programmable machine could make a fair fist of the 5 steps C to G. For instance, in the case of D it could be directed to search a large data-base for clues relevant to H_1 (a la Google-search). Step E requires judgement – which might prove tricky to program – but thanks to the PAW [Ch.10] there are only 5 crude categories of Weight to choose from – which enormously simplifies that task. The real challenge for the machine is set by steps B) and A):

Take B): in a closed game such as GO every hypothetical next move H can be listed and evaluated using algorithms (recipes) based on the rules and objectives of the game. Machines can be programmed to recognize patterns, and trained to exploit them using 'training sets'. Deep Mind apparently doesn't need such sets, improving its performance – by means not entirely transparent – with every game it plays. However the real world is OPEN – so presenting a very different challenge to any programmer: the hypothetical next moves are infinite in number and, as we have found, the required Weights require some cognizance of $\bar{H}$ as well as H [Appx. A8, Sect (10:6)]. Of course the human or animal mind must meet the same challenges too – and it may be that instincts, based on past evolutionary success, come to its aid at this point. Such instincts are likely to be unconscious – and therefore impossible to program. Many hypotheses pre-exist while others are obvious – like the Broad Street pump in the cholera outbreak. [(3:3)]. Even so others must be drawn out of the imagination [e.g. Faraday's illogical fixation that Magnetism can cause Electricity (3:18)] – like a story, a melody or a poem. And there be magic surely undefined – and un-definable.

The other challenge for the machine programmer comes with step A). How did Darwin come to recognize that the beaks of finches in the Galapagos Islands were significant (3:11)? Why was Kepler puzzled by the darkness of the night sky (3:2)? Why did Becquerel follow up the fogging of his photographic plates (3:7), or Joule persist and persist until he finally pinned down the Mechanical Equivalent of Heat (3:8)? Pasteur said; "In the fields of observation chance favours only the prepared mind." Nearly all the successful scientists we followed in Chapter 3 appear to have had expectations which were either confounded or confirmed by what they observed. It is beyond me to imagine how such a state of excited expectation (i.e. 'preparation') could be programmed into any machine. You would have to load it with human cravings and ambitions − not to mention the entire zeitgeist of the age in which those humans lived. If Darwin hadn't recently read Malthus, Lyle and his grandfather Erasmus's poem '*The Botanic Garden*' he probably wouldn't have bothered about the bloody finches.

One must concede that there are apparently creative tasks of a scientific kind that machines can be programmed to do − for instance to identify elusive but significant patterns in complex data-sets. Had such a machine existed in Mendeleev's day it might have tirelessly searched the chemical abstracts and identified the Periodic Table before he did (3:13). It is conceivable, though rather unlikely given that he simultaneously performed relevant experiments. It is more likely that a programmed machine might have beaten Hertzprung and Russel to the important correlation between the luminosities and colours of stars [Fig(3:1)]. Fig (A12/1) below shows a truly astonishing and suggestive set of correlations amongst the properties of galaxies, recently identified by a suitably programmed idiot-machine − though in this case only after humans had painfully discovered them first with pencil and paper.

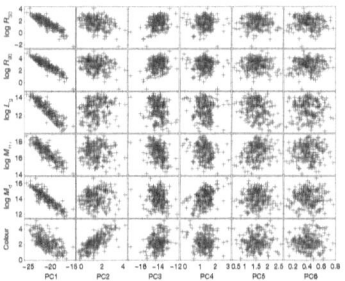

Fig A12/1. The properties of a sample of 200 galaxies found with the Parkes radio dish in Australia and then measured with the Sloan optical telescope in New

Mexico, are analysed by a computer to look for correlations. An astonishing, i.e. wholly unexpected regularity turned up—they all seem to be controlled by a single mysterious parameter (the one in the left hand column labelled PC1) – although they have a truly vast range of properties – for instance the Giants contain 10,000 times as many stars as the Dwarfs [box 3 down, left]. My son, who is not an astronomer, first discovered this when he chucked the original 3000 data into his computer. I simply didn't believe it; after all if you knew as much about galaxies as I did, such uniformity was incredible. So I ignored him and spent 2 years fumbling with pencil and paper. But he and his computer program turned out to be right eventually.[See *'Nature'*, 2008, vol 455, p1082.] So yes, computers can sometimes find patterns which might otherwise escape us. But that's not Thinking, it is Noticing.

So we shouldn't under-rate the potential of machines as scientific assistants. At the same time the burden of proof must lie entirely upon those who want to argue that machines will one day be able to think. In particular they must explain how they will first recognize significant problems (task A), and second formulate imaginative hypotheses to solve them (task B). For the moment I sense no earthly hope of progress in either direction.

It is important for all to recognize that there are powerful vested interests at work here. The US computer industry, backed up by the sickly (or sinister) US patent system – has greedily sequestrated huge funds to which it has no moral or sensible entitlement. Bill Gates for instance, who so far as I can see, contributed nothing substantive to computing beyond monopoly, cleared off with seventy thousand million bucks! Tycoons like him would like us to believe that they are so smart that they actually deserve to trouser fortunes which belong to the earlier gilded age of robber-barons like Rockefeller and Vanderbilt. And it might be easy for outsiders to imagine they *are* super-smart. Well they aren't. I can only illustrate that truth by anecdote.

Over the past 50 years computers have offered astronomy so much that astrophysicists of my generation (1965 –2015) have inevitably found themselves at the very cutting-edge of computing – where they were forced to innovate in order to progress. But those innovations were not the result of genius, merely the consequence of being the very first, or one of the first, to face a novel challenge. Thus in 1965 I found myself working on what was then the world's most powerful computer, the Ferranti Atlas at the University of London. I had written, *for its time*, an enormous program (6000 lines of code) to solve the equations of star-formation. The problem was that the computer stalled whenever it reached a trivial error in my code (every few lines) forcing me to wait a week before I could try the amended program again. I gradually came to understand why my predecessor in the enterprise had suffered a nervous breakdown. It took me another 12 months to recognize there was only one hope: to break the huge code down into thirty self-contained sub-routines, each of which could be tested separately each week, so that I could proceed at thirty times the previous speed. I had to start again from scratch but the stratagem eventually worked. My trick was novel at the time but it became the general philosophy for writing big programs. Was I first ? Certainly one of the first. Was I smart – definitely not. There was simply no other way to proceed.

After complaining to the Science Research Council about the abysmal computing facilities open to British observational astronomers I was appointed in 1977 to head up a mixed committee of astronomers and computer scientists to look into the matter. The sheer flood of new data coming from telescopes of many kinds (optical, satellite, radio) forced us to devise a novel network of computers – called '*Starlink*' – scattered about the globe, able to talk to one another, and able to display astronomical images on smart colour-terminals – which didn't exist then – until we designed and built the first. *Starlink* revolutionised astronomy for us – but more significantly it was copied all over the globe. Every multinational and bank on the planet wanted a copy of Starlink and, according to the chief executive of the mighty DEC corporation (now defunct, like Ferranti), they sold ten billion dollars worth of them. So far as I know Starlink spawned the civilian Internet (1978). I would like to claim the credit for that – but honestly I can't. We had never envisaged that our communication network would be used for anything but sharing software. But the bloody astronomers found they could send messages to one another – and off they went. And as for our uniquely clever image-display terminals, when a well known Silicon Valley expert saw mine he went down on his knees before it, raised his palms to heaven and prayed "Please Gahd I want one." And of course they came – from Silicon Valley. Poor old Britain got zero credit. But then everything would have come anyway – in the course of time. We just happened to be standing first in the queue.

One could go on – but I won't because I hope the point is made: that innovation in computing, as far back as Colossus in 1943, has been driven throughout by need rather than genius. Necessity has here truly been the mother of invention. Most of the really smart scientists I know hate computers, but realize they can't do without them. And the young scientists who do go into a career in the computer industry are most often the ones who find out they are either not smart enough or ambitious enough to do any better. So one mustn't treat computer gurus like gods – or pay them like gods either. Most computer firms are rightly terrified that they will soon go down the tubes like all their illustrious predecessors – which is why they are snatching everything they can while they can. In my opinion they are severely holding up progress by exploiting a totally compromised US patent system – they even tried to patent the curve on the corners of the i-Phone. If anyone uttered the words 'Microsoft' or 'Windows' while I was working on the Hubble Space Telescope, the whole room would burst out into jeers of execration and contempt.

In summary then computers can be turned into wonderful assistants—once a specific need for that assistance has been defined. But there is no evidence for now that they will ever be able to think: notice things yes, win games yes – but not think.

THINKING FOR OURSELVES (Mike Disney)
APPENDIX A13: A SKETCH-MAP OF THINKING

1.6 kw., version 10/9/18

When trying to understand a tricky argument it is sometimes helpful to make a little London-Underground map of how the different concepts relate to one another, which we do here. If you already accept that Inference Tables make sense then there's no need to bother with it because such tables already incorporate all such concepts; but if not, to make certain you have not been bamboozled, you may be willing to tackle some of its tricky symbolic arguments. Additionally it should convince you that we need to become familiar with *both* Odds and Probabilities, because both suffer from serious deficiencies – which the other can remedy. Had we recognized that back in the 17th century we might have pinned down the Detective's Equation, and with it Common Sense, back then.

Understandably, but unfortunately, Thinking has attracted far too many holier-than-thou Logic-Priests who try to browbeat one into following their own particular credos [e.g. {11/8} & {11/9}]. Their Holy Certainties stem partly from blindness to their own deficiencies,[principally that they can usually only deal with CLOSED i.e. largely uninteresting, systems], and partly from coming from the wrong kinds of universities (i.e. ex-seminaries) where priestliness was accounted not merely a virtue, but a necessity if you were employed to teach in them.

This appendix is not, I hasten to add, my own attempt at browbeating. I follow de Finetti's concept of the Probability of an event as the fraction of the total pot I would be prepared to bet on it occurring. Then The Product Rule, and all that follows therefrom (see postscript* at end here), are obvious – IF you are familiar with betting Odds, which too many scholars are not. And there's nothing fundamental to object to about a Prior – it is merely one's earlier (comparative) state of mind – yesterday's Posterior. De Finetti realised that serious i.e. Inductive Thinking is all about gambling – which is something all organisms constantly have to do. Priests cannot abide gambling – which is why they get so very confused when it comes to Thinking. But that is their problem – it doesn't have to be ours.

(a) Conversions between Odds and Probabilities:

Our definition of Odds was:

$$O(H) \equiv \frac{P(H)}{P(\bar{H})} \tag{1}$$

But $\quad P(H) + P(\bar{H}) = 1 \tag{2}$

Thus: $\quad P(\bar{H}) = 1 - P(H) \tag{3}$

So $\quad O(H) = \frac{P(H)}{1 - P(H)} \tag{4}$

But (4): $\quad O(H) - O(H).P(H) = P(H)$

Therefore: $P(H) = \frac{O(H)}{1 + O(H)} \tag{5}$

Equations (4) and (5) allow us to readily convert back and forth between the two concepts. [Note that they also work for Conditional Probabilities and Conditional Odds, i.e. for P(E|H)'s and O(E|H)'s.

(b) The Product Rule for Probabilities.

Is $\quad P(A \& B) = P(A|B) \times P(B) \tag{6}$

where P(A|B) is the 'Conditional Probability' or 'The Probability of A given that B is true'. Some authors, as I do, accept it as obvious or 'axiomatic' {11/8} whereas others {11/9} go to enormous (and to me unconvincing) lengths to prove it from more primitive considerations (but de Finetti gives a thoroughly convincing demonstration for Subjective Probabilities – see* below). (6) is at least very simple, and appears intuitive.

c) *Bayes' Rule in Probability*

follows straightforwardly from (6) because obviously P(B&A) = P(A&B) in which case:

If $P(H \& E) = P(H|E).P(E)$

and $P(E \& H) = P(E|H).P(H)$

Then $P(H|E) = P(E|H).P(H) / P(E)$ \hfill (7a)

the simplest form of Bayes' Rule. Rewriting it as:

$$\frac{P(H|E)}{P(H)} = \frac{P(E|H)}{P(E)} \quad (7b)$$

emphasises it as the way to update our certainty as to some hypothesis H [the LHS] in the light of new observational data E [the RHS] – surely what CST needs. The Prior P(H) is merely our earlier state of mind regarding H *before* incorporating a new clue E; i.e. it is comparative, not absolute.

But what is P(E)? It is the Probability of clue E whether H is true or not. If then $\bar{H}$ stands for "*All hypotheses which could explain E, apart from H*":

$P(E) = P(E \& H) + P(E \& \bar{H})$ \hfill (8)

which can be expanded using the Product Rule (6):

$P(E) = P(E|H).P(H) + P(E|\bar{H}).P(\bar{H})$ \hfill (8a)

and if we incorporate (8a) into (7a):

$$P(H|E) = \frac{P(E|H).P(H)}{P(E|H).P(H) + P(E|\bar{H}).P(\bar{H})} \quad (7c)$$

the form of Bayes' Rule, or Bayes' Theorem, found and used in most text-books. It's too clumsy to have evolved naturally, or be used by other animals, nor is it widely taught [Indeed I wasn't aware of it until I was 60] because, *on its own*, it is rarely much use because, for only one clue, it is weak.

(d) *The Detective's Equation in Probabilities.*

But suppose we have a second clue E_2 then (7c):

$$P(H|E_2) = \frac{P(E_2|H).P(H)}{P(E_2|H).P(H) + P(E_2|\bar{H}).P(\bar{H})}$$

But if P(H) is the Probability of H as a result of the *earlier* clue E_1 we could replace P(H) in the above equation with $P(H|E_1)$ using (7b) i.e.:

with $P(H|E_1) = \dfrac{P(E_1|H).P(H)}{P(E_1)}$

So: $P(H|E_2) = \dfrac{P(E_2|H).P(H|E_1)}{P(E_2|H).P(H|E_1) + P(E_2|\bar{H}).P(\bar{H}|E_1)}$

$$= \frac{P(E_2|H) \times \dfrac{P(E_1|H).P(H)}{P(E_1)}}{P(E_2|H) \times \dfrac{P(E_1|H)}{P(E_1)} + P(E_2|\bar{H}) \times \dfrac{P(E_1|\bar{H}).P(\bar{H})}{P(E_1)}}$$

Now the $P(E_1)$'s can be cancelled and we have the Detective's Equation:

$$P(H|E_1,E_2) = \frac{P(E_2|H).P(E_1|H).P(H)}{P(E_2|H).P(E_1|H).P(H) + P(E_2|\bar{H}).P(E_1|\bar{H}).P(\bar{H})} \quad (9)$$

in terms of Probabilities instead of Odds, but for 2 clues only. It is vastly unnatural, though one can see a sort of pattern emerging, which might allow one to guess at the form of $P(H|E_1,E_2,E_3)$. Surely it would never have evolved in animals. It is hardly surprising that (9) was never reached, or if reached never taught, as a practical tool for CST.

(e) *Bayes' Rule in Odds form.*

We *could* simply state it as a *definition* of the Weight:

$O(H|E) = W(E|H) \times O(H)$ \hfill (10)

because thus far W(E|H) is undefined, though it does seem intuitively obvious that:

$$W(E|H) = P(E|H)/P(E|\bar{H}) \tag{11}$$

However we can straightforwardly prove (11) because (10) has to be consistent with the Probability form of Bayes' Rule:

$$O(H|E) \equiv \frac{P(H|E)}{P(\bar{H}|E)} = \frac{P(E|H).P(H)/P(E)}{P(E|\bar{H}).P(\bar{H})/P(E)} \text{ using (7a)}$$

The P(E)s cancel leaving:

$$O(H|E) = \frac{P(E|H)}{P(E|\bar{H})} \times \frac{P(H)}{P(\bar{H})}$$

and as the last term is O(H) by definition [see (1)], comparison between (10) and this last equation proves (11). Although Dorothy Wrinch and Harold Jeffreys proved (11) back in 1921 they made nothing of it, and then Jeffreys went on to hang himself by insisting that all Probabilities be Objective, a form of religious mania. Thus they never reached the Detective's Equation, and neither did anyone else.

(f) The Detective's Equation in Odds

The form (10) of Bayes' Rule naturally generalises and translates into:

"Odds after New clue = New Weight × Odds after last clue"

Or $\quad O(H|E_2) = W(E_2|H) \times O(H|E_1)$

Which, replacing the last term with (10), becomes

$$O(H|E_1,E_2) = W(E_1|H) \times W(E_2|H) \times O(H) \tag{12}$$

which cycles (iterates) quite naturally up to any number of clues. Nature could certainly have exploited this simple algorithm many eons ago.

(g) The Product Rule in Odds

Thus far Odds have had it all over Probabilities when it comes to simplicity and power. But wait! What is the Product Rule for Odds?

Because $P(A \& B) = P(A|B).P(B)$

And $O(H) = \dfrac{P(H)}{1-P(H)}$

$O(A\&B) = \dfrac{P(A|B).P(B)}{1-P(A|B).P(B)}$

Now replace Probs. with Odds using (5):

$O(A\&B) = \dfrac{\dfrac{O(A|B)}{1+O(A|B)} \times \dfrac{O(B)}{1+O(B)}}{1 - \dfrac{O(A|B)}{1+O(A|B)} \times \dfrac{O(B)}{1+O(B)}}$

when, after multiplying out and cancelling:

$$O(A\&B) = \dfrac{O(A) \times O(A|B)}{1+O(B)+O(A|B)} \qquad (13)$$

which is far clumsier than the Probability form of the Product Rule (6) and not at all intuitive.

[You can check that (13) makes sense by considering two horses A and B running in separate races and calculating the Odds on a double, i.e. both winning. If $P(A) = 1/5$ and $P(B) = 1/2$ you should finds Odds on a double of 1/9 or 9 to 1 *against*. If you are a gambler (13) could be useful.]

The sheer clumsiness of (13) as compared with (6) was one of the main reasons scholars turned against Odds when it came to philosophical issues. I would have turned so myself. How could we have known that we would thereby be burying The Detective's Equation, and thus holding up human progress for 300 years?

(h) *The Probability of Alternatives: P(A or B)*

On reflection it is clear that:

$$P(A or B) = P(A) + P(B) - P(A|B) \qquad (14)$$

where the last term takes into account that A and B *may not be independent*, may overlap, in which case we could double-count the overlap region $P(A|B)$ if we didn't subtract it as in (14).

(i) The Odds on Alternatives: O(A or B)

Employing the same kind of jiggery-pokery we used to get (13) from its Probability form one can prove (it took me 15 hours, but then I'm not much cop at algebra) that;

$$O(A \text{ or } B) = \frac{O(A).O(B).O(A|B) + 2O(A).O(B). + O(A) + O(B) - O(A|B)}{1 + 2O(A|B) + O(B).O(A|B) + O(A).O(A|B) - O(A).O(B)} \quad (15)$$

which would surely be enough to put anyone off using Odds when they could use (14) instead.[Do a horse racing example to convince yourself of (15).]. I would like to bet animals don't use (15).

CONCLUSION

That such an apparently trivial choice between Odds and Probabilities could have held up human progress for 3 centuries seems astonishing, but straight thinking is hard enough without making it any clumsier. The situation reminds one of the Chinese choice of 2000 or more pictograms to write with instead of 26 phonetic symbols like the rest of us. But their choice wasn't as irrational as it seems—because the Chinese didn't have a common spoken language. Command and Control of a huge multilingual population demanded that the controlling mandarins used pictograms, not phonetic symbols, to communicate. It worked pretty well but unfortunately it precluded the vast majority of Chinese from becoming literate – with fatal, but unforeseeable consequences at the time. Further West our scholastic mandarinate buried the Detective's Equation from the rest of us, as well as from themselves.

Notation matters – in Thinking (Odds), in Language (phonemes), in Arithmetic (Arabic over Roman numerals), in Mathematics (Leibnitz's Calculus symbolism over Newton's) and in Mathematical Physics (Heaviside's Vector Calculus over Maxwell's Quaternions). But none of this was obvious at the time.

Perhaps somebody clever, or lucky, will come up with a better symbolic language in which to do our serious, our Common-Sense thinking. (See Appendix 9 on Animal Wisdom for a possible hint).

[***PS**; de Finetti begins by defining the Probability we attach to event Q as the fraction of the whole betting-pot we are prepared to bet on Q occurring. Suppose that Alice is prepared to bet an amount a against Bob that H will occur if E has happened already [i.e. on (H|E)], while Bob is prepared to bet b that it won't [i.e. on ($\bar{H}$|E)]. Then Alice's 'Expected gain' [i.e. the sum of her possible gains × the Probabilities of each], will be

$$-a \times P(E) + (a+b) \times P(H \& E) \equiv x_A$$

because she will lose a if only E turns up but gain (a+b) if both E and H do.

Therefore: $-P(E) + \frac{a+b}{a} \times P(H \& E) = \frac{x_A}{a}$

But if Alice has a rational expectation of generally winning, Bob must have a rational expectation of generally losing, so he won't accept the bet as it is (he could reduce his b however). For both to rationally play then x_A must $= x_B = 0$ (called "Coherence") and then as (a+b)/a = 1/P(H|E) – de Finetti's definition, then : $-P(E) + P(H \& E)/P(H|E) = 0$ which, rearranged with P(H&E) on the LHS, is the Product Rule. It says nothing about H, E or whether Alice and Bob's Probabilities coincide. It is all about the internal consistency of each's thinking [Bob would find, by the same argument $-P'(E) + P'(H \& E)/P'(H|E) = 0$ (where primes denote his Probabilities) which is simply The Product Rule again.]

[**PSS**:For Cognoscenti: Odds are so much simpler than Probs for compounding evidence because (8a) is symmetric in H and $\bar{H}$ thus P(E) cancels in (11a) whereas it has to be calculated in Probs.; often very awkward; textbooks struggle

INDEX

NB; entries in **Bold** are in the Glossary, in CAPS are Appendices.

Affirming the Consequent' 377
Albert Einstein 34
Algorithm 207, 513
Analogy 63, 397
Argument by Design 192
Aristarchus 60
Aristotle 269, 413
Ascent of Man 494
Association of ideas 477
Categorical Inference 510
bad-simple 325
Bayes, Thomas 423
Bayes Theorem 425
Bayes' Rule 18, 102
Bayes' Rule (or 'Theorem') 513
Bayes' Theorem' 102
Bayesians 431
Becquerel 56
Behaviourist 144
Berlin Isiah 18, 457
Bernoulli, Jacob 307
Big Bang 50, 229
Big Bang Cosmology 194
Birmingham University 418
Black, Joseph 66
Blackstone Ratio 513
Blackstone's Ratio' 209
Boorstin, Daniel 82, 192, 272
Bosma, Albert 74
Boy/Girl problem (& Appendix 4 p554).. 298
Bradford Hill 302
Bradley 63
Bruno de Finetti 428
burden of proof 83, 358
Careerism 393
Carl Sagan 357
CATEGORICAL INFERENCE 581
Categorical Inference' 207
Cavity Magnetron 418
Central Limit Theorem 514
Central Limit Theorem' 315
Certainty 169, 293, 448
Childbed Fever 273
Cleverness 358
Combinations and Permutations 308
Common Sense 513
Common Sense' 32
Comparative Logic 32
Complex hypotheses' 178
Complexity of an hypothesis 197
Conditional Odds $O(H|E)$ 514
Conditional Probability $P(H|E)$ 514

Confirmation Bias 358
Conservation of Energy 73
Continental Drift 227
Copernicus 184
correlation 71
Cosmology 229
Counting 167
Counting Errors 239, 514
COWDUNG 227
Cox, Richard   205
Criticism of Statistics 333
Cyril Burt 372
Dark Galaxy 224
Dark Matter 76
Darwin 62, 186
De Finetti, Bruno 428
de Moivre, Abraham 248
Decisiveness 515
Decisiveness of Evidence 481
Decisiveness Rule' 482
Deduction 26, 512
Deep Time 273
Demarcation Problem 358
Descartes 144, 198, 295
Detective's Equation 132, 432
Detective's Equation (DE) 513
Detectives Equation 151
Differential Equations 472
discrepancy 158
Doll., Richard   302, , 304
Economics 365
Eddington, A,S 210
Education 486
Einstein 80, 185, 460
Eratosthenes 270
Error Analysis' 236
Evolution by Natural Selection 186
Expected Value E() 516
experiment 89
experiments 456
Fallacy of Falsifiability 357
Falsifiability 169
Faraday 52, 81
Fisher, R A 296
Flying Saucers 124
Forgetting the Tarantula 358, 398
Four Elements 190
Free parameters 194, 396, 516
Freud 372
Friedman Milton 369
Galaxies 106
Galileo 49, 418, 458
Galton, Francis 296
Gambler's Secret 515
Gambler's Secret' 298
Gambler's Theorem 310
Gambling Theory 306
Gauch, Hugh Jr. 170, 443
Gaussian   518
genius 90
George W Corner 409
Giacconi, Riccardo 61
Global Warming 390

good-simple	323, 325
governance	491
Grassi, Giovanni	69
Graunt, John	299
Group Think	358
Groupthink	395
Heaviside, Oliver	276
Heliocentric hypothesis	184
Henry Marsh	371
Herring Gulls	143, 447
Hertz, Heinrich	275, 459
Hidden galaxies	111, 260, 283, 499, 592
Hidden Galaxies'	219
History Equation	471
Horton	351
Howson, Colin	334
Hoyle, Fred	380
Hubble, Edwin	176
Hubble Space Telescope	114, 225, 389
Hughes, David	275
Human Thermodynamics	485
Hume, David	95, 143, 422
Huxley, Thomas	316
Huyghens Christiaan	420
Hypothesis	39
Hypothesis Testing	340
Induction	27, 95, 514
Inductive Certainty	212
Inductive Conviction	152
Inference	514
Inference Table	137, 199, 257, 514
Inference Tables	204
Information Theory	320
Instrument builders	455
instruments	452
Ioannidis	335
J.J. Thompson	277
Jane Goodall	145
Jared Diamond	451
Jaynes, Edwin	333, 334
Jeffreys, Harold	55, 379
Joseph, Black	191, 263
Joule, James	58
Judson, Freeman	46
JURY DUTY	216
Kelvin, Lord	276, 278
Kendall M G	294
Kepler, Johannes	48, 262
Keynes, John Maynard	154, 502
Lactantis	269
Leeuwenhoek, Antoon	60, 274
Leonard Digges	415
Leveraging	466
Lewis Fry Richardson	468
Lind, James	300
Lorentz, Antoon	78
LTCM	369
Ludic Fallacy	400
Lunar Society	191
SALLY CLARK CASE	562
Malthus, Thomas	61
Marconi	64, 275
Mark Twain	476
mathematics	464
Maxwell, James Clerk	54, 459, 462
Maxwell's Equations	459
MCO	343
Measurement Error	314
Measurement Errors	265, 319, 517
measurements	261
Medawar, Peter	46, 360, 494
Mendeleev	67
Minimum Consensual Odds (MCO)	518
Minimum Consensual Odds'	343
Misplaced Deference	358, 386
Newton	26, 53, 411, 419
Nicholas Nassim Taleb	400
Nightingale	296
Nikolaus Steno	478
noise	188
Normal Curve of Errors	248
Normal Distribution	518, 559
Normal Weights	518
Null Hypothesis	327, 328, 331, 332
Null Prior	222
Objective Bayesians	430
Ockham's Razor	171, 183, 439, 442, 519
Odds	98
Odds on H, i.e. O(H)	517
Oersted Christian	52
Open University	320
Oreskes, Naomi	390
'Origin of Species' by Darwin	186
Orwell, George	268
Oyster Club	483
P(E|)	165
Palla Strozzi	271
Parsimony	184, 196
Pasteur, Louis	275
Pauli	74
PAW	292
Pearson, Karl	296
PEER REVIEW	347
Periodic Table	67
Perlmutter, Saul	211, 236
Philosophy	502, 566
Plausible Inference	32, 152
Poincaré, Henri	79
Popper, Karl	360
Posterior Odds'	151
Principle of Animal Wisdom (PAW)	520
Principle of Limited (Independent) Variety (PLV)	520
Principle of Limited Variety	358
Principle of Relativity'	77
Principle of Uniformity	521
Printing	417
Prior	213
Prior Odds	103
Prior Odds O(H)	521
Probability	97, 241, 313
Probability P(H)	522
Probability Theory	305, 313
Problem of Induction'	28
Prosecutor's Fallacy	522
Provisionality	497
Psychiatry	369
Psychology	369
Ptolemy	271
Pygmalion	401
Pygmalion Complex	358
Quine, R van Ormond	362

R.V. Jones	442
RAF Bomber Command	306, 311, 326
Refereeing	349
REFERENCES	526
religion	493
Resurrective Thinking	358
Richard Price	423
Russel, Bertrand	291
Sagan, Carl	404
Saint Augustine	416
Sally Clark	166, 337
Scatter,(Appendix 3)	239
Scientist's Approximation'	245
Scientists Approximation	523
Second Law of Thermodynamics	85
Second Law of Thermodynamics [SLOT]	484
Semmelweiss, Auguste	273
Shannon, Claude	320
Slipher, Vesto	164
Snake Oil Delusion (SNOD)	523
SNOD	128
Snow, John	51
Special Theory of Relativity'	80
Squared Knowledge Theorem'	478
Standard deviation'	240
Standard Model	193
Stapel, Diedricht	374
Statistics	**294**
Statistics, the Central Fallacy of	524
Subjective Bayesian	428
Systematic Error	268
Systematic Errors	521
Theory of Probability	524
Thomas Young	478
Time, Clock & Human	486
Tommy Flowers	469
Tunnel Vision	358
Turing, Alan	316
Tycho Brahe	272
Uniformity	422
universe is expanding'	159
unknown unknowns	228
Urbach, Peter	334
vertex-numbers	243
Wallace, Alfred Russel	62, 186
Wegener, Alfred	55, 381
Weight	103, 267
Weight W(E|H)	524
Weights	200, 212, 286
William James	412
William Sargant	373
woolly thinking	189